The Anatomy of Murder

THE ANATOMY OF MURDER

*Ethical Transgressions and Anatomical Science
during the Third Reich*

SABINE HILDEBRANDT

berghahn
NEW YORK · OXFORD
www.berghahnbooks.com

First published in 2016 by

Berghahn Books

www.berghahnbooks.com

© 2016, 2017 Sabine Hildebrandt
First paperback edition published in 2017

Library of Congress Cataloging-in-Publication Data

Names: Hildebrandt, Sabine, author.
Title: The anatomy of murder : ethical transgressions and anatomical science during the
Third Reich / Sabine Hildebrandt.
Description: New York : Berghahn Books, 2016. | Includes index.
Identifiers: LCCN 2015026607| ISBN 9781785330674 (hardback : alkaline paper) |
ISBN 978-1-78533-732-1 paperback | ISBN 9781785330681 (e-book)
Subjects: LCSH: Human anatomy—Study and teaching—Germany—History—20th
century. | Human anatomy—Research—Germany—History—20th century. |
Anatomists—Germany. | Medical ethics—Germany—History—20th century. |
Human experimentation in medicine—Germany—History—20th century. |
National socialism—Moral and ethical aspects. | World War, 1939-1945—Atrocities.
| Germany—History—1933-1945.
Classification: LCC QM33.3.G3 H55 2016 | DDC 612.0071/143—dc23
LC record available at http://lccn.loc.gov/2015026607

British Library Cataloguing in Publication Data

A catalogue record for this book is available from the British Library.

ISBN 978-1-78533-067-4 hardback
ISBN 978-1-78533-732-1 paperback
ISBN 978-1-78533-068-1 ebook

To

Libertas Schulze-Boysen
executed at Berlin-Plötzensee on 22 December 1942

and

Charlotte Pommer
who ended her career in anatomy after finding
Libertas' body on a dissection table

CONTENTS

Appendix

FOREWORD

In 1873, a young Canadian physician by the name of William Osler traveled from Montreal to Berlin and Vienna to advance his medical studies.[1] Along with many other young doctors from North America and Great Britain, Osler saw Europe, the German-language universities in particular, as being the most advanced in the world. It was to these institutions the young graduates flocked to study with leading professors and scientists, many of whom achieved immortality through the eponymous designation of diseases and organs that carry their names to this day.

The institutions and great professors of Germany and Austria so admired by Osler and his peers represented the principal source of modern medical science as we know it today. It was in the hospitals and clinics of Germany and Austria that, among others, Robert Koch discovered the tubercle and cholera bacilli, thus establishing the basis of infection and microbiology; Paul Ehrlich discovered chemo/antimicrobial therapy and the basis of immunology; Rudolf Virchow pioneered cellular pathology; the Vienna pathologist Carl von Rokitansky established the correlates of disease and anatomic pathology. The German university in general served as the prototype for Abraham Flexner's new paradigm for medical education that merged clinical medicine with advanced education and research as exemplified by the German university. Academic medicine and modern medical science today are the inheritors of that legacy.

Osler's exuberant chronicle of his encounters with great physicians and scientists was offset by a somber description of anti-Semitic sentiment within the Berlin medical community and a prescient prediction of what might happen should there be an exodus of Jewish physicians from Germany. Not even a person with the intelligence and intellectual acumen of William Osler could have predicted the calamity that would ensue.

Sabine Hildebrandt's fascinating and important book, *The Anatomy of Murder: Ethical Transgressions and Anatomical Science during the Third Reich,* is the first detailed history of the role played by the universities and medical faculties of Germany and Austria during the Third Reich—the same institutions once so admired by Osler and his peers. It is a story of academia's exploitation of the helpless victims of Nazi terror, in life and death. It is a profoundly moving account of evil and suffering. It includes not only documentation of the exploitation of the living and dead victims, including a chilling description of the exploitation of the "future dead," but also the travails and suffering of those few who attempted to resist. As well it includes accounts of academics who suffered and died because of their "race/ethnicity" and the stories of some who survived in exile.

The potency of this narrative cannot be understated. The malevolence of distinguished academics and scientists is revealed and the suffering of the victims and their families documented. Heretofore anonymous victims are given a biography and a face. Students who experienced the consequences of Nazi terror in the dissecting room are given voice through their accounts. The agony of exile and the ambivalence of admiration for the language and culture of the country of his birth, achievement and torment, are movingly illustrated in the story of the expatriate anatomist/artist Hans Elias.

The subject of anatomy provides a critical analytic perspective on the relationship of distinguished professors and institutions with the National Socialist regime and how many were able to exploit mass murder by the Hitler state to enhance their research, teaching, and professional status. The victims of Hitler's program of terror were the source of valuable "fresh" specimens from otherwise "healthy" young victims. Hildebrandt documents the research of the internationally renowned Professor Hermann Stieve of the Charité Hospital and the University of Berlin who exploited the execution of young female political prisoners at Plötzensee Prison in Berlin to advance his studies on the influence of psychic stress on the female genital tract. Stieve's students were the hapless benefactors of the death of female victims, cadavers from whom had heretofore been rarely available to anatomical institutes. Stieve's professional reputation was based in no small part on these studies on the doomed women.

As important as Hildebrandt's account of the Hitler period may be, her story of events after the war is of equal importance. She appropriately and dramatically refers to the postwar period as "The Great Silence." The silence encompassed not only anatomy but all of German medicine, including academia. Hildebrandt's studies comprised not just individual anatomists and university departments but also the professional society of anatomists, the Anatomische Gesellschaft. They include the fate of professors, a few of whom were held to account for their actions during the Third Reich. Most evaded judgment and rose to distinguished postwar careers. Hildebrandt's discussion of the case of Dr. Robert Herrlinger is an especially important example of the ambivalent hypocrisy of the academic community that, despite its awareness of Herrlinger's nefarious conduct as an anatomist at the SS-linked Reichsuniversität at Posen in occupied Poland, appointed him to the position of professor of medical history at the University of Würzburg. Thus a person responsible for acts of inhumanity became indirectly responsible for teaching medical students in a subject area that is considered to be within the domain of the medical humanities.

The story of the postwar response of the German academic medical community to their own history is one of denial, deceit, and deception. It is about the assertion of power for protection of self, protection of colleagues, and protection of institutions. The "Great Silence" persisted for decades after the war. It took almost half a century for the first voices to be heard. The initial sounds were whispers, but gradually they were joined in a collective chorus by a new generation of scholars exemplified by Dr. Hildebrandt and her peers. They asked questions, discovered and explored heretofore inaccessible archives, and allowed previously silent voices to be heard.

Dr. Hildebrandt enriches this saga with a history of human anatomy and the moral challenges of the discipline before the Hitler period. She concludes with an insightful analysis of the implications for the future in the context of the ethics of medicine and the implications of the teaching of anatomy for the moral development of future health professionals and their sensitivity to the human experience of the deceased subject as a person. Hildebrandt's research, documentation and writing reveal the talent of a skilled historian and storyteller.

The Anatomy of Murder: Ethical Transgressions and Anatomical Science during the Third Reich is a dramatic and eloquent telling of the universal story of the pathology of power—political, professional, personal. It is a powerful cautionary tale not just for anatomists but for every person involved in the education and training of future health professionals, a group that encompasses curriculum planners, clinical teachers, academic and departmental administrators including deans and department heads, health-care policy analysts, university presidents, government officials, and students. It should be compulsory reading for every member of ethics committees and institutional review boards.

This book is not only about perpetrators and victims of the past, it is also about the implications for the present and the future. It is about us and our responsibility for future generations of health professionals.

William E. Seidelman, MD
Emeritus Professor
Department of Family and Community Medicine
Faculty of Medicine, University of Toronto

Beer-Sheva, Israel
September, 2014

Note

1. Cushing, *Life of Osler,* 106–19 and 214–15; Bliss, *William Osler.*

Bibliography

Bliss, Michael. 1999. *William Osler: A Life in Medicine.* U. Toronto: Toronto.
Cushing, Harvey. 1925. *The Life of Sir William Osler.* Vol. 2. Oxford: Oxford University Press.

ACKNOWLEDGMENTS

This book is the result of several years of discussions with colleagues and students at the University of Michigan, Boston Children's Hospital/ Harvard Medical School, and in Germany. They all have contributed in different ways.

Dr. John DeLancey started it all by asking about the Pernkopf atlas, Dr. Tom Gest gave me the freedom to follow my own ideas in anatomy, and Dr. Alex Stern was the first medical historian to acknowledge my work. Dr. Paul Weindling and Dr. Michael Grodin encouraged me to delve deeper into the world of medical history, and Dr. Stephen Carmichael and Dr. Shane Tubbs were ceaseless supporters of my writing. Dr. Christoph Redies was a forceful collaborator, and Dr. Andreas Winkelmann and Dr. Gareth Jones remain fellow thinkers on the paths of history and ethics in anatomy. Dr. Gary Fleisher and Dr. Mark Schuster have graciously given my work a home at Boston Children's Hospital. There are many more who will find their names and influence in these pages. The "gray eminence" behind all of it is Dr. William Seidelman, whom I cannot thank enough.

At the heart is my family: Friedhelm, Clara, Jakob.

Abbreviations and German Terms

Abbreviations

AG: *Anatomische Gesellschaft*, international organizational body of German-language anatomists

BA: *Bundesarchiv*, federal archives

BÄK: *Bundesärztekammer*, German physicians' association

BayHst: *Bayerisches Hauptstaatsarchiv*

BEG: *Bundesgesetz zur Entschädigung für Opfer der nationalsozialistischen Verfolgung*, Federal Law for the Compensation for Victims of National Socialist Persecution

DÄ: *Deutsches Ärzteblatt*, official journal of the BÄK

DFG: *Deutsche Forschungsgemeinschaft*, German research association

FRG: Federal Republic of Germany

GeStapo: *Geheime Staatspolizei*, secret state police

GDR: German Democratic Republic

GDW: *Gedenkstätte Deutscher Widerstand*, memorial site of the German resistance

Hschf: *Hauptscharführer*

KWI: Kaiser Wilhelm Institute

MPI: Max Planck Institute

NS: National Socialism

NSDAP: *Nationalsozialistische Deutsche Arbeiterpartei*, National Socialist German Workers' Party

NSDDB: *Nationalsozialistischer Deutscher Dozentenbund*, National Socialist German Lecturers' League

NSDStB: *Nationalsozialistischer Deutscher Studentenbund*, National Socialist German Students' League

REM: *Reichserziehungsministerium*, short for *Reichsministerium für Wissenschaft, Erziehung und Volksbildung*, Reich Ministry of Science, Education and Culture

RFR: *Reichsforschungsrat*, Reich research council

RJM: *Reichsjustizministerium*, Reich ministry of justice

SA: *Sturmabteilung*, brown shirts

SB-BB: *Staatsbibliothek Berlin Preussischer Kulturbesitz*, Estate Elias

SD: SS *Sicherheitsdienst*, security service

SS: *Sturmstaffel*, Nazi elite troupe

UA: University Archive

Erlangen (UAE)

Freiburg (UAF)

Greifswald (UAG)

Jena (UAJ)

Leipzig (UAL)

Marburg (UAM-EB), Estate (*Nachlass*) Benninghoff (EB)

Münster (UAM)

Rostock (UAR)

Wien (UAW)

Würzburg (UWü)

UK: United Kingdom

WWI: World War I

WWII: World War II

German Terms

Ahnenerbe: Forschungs- und Lehrgemeinschaft das Ahnenerbe e.V., organization for the research on "race" and heredity under the patronage of Heinrich Himmler, head of the SS

Anatomische Anzeiger: journal of the *Anatomische Gesellschaft*

Ausserplanmässiger Professor: professor extraordinarius who is not a civil servant

Ausserordentlicher Professor: professor extraordinarius, full professor without a chair

Baum-Gruppe: group of Jewish German dissidents associated with the *Rote Kapelle*

Gedenkstätte Deutscher Widerstand: Memorial Center for the German Resistance in Berlin

Privatdozent: senior lecturer

Professor ordinarius: full professor with chair

Prosektor: senior assistant of professor of anatomy, responsible for body procurement

Reichsführer: leader of the *Reich*, i.e. Hitler

Reichsmark: German currency from 1924 to 1948

Rote Kapelle: "red orchestra," resistance group of mostly German political dissidents

Sondergericht: special court

Sondervotum: dissenting opinion

Venia legendi: license to teach at an academic institution

Volksgerichtshof: People's Court, highest court in the Third Reich

Weisse Rose: "white rose" resistance group with a majority of students

INTRODUCTION

The difficulty is, you see, that our imagination
cannot count [...]
And if I say one died - a man I have made you know
and understand [...] then perhaps I have told you
something that you should know about the Nazis.
                                    Erich Maria Remarque, 1946[1]

The connection between Erich Maria Remarque, author of the World
War I novel *All Quiet on the Western Front*, and anatomy in National
Socialism was a woman named Elfriede Scholz. She was a seamstress, born
on 25 March 1903, and was executed in Berlin-Plötzensee at noon on
16 December 1943. Her execution had followed a verdict of "aiding and
abetting the enemy"; she had criticized Hitler in conversations with her
neighbors, who then denounced her to the authorities. Directly after the
execution her body was taken to the anatomical department of the Fried-
rich-Wilhelm University of Berlin and used by Hermann Stieve, chairman
of anatomy, for his research. Her maiden name was Remark, and she was
Erich Maria Remarque's sister.[2]

I had never heard of fates like that of Elfriede Scholz, or of any involvement of anatomists with the National Socialist regime and its victims, despite my interest in this period of German history. Born not that
long after World War II in a small town in Western Germany, I always felt
the silent presence of the aftermath of the Third Reich around me. I grew
up without Jewish neighbors and knew synagogues only as empty spaces
in urban landscapes. My first school was called *Geschwister-Scholl-Schule*
after Sophie and Hans Scholl, university students who had been executed
because of their resistance against Hitler. Even as a six-year-old I understood the enormity of their story. The voices of the adults around me grew
unusually quiet and halting when they spoke about such matters, while
they tended to turn louder and exuberant for many other reminiscences

of past times. Anne Frank's diary and Wolfgang Borchert's and Heinrich Böll's pacifist writings became part of my world of ideas, and schoolteachers introduced me to the facts of National Socialism. I felt drawn to this history, as I needed to know why the people I lived with had become part of a clearly atrocious past. When I began my medical studies in the early 1980s, the first comprehensive investigations on the involvement of medicine in National Socialist crimes were published, which I followed avidly. I felt reasonably well informed. However, it was only when I started my career in anatomy that I first heard of a connection between this field and victims of the National Socialist regime. A colleague asked me about the Nazi origins of Eduard Pernkopf's atlas of anatomy, and I had no answer for him. His inquiry sent me on a quest for information, and I soon realized that the subject was much more complex than I had anticipated and needed further exploration. While many scattered facts were available, there existed no overall narrative of this past. Fortunately, quite a few of my German colleagues had by then performed more systematic investigations into the history of their anatomical departments. Also, ventures into various German archives revealed that many documents had not yet been investigated. Based on the work of recent years, a first account of the subject is possible now. This book endeavors to follow Michael Kater's demand to be "neither apologetic nor demonological, but factual, precise, and to the point."[3]

To understand what happened to Elfriede Scholz, it is necessary to study the many aspects of the history of anatomy in the Third Reich. The first is that this period of our past has remained unexplored for many years. The great silence in postwar Germany regarding the National Socialist (NS) involvement of medicine lasted particularly long,[4] and even longer for anatomy. It took the passing of the last anatomists who had been active during this time and the retirement of their loyal students before serious investigations of the individual institutions became possible.[5]

Based on the general background of the National Socialists' efforts to control the universities, the relationship between the new government and the anatomical departments and anatomists will be explored. Apart from the general academic administration, the NS government was involved in body procurement, recruitment of personnel, and research funding. It also attempted to control the professional societies, often with great success. However, the *Anatomische Gesellschaft*, an international anatomical society based in Germany, managed to retain its international character and its autonomy to a certain extent. At the same time, the political spectrum of the anatomists at the universities reached from political dissidents, over those persecuted for so-called "racial" reasons to the politically vague, and finally to the convinced ideologues. Many anatomists lost their positions, and their fates ranged from professional success after forced emigration to suicide and death in concentration camps

All of the anatomists remaining in Germany used bodies of victims of the National Socialist system for medical education or research. The victims were part of the traditional body procurement and included political dissidents from inside and outside Germany, petty criminals, psychiatric patients, deserters, and many others. They served as "material" for anatomical scientific studies. Research on bodies of executed persons, which had been an established practice in German anatomy before 1933, soared during the war. It will be shown that anatomists started to realize entirely new "opportunities" of research, which, after going through distinct stages of escalation, transgressed the traditional anatomical paradigm of work with the dead: three anatomists started to include the expected death of a person, a prisoner, in their research design, thus entering the field of human experimentation.

There was only one anatomist who refused to work with the bodies of executed political prisoners, Hermann Stieve's young assistant Charlotte Pommer. When she recognized the bodies of prominent dissidents on the dissection tables of the anatomical department in Berlin, she decided to abandon her career in anatomy. She remained the only anatomist who protested the use of NS victims in the dissection room by changing careers; all others remained silent, even if they might disapprove of the regime. They knew their work was sanctioned by law, and as long as their methods were scientifically correct, most had no other ethical or professional worries. They maintained this attitude for the rest of their lives, as the postwar history of German anatomy shows. However, distinct changes in the professional ethics of anatomy since the war have led to a change of heart in a new generation of anatomists, who realize that the history of anatomy in National Socialism has implications for contemporary medicine.

This book is meant to lay the foundation for the remaining work still to be done in the field of anatomy. Other related medical fields, which also exploited the bodies of victims of the NS system for research purposes, can only be mentioned in passing to compare them, as a full presentation of the history of disciplines like pathology, neuropathology, or forensic medicine would go beyond the scope of this book. However, the most relevant studies are quoted for those interested in further information. Among the future areas of historical investigations in anatomy are the full reconstruction of the biographies of the anatomists, regardless whether their careers were disrupted by the NS regime or whether they thrived, and the documentation of the names and lives of the victims whose bodies were used for anatomical purposes. Such documentation will become the keystone for the necessary memorialization of the victims. As Hans-Joachim Lang said, "Forgetting them would be the victims' final anihilation."[6]

After studying this history for several years now, I believe that it represents an example of the ethical dangers inherent to a medicine that believes itself on secure moral ground and has ceased to reflect on the paradigmatic basis of its methods of gaining knowledge. This lack of doubt concerning one's own moral convictions, the institutional culture of medicine, and the current political environment is not specific to the Third Reich. Rather, it can be seen in many other periods of the history of medicine and certainly in current medical research, education and practice.

Notes

1. Remarque, quoted after Gelder, *An Interview with Remarque.*
2. Glunz and Schneider, *Elfriede Scholz.*
3. Kater, "Burden of the Past," 56.
4. Caplan, "The Stain of Silence."
5. Hildebrandt and Redies, "Anatomy in the Third Reich."
6. Lang, *Die Namen der Nummern,* 13.

Bibliography

Bliss, Michael. 1999. *William Osler: A Life in Medicine.* Toronto: University of Toronto Press.

Caplan, Arthur L. 2010. "The Stain of Silence: Nazi Ethics and Bioethics. In: *Medicine after the Holocaust: From the Master Race to the Human Genome and Beyond,* edited by Sheldon Rubenfeld, 83–99. New York: Palgrave Macmillan.

Cushing, Harvey. 1925. *The Life of Sir William Osler.* Vol. 2. Oxford: Oxford University Press.

Gelder, Robert van. 1946. "An Interview with Erich Maria Remarque." In: *Writers and Writing,* by Robert van Gelder, 377–81. New York: Charles Scribner's Sons.

Glunz, Claudia, and Thomas Schneider. 1997. *Elfriede Scholz, geb. Remark: Im Namen des deutschen Volkes. Dokumente einer Justiziellen Ermordung.* Osnabrück: Universitätsverlag Rasch.

Hildebrandt, Sabine, and Christoph Redies. 2012. "Anatomy in the Third Reich." *Annals of Anatomy* 194:225–27.

Kater, Michael H. 1987. "The Burden of the Past: Problems of a Modern Historiography of Physicians and Medicine in Nazi Germany." *German Studies Review* 10:31–56.

Lang, Hans-Joachim. 2007. *Die Namen der Nummern: Wie es gelang, die 86 Opfer eines NS-Verbrechens zu identifizieren.* Überarbeitung der Originalausgabe von 2004. Frankfurt am Main: S. Fischer Verlag.

Chapter 1

HISTORY OF RESEARCH ON MEDICINE AND ANATOMY IN NATIONAL SOCIALISM

> I realized … that there was a sharp distinction between what
> was remembered, what was told, and what was true.
>
> Kevin Powers[1]

The history of research on medicine and anatomy in National Socialism resembles in some ways the progress of investigations into the
overall history of the Third Reich, which after early general studies have
shown new and varied scholarly approaches since the 1970s and thereafter.[2] These developments moved in parallel with "specific stages of postwar
German society."[3]

When, in 1987, Canadian historian Michael H. Kater reviewed the status of research on the role of physicians in National Socialism, he commented on the scarcity of such studies up to that date.[4] This was all the
more remarkable as directly after the war, the newly reconstituted *Ärztekammern* (regional professional associations of physicians) of West Germany were actively involved in understanding the participation of German
physicians in NS atrocities. The *Ärztekammern* (West) declared an interest
in the publication of a report on the 1946–47 Nuremberg Doctors' Trial
from a physician's point of view, and commissioned neurologist Alexander Mitscherlich and physician Fred Mielke among others for this task.
They produced a preliminary report in 1947,[5] which initiated acerbic controversies among the authors and some prominent physicians who were
named as collaborators of the NS regime. Among those named in the report was surgeon Ferdinand Sauerbruch, who, along with other physicians,
saw Mitscherlich and Mielke's frank report as undermining the newly
re-growing trust between patients and doctors in Germany.[6] Nevertheless,

the Ärztekammern (West) released a declaration in 1947 stating German physicians' sympathy for "the victims of the NS tyranny, which employed science as a means for its deeds," and their "sorrow about the fact that men out of our own ranks, have committed crimes that elicited revulsion in the whole world."[7] Apparently Mielke himself was of the opinion, voiced in a speech delivered at the 51. *Deutsche Ärztetag,* the first postwar meeting of the Ärztekammern in 1948, that among the 90,000 doctors practicing in NS Germany, only three to four hundred had been involved in NS crimes.[8] Most German physicians adopted this assessment at the time, as it conveniently allowed them to declare themselves not responsible for NS medical atrocities committed by only a few perverted psychopaths, and to detach themselves completely from these deeds.

The explanation of the culpable few and the innocent majority also seems to have been accepted by the World Medical Association, which initiated the admission of the German physicians' association into this international body in 1950. For many German and international physicians the question of medicine's involvement in the NS policies was satisfactorily resolved with the report.[9] Mitscherlich and Mielke's final account of the Nuremberg Doctors' Trial, as well as Alice Platen-Hallermund's report on the murder of the mentally ill and other writings on the subject like Werner Leibbrand's in 1946,[10] were largely ignored by the German medical establishment. No further investigations were pursued and a long-lasting silence on the topic followed.[11]

Apart from the generally prevalent German apathy concerning its NS history,[12] this silence had its root in the widespread denial of any personal responsibility that existed in all parts of the population, not only among physicians. The latter felt that they had labored under a regime of injustice which had fatefully and forcefully overtaken an otherwise rational medicine.[13] In addition, doctors had become "doubly homogenized",[14] starting with the beginning of the Third Reich, when they were silent about the expulsion of their Jewish and dissident colleagues and the usurpation of the universities by the NS regime. The second homogenization occurred later, after the collapse of the NS regime, when they passed through denazification and reclaimed their positions in postwar Germany, often adhering to the ideological thinking from their past.[15] This homogenization led to a silence of mutual solidarity, which was only broken with the advent of a new generation of physicians who felt the need to inquire more deeply into the history of medicine in the Third Reich.[16] The silence reigning in German historical research on the topic of NS doctors, or in any publications in which the authors distanced themselves from NS perpetrators and their science,[17] was all the more notable as very public trials of war criminals, including physicians, were pursued in the 1950s and 1960s. Among those being prosecuted were the gynecologist Carl Clauberg, accused of medical experiments on women in Auschwitz in 1956, and the anatomist Johann Paul Kremer in 1960, who was charged with assistance in the death of prisoners in Auschwitz.[18] In 1956 the BÄK (*Bundesärztekammer,* a later name for *Ärztekammern* West) refused to retract Clauberg's membership for formal reasons pending the outcome of the trial, but issued a preliminary withdrawal of Clauberg's professional privileges when he was charged with several war crimes in 1957.[19] In 1958 the BÄK officially distanced itself from individual doctors who had committed crimes in concentration camps and recommended the preliminary removal of BÄK membership in all such cases.[20]

The extent to which active denial, a possible lack of intrapersonal insight, or a persistence of ideological thinking patterns existed in German academic physicians of the first postwar decades is well illustrated by the person of August Mayer, emeritus chair of gynecology at the University of Tübingen. He was one of the very few leaders of the medical establishment who published reflections on medical ethics in the 1960s, including ethical failures of medicine in the Third Reich.[21] While he named "euthanasia," forced abortions, and infanticide as crimes of the NS regime, he distanced himself from such activities in a clear manner. Mayer seemed to have overlooked the fact that he himself had taken part in such crimes. He had joined the NSDAP and SA, was an early defender of forced sterilization, and, in his clinic, 740 forced sterilizations were carried out, as well as abortions on two forced-laborer women. Interestingly, Tübingen was one of the first universities to offer a lecture series on the involvement of German academia in National Socialism in 1964–65. While sterilizations and "euthanasia" were mentioned, the involvement of Mayer was not.[22]

Other sporadic German publications in the 1940s, 1950s, and 1960s focused on medical atrocities and the ideological roots of medicine during National Socialism.[23] Some international authors approached questions of the involvement of doctors in NS crimes in various publications. Directly after the war, occupying military forces gathered information on German science, e.g. in the FIAT reports.[24] In 1946 Czech internist Josef Charvát published a compilation of his own and colleagues' testimonies of abuses of medical science during the NS regime. These included some of the earliest firsthand descriptions of medical experiments in concentration camps,[25] and other autobiographical reports by former prisoners like Eugen Kogon and Miklos Nyiszli in 1946.[26] In 1949 neurologist Leo Alexander reported on "medical science under dictatorship" in a review in the *New England Journal of Medicine,* as well as in a three-part series of psychological reflections, following his work as advisor during the Nuremberg Doctors' Trial.[27] Alexander drew lessons for American medicine from his experiences, reminding his colleagues of the dangers of a technicalized medicine and concluding his essay with the reminder, "Yes, we are our brothers' keepers".[28] In 1950 Francois Bayle published a detailed account of the Nuremberg Doctors' Trial in French, Carlos C. Blacker condemned the NS human experiments in the *Eugenics Review* in 1952, and in 1959 the Museum of the Memorial Site at Auschwitz started a journal series called "*Zeszyty Osięcimskie*" ("Papers from Auschwitz"), which was also translated into German. One of the first issues contained Polish lawyer Jan Sehn's detailed account of Carl Clauberg's medical experiments on women in Auschwitz.[29] A first sign of the changes to come may have been a 1976 conference organized by the Hastings Center, which explored the "proper use of the Nazi analogy in ethical debate" under the title "Biomedical ethics in the shadow of Nazism".[30] Medical crimes and the interrelationship between the medical community and the NS government, as well as the Nuremberg Doctors' Trial, were discussed in order to highlight lessons for modern bioethics. Overall, however, there were only very few public discussions or serious research efforts on the subject in Germany and internationally, so that Kater found a "deplorable dearth of sources" concerning the history of medicine in the Third Reich in 1987.[31]

The "Turn of the Tide," 1980

As physician Christian Pross noted years later, "the tide was turned" at the *Gesundheitstag* in Berlin in 1980, an alternative meeting held at the same time as the *Bundesärztetag*, the official conference organized by the BÄK.[32] This assessment was shared by historians Paul Weindling and Walter Wuttke.[33] What had happened? During the 1970s and 1980s German and international medical historians, together with politically left-leaning physicians and students in Germany, had started to explore the field of medicine in the Third Reich.[34] Foremost among them was a group of psychiatrists who disagreed with their colleagues' previous assessment that they had nothing to do with NS atrocities in psychiatry; instead, the group identified continuities in ideology and personnel from the NS period to postwar Germany, thereby implicating these colleagues.[35] Weindling traced the roots of this new interest "in part to a heightened critical awareness of scientific knowledge and the associated structures of social power in modern society" and saw the impetus coming from radical historians "outside the academic discipline of the history of medicine in the Federal Republic of Germany". These historians were subsequently accepted by "the more liberal-minded historians of medicine, and professional medical organizations."[36] During the *Gesundheitstag*, medicine in National Socialism became a main topic of discussion between German physicians, allied health workers, patients, and Jewish physicians who had suffered persecution.[37] This development in the field of medicine was accompanied by a generally enhanced awareness of the public concerning NS atrocities following the release of a television series on the Holocaust and Claude Lanzman's film *Shoah*.[38]

Among the first of the German publications coming out of this movement was Tübingen historian Walter Wuttke-Groneberg's collection of documents on medicine and National Socialism[39] and the documentation of the *Gesundheitstag* 1980, edited by medical historian Gerhard Baader.[40] Early international scholars in the area included Michael Kater, who had written his dissertation on the *Ahnenerbe* in Heidelberg in 1966. The SS *Ahnenerbe* was a research foundation headed by Heinrich Himmler promoting the "racial" supremacy of Germans. Out of this research Kater developed an inquiry into the involvement of various professional groups with NS politics, among them students, teachers, and physicians.[41] His 1989 book *Doctors under Hitler* belongs now to the standard texts in the field. Fridolf Kudlien, professor of history of medicine at the University of Kiel, published the first comprehensive German overview on the topic in 1985, *Ärzte im Nationalsozialismus* (Doctors during National Socialism). Among his coauthors were some of the most active researchers on medicine in the Third Reich at the time, Michael Kater and Gerhard Baader, medical historian Rolf Winau from Freie Universität Berlin, and physician-activist Karl-Heinz Roth.[42] An international meeting that included scholars from the Federal German Republic and the German Democratic Republic (GDR) was held in 1988 in Erfurt and Weimar under the auspices of the International Physicians for the Prevention of Nuclear War (IPPNW),[43] and a first review of the topic from the perspective of a GDR researcher was presented by medical historian Achim Thom in 1989.[44] Other early landmark publications were Cologne geneticist Benno Müller-Hill's *Tödliche Wissenschaft* (Deadly Science), a look at German eugenicists and anthropologists; historian Gisela Bock's study of forced sterilizations, *Zwangssterilisation im Nationalsozialismus*; Hans-Walter Schmuhl's *Rassenhygiene, Nationalsozialismus, Euthanasie*, an early analysis of racial hygiene; *Rasse, Blut und Gene* by sociologists Peter Weingart, Jürgen Kroll, and Kurt Bayertz's, a history of German eugenics and racial hygiene; American psychiatrist Robert Lifton's *The Nazi Doctors*, a first approach to understanding the psyche of NS physicians, which received much criticism; American historian Robert Proctor's *Racial Hygiene: Medicine under the Nazis*, an analysis of physicians' involvement in NS ideology; British historian Paul Weindling's *Health, Race and German Politics between National Unification and Nazism 1870–1945*, which traces the roots of NS policy and ideology back to the nineteenth century; American bioethicist Arthur L. Caplan's

When Medicine Went Mad, a documentation of the first conference on the meaning of the Holocaust for bioethics; bioethicists George J. Annas and Michael A. Grodin's *The Nazi Doctors and the Nuremberg Code,* a first analysis of the relationship between NS medicine and postwar codes of bioethics; and the historians Manfred Berg and Geoffrey Cocks's international essay collection *Medicine and Modernity,* which analyzed the historical roots of public health and medical care in National Socialism.[45] In Germany, early essential and persistent contributors to the field included authors Ernst Klee and Götz Aly, who over the years pursued among other topics the continued careers of academics active during the NS period and after the war.[46]

In 1987, the still-existing contentiousness of publishing data on the involvement of physicians in NS policies was demonstrated by the harsh reaction of the BÄK to a paper by physician Hartmut Hanauske-Abel. In a 1986 article for the British medical journal *Lancet* he reported on the involvement of doctors in the inequities of the NS state and asked whether modern medicine became guilty again by acquiescing to governments preparing for nuclear warfare.[47] The BÄK openly criticized Hanauske-Abel for what they considered a denigration of German physicians and tried to seriously compromise his professional standing.[48] Since then the BÄK has officially changed its position, but remains slow in responding adequately to the issues at hand.[49] At the *Bundesärztetag* in 1989 medical historian Richard Toellner gave a keynote speech on "Physicians in the Third Reich," which was printed in *Deutsches Ärzteblatt* (DÄ),[50] the official journal of the BÄK. The BÄK has also financially supported some limited investigations on the history of medicine in the Third Reich[51] as well as creating an award for studies on the role of physicians during the NS period.[52] However, the BÄK failed to significantly assist the landmark project of the complete documentation of the Nuremberg Doctors' Trial in the 1990s, and the leaders of this effort, Klaus Dörner and Angelika Ebbinghaus, resorted to crowd-funding from their physician colleagues to finance the work.[53] A further lapse occurred with the publication of an uncritical obituary in the DÄ on Hans Sewering, the former member of the board of directors of the BÄK.[54] The article failed to mention Sewering's past as a member of NSDAP and SS, and his contribution to the death of several "euthanasia" patients by signing their transfer papers from a psychiatric hospital to one of the NS extermination centers.[55] The obituary was disputed by a letter of protest endorsed by the majority of German medical historians, which was published in the DÄ.[56] Finally in 2012, on the basis of a petition by physician activists, the *Bundesärztetag* issued the so-called Nuremberg Declaration, in which it acknowledged and apologized for the role German medicine played in National Socialism.[57]

Meanwhile, the different areas of research into the role of physicians in the Third Reich have proliferated immensely since the 1980s. They focus on such disparate subjects as the systematic political and ideological involvement of physicians and allied health workers in NS government and crime; biographies of victims and perpetrators of NS policies; histories of individual universities, their faculties, and departments; as well as analyses of medical disciplines and professional medical associations. Two recent publications can serve as introductions for new readers to the field: a bibliography assembled by Robert Jütte, Hans-Walter Schmuhl, Wolfgang Uwe Eckart and Winfried Süss, and a monograph by Wolfgang Uwe Eckart. Neither of these books provide a complete overview of the fast growing field.[58] The latest additions of in-depth analyses include Sascha Topp's study of the meaning of the rise in historical research activity and its results for German postwar medicine, a collection of essays on the impact of NS medicine on bioethics by Roelcke and coauthors, and a first comprehensive account on the victims of human experimentation by Weindling.[59]

Research on the History of Anatomy in the Third Reich

If research on the history of medicine during National Socialism in general had a slow start after the war, it was even slower for the medical discipline of anatomy. The causes of this phenomenon have yet to be investigated, but may partially lie in the high NSDAP membership of anatomists and their complex connections with the NS authorities through the use of bodies from NS victims for anatomical purposes. First historical studies of anatomical institutes were not published before 1990 and a preliminary overview of the field appeared only twenty years later.[60] The apparent "lack of interest" in anatomy's NS history is all the more striking as anatomist August Hirt's name came up during the Nuremberg Doctors' Trial in 1947 and thus the involvement of at least one anatomist in war crimes was known to a wider public.[61] Even before then in 1945, histological specimens from Hirt's research on bodies of NS victims had been discovered and discussed in a French scientific journal.[62] At that time the military authorities of the Allied Forces and families of political dissidents and other NS victims were looking for the bodies of foreign nationals or relatives in German anatomical institutes, thus testifying to the fact that the public was well aware of the anatomical departments' involvement with NS institutions.[63] And while Johann Paul Kremer's Auschwitz diary had been extensively quoted during his trial in 1960, and indeed had been historically the first document providing proof for medical experiments in concentration camps, Kremer's deeds[64] had no greater echo within German historiography.

On the other hand, the specter of the "sinister anatomist" could be found in various works of fiction after the war.[65] Polish author Zofia Nał-kowska and German author Rolf Hochhuth achieved a greater societal resonance with this topic based on their personal experience with anatomy in the Third Reich. In 1946 Nałkowska published the popular essay collection *Medaliony*, in which she reflected on her experiences as a member of the Commission for the Investigation of Nazi War Crimes in Poland. This included a graphic description of her encounter with the remains of NS victims during a visit to the anatomical institute in Danzig/Gdańsk. Hochhuth published two works in 1963 that referred to the involvement of German anatomists in NS iniquities. His play *Der Stellvertreter* (*The Representative*) accused Pope Pius XII of a lack of support for those persecuted by the NS regime, and one of the minor characters in this drama was called Hirt. In the play, Hochhuth reveals detailed knowledge of the anatomists' interest in a skull collection and Hirt's close connection with Himmler.[66] In his novella *Die Berliner Antigone*, set during WWII, Hochhuth's main character is a medical student who steals the body of her brother, an officer executed for treason, from an anatomical institute and buries him in a cemetery.[67] The story shows Hochhuth's insight into the connection between execution sites and anatomical institutes. His first wife Marianne Heinemann Sideri was the daughter of Rose Schlösinger, one of the women of the dissident group *Rote Kapelle*. Schlösinger was executed on 5 August 1943 and dissected by the anatomist Hermann Stieve.[68]

Postwar historiography in anatomy rarely mentions the period of the Third Reich, except for a brief paragraph on the topic by Wolf-Heidegger and Cetto in their history of artistic depiction of anatomical dissection.[69] Gerhard Wolf-Heidegger may have been more aware of the subject than other anatomists because he had been dismissed from his position in Bonn for so-called "racial" reasons in 1933.[70] A 1965 history of the *Anatomische Gesellschaft* by Robert Herrlinger did not mention politics or the dismissal of colleagues. This finally changed in 1986, when Schierhorn published an article on the fate of dismissed anatomical colleagues, and Kühnel included comments on the NS period in his reflections on the hundredth anniversary of the society in 1989.[71]

The first incentive to take a closer look at the anatomical institutes came in the 1980s, when Aly's and Müller-Hill's inquiries into the history of the Max Planck Institute (MPI) revealed the connection of the MPI's predecessor, the Kaiser Wilhelm Institute (KWI), with NS atrocities and the continued presence of tissue specimens that stemmed from NS victims in MPI collections. A contentious argument ensued between Aly and the MPI, first about access to the MPI archives and then concerning Aly's documentation of specimens deriving from thirty-three NS victims in a collection assembled by the neuropathologist Julius Hallervorden.[72] The very public nature of this controversy contributed to an increasing interest by students and others concerning the possible continued use of specimens deriving from NS victims in German universities, especially in departments of anatomy. Aly also published the war diary of anatomist Hermann Voss, chairman of anatomy at the University of Posen/Poznan. Voss' journal revealed him as an anti-Semite who hated Poles and used bodies of executed Poles and Jews not only for teaching and research purposes, but also for the commercial production of bone specimens and death masks.[73] The diaries mention his assistant Robert Herrlinger, with whom Voss had published an abbreviated manual of anatomy for medical students, which was still well known in the 1980s. The connection of these two anatomists' names with NS atrocities may have contributed to students' interest in finding out whether they were still learning not only from textbooks authored by men with a questionable biography, but whether they were also studying anatomical specimens derived from NS victims. Furthermore during these years, an international discussion about the use of data derived from NS coercive medical experiments led to an inquiry into the origin of brain specimens from the Vogt archive, a collection of neuroanatomical specimens assembled by Oskar and Cécile Vogt, now housed at the Institute for Brain Research at the University of Düsseldorf. The possibility that two of these tissue samples originated from NS victims could not be excluded.[74]

Tübingen became one of the first focal points of research in the history of anatomy in the Third Reich, starting with an inquiry into the provenance of a mass burial called *Gräberfeld X* in the municipal cemetery. Shortly after the war the public had been made aware that the remains of NS victims whose bodies had been used for anatomical purposes were buried in this grave.[75] Following requests by the *Vereinigung der Verfolgten des Naziregimes* (association of persons persecuted by the NS regime), a memorial stone was erected at the site in 1952. This memorial held no reference to the individual victims or their history. In 1963 a plate was added that declared this place to be the burial of "several hundred persons from various countries, who died in camps and institutions."[76] When in 1980 the city started to refurbish the grave, the public asked questions about the identity of those who were buried there. In a subsequent search, the names of more than one thousand persons whose bodies had been delivered to the anatomical department were found in the body registers of the anatomical institute, among them more than five hundred NS victims. All these names were engraved on six new bronze markers.[77] Finally, in 1986, the city of Tübingen engaged historian Benigna Schönhagen for a detailed investigation of the connection between the grave, the origin of the bodies, and their use at the anatomical institute. Schönhagen's study became the first of its kind and is still among the most detailed. However, she found that many documents were missing, probably destroyed after the war or held in inaccessible archives, and that in many cases the history of certain victim groups—e.g., those who died in psychiatric institutions outside the so-called "euthanasia" program—had never been investigated.[78] In 1988, Tübingen medical students organized a lecture series, in which anatomist Ulrich Drews addressed the problem of body procurement during National Socialism. They also demanded more information about this history from the university leadership. A constant observer and commentator of these controversies was the author Hans-Joachim Lang.[79]

Late in 1988 the subject of the potential continued existence of specimens from NS victims in anatomical institutes was taken up first by local and later national television and led to a strong nationwide and international response. An inquiry by the Israeli minister for religion, Zevelun Hammer, induced German chancellor Helmut Kohl to order all state ministries to instruct the anatomical institutes to inspect their collections and to remove any specimens derived from NS victims.[80] Hammer also demanded the return of all remains of NS victims to the Jewish community for proper burial.[81] Thus in January 1989, the Standing Conference of the Ministers of Culture in the German States (*Kultusministerkonferenz*) sent inquiries to all German universities concerning the existence of specimens from NS victims in their collections. Later, the same question was put to the MPI. The final report by the *Kultusministerkonferenz* from 1994 lists the replies from the universities in all states of the newly reunited Germany and reveals a highly variable level of compliance by the individual institutes.[82] Some launched closer inspections of their collections, whereas others just pointed to the fact that their buildings had been destroyed during the war or that no documentation was available. According to the report, two states still held specimens from NS victims, and seven more owned specimens that were not clearly identifiable as such but possibly belonged to this group. The MPI also held such remains. These specimens were removed from the collections and buried without further study. Weindling and others criticized this course of action, as no priority had been given to a detailed historical investigation and identification of the specimens, which were usually interred anonymously.[83] Despite existing guidelines,[84] there are still problems with the handling of historic human remains, as the events around the discovery of human bones found in the vicinity of the building that used to house the KWI anthropology in Berlin Dahlem in 2014 show. The potential historic roots of these remains were not communicated, and the bones were cremated without an attempt at possible identification and appropriate burial.[85] Furthermore, former German anatomical departments in the occupied territories were not included in this inquiry and some remain uninvestigated. Recent research by Raphael Toledano has led to the discovery of tissues from Hirt's victims at the University of Strasbourg.[86]

The earlier historical studies of anatomical departments were mostly initiated by outside agencies, including international ones, not by the anatomists themselves. Indeed, older anatomists often still frowned upon colleagues who launched such investigations. In the 1990s Gerhard Aumüller, chairman of anatomy at the University of Marburg, was still chided by colleagues as "fouling his own nest" when he spoke of first research results gathered with historian Cornelia Grundmann on the history of anatomy during NS times.[87] In anatomy, the tide only truly turned with the advent of the new millennium, more than fifty years after the end of the war. Since then, studies of individual anatomists and anatomical institutions have been published in increasing numbers, information on professional medical societies has become available, and lately the focus has turned on the two groups of victims in NS anatomy: the anatomists, whose careers were disrupted by NS policies, and the NS victims, whose bodies were used for anatomical purposes.

Key Terms in the History of Anatomy in National Socialism

Starting with the first systematic study of the practices of an anatomical institute during the Third Reich, the university in Tübingen, it emerged that key terms used in the description of this history had to be clearly defined in order to understand the impact and extent of the influence of the NS regime on the field of anatomy.

The term "victim of the National Socialist regime" is often discussed controversially; for some it implies a weakness of the person who suffered grievous harm.[88] Some "victims" prefer to call themselves "survivors," a term that connotes strength rather than weakness. The US Holocaust Memorial Museum has accepted both terms in its communications with the public.[89] In the history of anatomy in the Third Reich, however, there are no survivors, as the bodies of persons killed by the regime were used for anatomical purposes. The only survivors in this history are the families of the victims. Thus the term "victim" is the only appropriate one to use in this context. During the discussions between the external commission and the anatomical department in Tübingen it became obvious that they used different definitions of the term "victim of the National Socialist regime".[90] The anatomical department had originally worked with the official definition employed by the Federal Republic of Germany in its Federal Law for the Compensation for Victims of National Socialist Persecution

(BEG). This law holds that a "victim of NS persecution" must have been persecuted because of political dissent against National Socialism, "race", religion or personal beliefs and have suffered harm through this persecution to body, health, freedom, property, personal assets, or to the person's professional or economical career.[91] The external commission overseeing the investigation in Tübingen, however, pointed out that, whereas the BEG definition applied to most victims whose bodies were used for anatomical purposes, it left out several other groups of victims with questionably "unnatural" deaths: suicides, victims of the fast-acting special courts (*Sondergerichte*), inmates of special camps (prisoners of war, forced laborers), psychiatric institutions, and nursing centers. Thus the narrow BEG definition was extended by the Tübingen commission to include bodies of persons with questionable causes of death from these particular sources. Other authors[92] have excluded persons executed for so-called "serious crimes" (*Schwerverbrecher*) from their definition of NS victim, thus differentiating them from persons who were executed for political reasons. However, given the fact that execution rates for capital offenses like murder had dropped to very low rates in the Weimar Republic and women were not executed at all, it is reasonable to view those executed for "serious crimes" also as victims of the NS regime, as most of them would not have been executed before 1933. During the Weimar Republic capital punishment verdicts were handed down in 1,061 cases[93] and led to a total of 184 executions of men.[94] Executions were suspended from 1928 to 1929 due to an active public discussion about the abolition of the death penalty, and were still rare thereafter.[95] After 1933 the death penalty rose exponentially. Estimates of total numbers vary from 16,000 civilian death sentences of which 90 percent were carried out[96] to 16,569 with more than 12,000 executed[97] and 17,383 with 11,881 executed by 1944.[98] In addition there were a minimum of 16,000 military cases and possibly as many as 25,000, with an execution rate around 90 percent.[99] The wider perception of the term "NS victim" seems particularly advisable as the meaning of "serious crime" was constantly extended by the NS regime throughout its tenure, to include lesser crimes such as looting and mail theft, i.e., crimes that only under the NS regime led to executions.[100] Murderers made up only a small fraction of the persons whose bodies were delivered to the anatomical departments during the Third Reich, and even then some of the murders had a political background.[101] Several of the memorial centers at former NS execution sites in Germany share the wider definition of the term,[102] as they base their definition on the commonly held modern perception that capital punishment is inhumane.

It is necessary to define the words "race" and "racial" to understand their impact on the lives of many victims. "Race" is a multifaceted term and has meant different things to different people throughout the course of history.[103] In the discussion of the history of anatomy in the Third Reich, "race" will be used exclusively in the context of the NS biologistic and racist terminology, which is distinct from other possible meanings of the term, e.g., "ethnicity".[104] To mark the author's distance to the racist and discriminatory use of the term, the words "race" and "racial" will be used in quotation marks throughout, except in the terms racial hygiene, racial anthropology, racial genetics and racial biology. The term "Jew" in the National Socialist context did not refer to a religious, "ethnic" or "biological" entity. Individuals identified as Jews by the National Socialists did not share a common biology, ethnicity, or religion. At least since the ratification of the law of 11 April 1933 and its specification on 14 November 1935, the NS official definition of who was considered to be a Jew was ultimately a biologistic and administrative one, and based on ancestry: whoever had a certain number of documented ancestors of Jewish religion was called a Jew (or "non-Aryan"), whatever the practiced religion or political and national affiliations of the person in question were. Included in the group of the persecuted were also persons with spouses of Jewish descent. Decisions concerning employment of Jews and their spouses were based on the 1933 *Gesetz zur Wiederherstellung des Berufsbeamtentums* (law on the restoration of the civil service) and the 1935 *Nürnberger Gesetze* (Nuremberg race laws).[105]

The word "euthanasia" was coined in the fifth century BCE and described a "good death" or "easy death," sometimes also a "quick" or "honorable death".[106] In the language of National Socialism the term was associated with the so-called "mercy-killing" of all those deemed unfit to continue living as part of the German people. Since the end of WWI economical arguments were combined with ideological-biologistic ones and led to plans ranging from forced sterilization to the NS "euthanasia" program.[107] This killing program was authorized by Adolf Hitler in October 1939 and included the murder of children and adults hospitalized in German psychiatric institutions. Jewish patients were among those specially targeted, as were concentration camp inmates. They died through shootings, gas poisoning, over- and undermedication, starvation, and general neglect. Even after the official program was halted in 1941, the murders continued. Careful estimates arrive at a number of 200,000 to 300,000 victims.[108] These deaths were anything but "good" or "easy," they were not voluntary endings of life but murders executed by a merciless regime. For this reason many historians use the term "euthanasia" in reference to the NS program in quotation marks and this practice is adopted here.[109]

The term "material" for studies of tissues from animals and humans is common and exact phrasing in scientific anatomical literature, and generally used without quotation marks. However, in the context of executions by a criminal regime like the National Socialists, addressing the tissues of the executed as "material" or "*Werkstoff*" (literally: stuff to work with) has been interpreted as acquiescence of the researcher with the demeaning purposes of a regime that aimed at the depersonalization of the victim.[110] Thus the term "material" is used in quotation marks throughout this book to mark a distancing to the depersonalization potentially implied in the word, and to demonstrate respect for the persons whose tissues were used for research without their voluntary consent.

The phrase "German anatomical department during National Socialism" is used for all functional anatomical departments of universities that fell under German jurisdiction at any time between 1933 and 1945. These include not only the institutes existing in Germany in 1933, but also the ones in the occupied and annexed territories and states. As several cities have different names in different languages, the more common versions of the names in the literature and the historical documents or their English versions are used here: Posen for Poznan, Danzig for Gdańsk, Breslau for Wrocław, Dorpat for Tartu, Strasbourg for Strassburg, Cologne for Köln, Munich for München, Nuremberg for Nürnberg, Brünn for Brno, Prague for Prag, Pressburg for Bratislava, Terezin for Theresienstadt, Crakow for Kraków, Warsaw for Warszawa, Königsberg for Kaliningrad. Any quotes from German originals were translated by the author.

Notes

1. Powers 2012, 60.
2. For a review see: Eley 2003a,b.
3. Roelcke 2007, 226.
4. Kater 1987a; Kater 1987b, 300.
5. Mitscherlich and Mielke 1947; Gerst 1994; Peter 1994.
6. Pross 1991.
7. Quoted in Sehn 1959, 30; see also: Peter 1994, 245.
8. Gerst 1994, A-1617.
9. Wuttke-Groneberg 1980.
10. Mitscherlich and Mielke 1960; Platen-Hallermund 1948; Leibbrand 1946.
11. Kater 1987b; Zimmermann; Baader 1999; Beushausen et al. 1998; Caplan 2010; Roelcke 2014a.
12. Mitscherlich and Mitscherlich 1977.
13. Esch 1951; Kater 1989, 222–23; Hofer and Leven 2003.
14. Walther 2008, 49.
15. Baader and Schultz 1987, 17.
16. Pross and Aly 1989; Bleker and Jachertz 1989; Kolb and Seithe 1998.
17. Roelcke 2007, 229–32.
18. Weinberger 2009; Landgericht Münster 1960.
19. Sehn 1959, 31; Peter 1994, 263.

20. Peter 1994, 259.
21. Mayer 1961; Mayer 1966.
22. Doneith 2007, 164.
23. Conrad-Martius 1955; Honolka 1961; Saller 1961; Dörner 1967; Kaul 1968.
24. E.g., Stöhr 1947.
25. Charvát 1946.
26. Kogon 2006; Nyiszli 1993.
27. Alexander 1949a; Alexander 1949b.
28. Alexander 1949a, 47.
29. Bayle 1950; Blacker 1952; Sehn 1959.
30. Callahan et al. 1976.
31. Kater 1987a, 31.
32. Pross 1991, 14.
33. Weindling 1991; Wuttke 1989.
34. Kater 1987b, 301; Jütte et al. 2011, 313.
35. Roelcke 2007, 232–34.
36. Weindling 1991, 416.
37. Baader and Schultz 1987, 9.
38. Roelcke 2014b, 259.
39. Wuttke-Groneberg 1980.
40. Baader and Schultz 1987.
41. Kater 1966; Kater 1974; Kater 1981; Kater 1983; Kater 1987b, Kater 1989.
42. Kudlien 1985; Bader and Schultz 1987; Winau and Hafner 1974; Roth 1981; Roth 1984a; Roth, ed., 1984b.
43. Rapoport and Thom 1989.
44. Thom and Caregorodcev 1989.
45. Müller-Hill 1984; Bock 1986; Schmuhl 1987; Weingart et al. 1992; Lifton 1986; Proctor 1988; Weindling 1989; Caplan 1992; Annas and Grodin 1992; Berg and Cocks 1997.
46. E.g., Klee 1983; Klee 1985; Klee 2003; Aly 1985; Aly 1987; Pross and Aly 1989.
47. Hanauske-Abel 1986.
48. Stock 1987; Peter 1994.
49. Roelcke 2014a.
50. Toellner 1989.
51. Jütte et al. 2011, 10.
52. Deutsches Ärzteblatt 2012.
53. Ebbinghaus and Dörner 2001; Roelcke 2014a.
54. Hoppe and Vilmar 2010.
55. Seidelman 1996; Seidelman 2014; Kater 1997.
56. Hohendorf 2010; Hohendorf 2014.
57. In Remembrance 2012; Reis 2012; Roelcke 2014a.
58. Jütte et al. 2011; Eckart 2012.
59. Topp 2013; Roelcke et al. 2014; Weindling 2015.
60. Hildebrandt 2009a; Hildebrandt 2009b; Hildebrandt 2009c.
61. Mitscherlich and Mielke 1947.
62. Champy and Risler 1945.
63. Hildebrandt 2013b.
64. Landgericht Münster 1960; Höss et al 1984.
65. E.g., Nałkowska 2000; Hochhuth 1967.
66. Hochhuth 1967, 43–44.
67. Hochhuth 2011.
68. Fornaro 2012; Hildebrandt 2013a.

69. Wolf-Heidegger and Cetto 1967.
70. Höpfner 1999.
71. Herrlinger 1965; Schierhorn 1986; Kühnel 1989.
72. Aly 1985; Aly 1988; Aly 1989; Müller-Hill 1984; Weindling 2012.
73. Aly 1987 and 1994; Aly 2003.
74. Bogerts et al 1988.
75. Schönhagen 1987, 9.
76. Ibid., 13.
77. Ibid., 16.
78. Ibid., 20–21.
79. Schwäbisches Tageblatt 1989; Weindling 2012.
80. Weindling 2012; Seidelman 2012.
81. Dickman 1989.
82. Kultusministerkonferenz 1994.
83. Weindling 2012.
84. E.g., Bundesärztekammer 2003; World Archeological Congress 2009.
85. Campus 2015; Kühne 2015; Aly 2015.
86. Bever 2015
87. Personal communication, Gerhard Aumüller; Aumüller 1991; Grundmann and Aumüller 1996.
88. Fangerau and Krischel 1996.
89. USHMM 2014.
90. Universität Tübingen 1990.
91. BEG 2009.
92. E.g., Schultka and Viebig 2012.
93. Evans 1996, 915–16; 1,141 according to Siebenpfeiffer 2005.
94. Evans 1996, 915–16.
95. Ibid., 561.
96. Wagner 1974; Messerschmidt and Wüllner 1987.
97. Oleschinski 2002, 7.
98. Seeger 1998.
99. Evans 1996; Wagner 1974; Messerschmidt and Wüllner 1987.
100. Hildebrandt 2009b.
101. Hildebrandt 2013a.
102. Keller 2011.
103. Fuller explorations of the topic can be found, for example, in Banton 1998, and Hutton 2005.
104. Weingart et al. 1992; Krohn et al. 1998; Hutton 2005; for a discussion of the terms "race" and "ethnicity" see Coker 2001.
105. Mendes-Flohr and Reinharz 2011, 722, 734.
106. Benzenhöfer 2009, 13–19
107. Binding and Hoche 1920; Friedlander 1995.
108. Faulstich 2000.
109. E.g., Klee 1985; Roelcke et al. 2001.
110. E.g., Klee 2004.

Bibliography

Alexander, Leo. 1949a. "Medical Science under Dictatorship." *The New England Journal of Medicine* 241(2): 39–47.

————. 1949b. "The Molding of Personality under Dictatorship: The Importance of the Destructive Drives in the Socio-Psychological Structure of Nazism." *Journal of Criminal Law and Criminology* (1931–51), 40(1): 3–27.

Aly, Götz. 1985. "Der saubere und der schmutzige Fortschritt." In: *Reform und Gewissen: "Euthanasie" im Dienst des Fortschritts*, ed. Götz Aly, 9–78. Berlin: Rotbuch Verlag.

————. 1987. "Das Posener Tagebuch des Anatomen Hermann Voss." In: *Biedermann und Schreibtischtäter: Materialien zur deutschen Täter-Biographie*, ed. Götz Aly, Peter Chroust, and Christian Pross, 15–66. Berlin: Rotbuch Verlag.

————. 1988. "Forschen an Opfern: Das Kaiser-Wilhelm-Institut für Hirnforschung und die 'T4.'" In: *Aktion T4 1939–1945. Die "Euthanasie"-Zentrale in der Tiergartenstrasse 4*, ed. Götz Aly, 153–59. Berlin: Edition Hentrich.

————. 1989. "Je mehr, desto lieber." *Die Zeit*, 3 February. Accessed 26 September 2013. http://www.zeit.de/1989/06/je-mehr-desto-lieber.

————. 1994. "The Posen Diaries of the Anatomist Hermann Voss." In: *Cleansing the Fatherland: Nazi Medicine and Racial Hygiene*, ed. Götz Aly, Peter Chroust, and Christian Pross, 99–155. Baltimore: The Johns Hopkins University Press.

————. 2003. *Rasse und Klasse: Nachforschungen zum deutschen Wesen*. Frankfurt am Main: S. Fischer Verlag GmbH.

————. 2015. "Geistlos und roh." *Berliner Zeitung*, 2 February. Accessed 2 February 2015. http://www.berliner-zeitung.de/meinung/kolumne-zur-freien-universitaet-geistlos-und-roh-an-der-fu-berlin,10808020,29728282.html .

Annas, George J. D., and Michael A. Grodin. 1992. *The Nazi Doctors and the Nuremberg Code: Human Rights in Human Experimentation*. New York, Oxford: Oxford University Press.

Aumüller, Gerhard. 1991. "Die Anatomie in der NS-Zeit." In: *"bis der langersehnte Umschwung kam": Von der Verantwortung der Medizin unter dem Nationalsozialismus*, ed. Fachschaft Medizin der Philipps-Universität Marburg, 87–112. Marburg: Schuren.

Baader, Gerhard. 1999. "Die Erforschung der Medizin im Nationalsozialismus als Fallbeispiel einer kritischen Medizingeschichte." In: *Eine Wissenschaft emanzipiert sich: Die Medizinhistoriographie von der Aufklärung bis zur Postmoderne*, ed. Ralf Bröer, 113–20. Pfaffenweiler: Centaurus-Verlagsgesellschaft.

Baader, Gerhard, and Ulrich Schultz. 1987. *Medizin und Nationalsozialismus: Tabuisierte Vergangenheit- Ungebrochene Tradition? Dokumentation des Gesundheitstages Berlin 1980*. 3rd ed. Frankfurt: Dr. Med. Mabuse.

Banton, Michael. 1998. *Racial Theories*. 2nd ed. Cambridge: Cambridge University Press.

Bayle, Francois. 1950. *Croix Gammée contre Caducée: Les Expériences Humaines en Allemagne pendant la Deuxième Guerre Mondiale*. Neustadt: Centre De L'Imprimerie Nationale a Neustadt (Palatinat).

BEG. 2009. "Bundesgesetz zur Entschädigung für Opfer der nationalsozialistischen Verfolgung." Accessed 7 March 2013. http://www.gesetze-im-internet.de/beg/BJNR01387 0953.html#BJNR013870953BJNG000100328.

Benzenhöfer, Udo. 2009. *Der Gute Tod? Geschichte der Euthanasie und Sterbehilfe*. Göttingen: Vandenhoeck & Rupprecht.

Berg, Manfred, and Geoffrey Cocks, eds. 1997. *Medicine and Modernity: Public Health and Medical Care in Nineteenth- and Twentieth-Century Germany*. Cambridge: Cambridge University Press.

Beushausen, Ulrich, Hans-Joachim Dahms, Thomas Koch, Almut Massing, and Konrad Obermann. 1998. "Die medizinische Fakultät im Dritten Reich." In: *Die Universität Göttingen unter dem Nationalsozialismus*, ed. Heinrich Becker, Hand Joachim Dahms, and Cornelia Wegeler, 183–286. München: K.G. Saur.

Bever, Lindsey. 2015. "Remains of Holocaust experiment victims found at French forensic institute." The Washington Post July 22 2015. http://www.washingtonpost.com/news/

morning-mix/wp/2015/07/22/remains-of-holocaust-victims-used-as-guinea-pigs-found-at-french-forensic-institute/. Accessed 12 August 2015.

Binding, Karl, and Alfred Hoche. 1920. *Die Freigabe der Vernichtung unwerten Lebens.* Leipzig: Verlag Felix Meiner. Accessed 3 April 2014. http://www.staff.uni-marburg.de/~rohrmann/Literatur/binding.html English excerpts. Accessed 3 April 2014. http://germanhistorydocs.ghi-dc.org/sub_document.cfm?document_id=4496 .

Blacker, Carlos Paton. 1952. "'Eugenic' Experiments Conducted by the Nazis on Human Subjects." *The Eugenics Review* 44: 9–19.

Bleker, Johanna, and Norbert Jachertz, eds. 1989. *Medizin im "Dritten Reich."* Köln: Deutscher Ärzteverlag.

Bock, Gisela. 1986. *Zwangssterilisation im Nationalsozialismus: Studien zur Rassenpolitik und Frauenpolitik.* Opladen: Westdeutscher Verlag.

Bogerts, Bernhard, Bernd-M. Becker, Martina Krüger, Elliot S. Gershon, and Margaret R. Hoehe. 1988. "Letters to the Editor: the Brains of the Vogt Collection." *Archives of Genetics and Psychiatry* 45: 774–76.

Bundesärztekammer. 2003. "Arbeitskreis 'Menschliche Präparatesammlungen': Empfehlungen zum Umgang mit Präparaten aus menschlichem Gewebe in Sammlungen, Museen und öffentlichen Räumen." *Deutsches Ärzteblatt* 8: 378–83, English translation. Accessed 5 February 2015. http://www.aemhsm.net/ressources/actus/TranslationGuidelines_final.doc.

Callahan, Daniel, Arthur Caplan, Harold Edgar, Laurence McCullough, Tabitha Powledge, Margaret Steinfels, Peter Steinfels, Robert M. Veatch, Joseph Walsh, Joel Colton, Lucy C. Davidovitcz, Milton Himmelfarb, and Telford Taylor. 1976. "Biomedical Ethics and the Shadow of Nazism: A Conference on the Proper Use of the Nazi Analogy in Ethical Debate, April 8, 1976." *Hastings Center Report* 6 (4 August 1976): 1–19.

Campus. 2015. "Unhaltbare Vorwürfe." *Campus.leben,* 4 February 2015, accessed 4 February 2015. http://www.fu-berlin.de/campusleben/campus/2015/150204_interview-alt-knochen/index.html.

Caplan, Arthur L, ed.. 1992. *When Medicine Went Mad: Bioethics and the Holocaust.* Totowa, NJ: Humana Press.

Caplan, Arthur L. 2010. "The Stain of Silence: Nazi Ethics and Bioethics." In: *Medicine after the Holocaust. From the Master Race to the Human Genome and Beyond,* ed. Sheldon Rubenfeld, 83–99. New York: Palgrave MacMillan.

Champy, Christian, and Dr. Risler. 1945. "Sur une Série de Préparations Histologiques Trouvées das la Laboratoire d'un Professeur Allemand: Expériences Faites sur L'Homme au Camp de Struthof." *Bulletin de l'Académie de Médicine* 129: 263–65.

Charvát, Josef. 1946. *Medical Science Abused: German Medical Science as Practised in Concentration Camps and in the So-Called Protectorate.* Prague: Orbis.

Coker, Naaz. 2001. "Understanding Race and Racism." In: *Racism in Medicine: An Agenda for Change,* ed. Naaz Coker, Miachel Mansfield, and Julia Neuberger, 1–21. London: King's Fund Publishing.

Conrad-Martius, Hedwig. 1955. *Utopien der Menschenzüchtung. Der Sozialdarwinismus und seine Folgen.* Munich: Köselverlag.

Deutsches Ärzteblatt. 2012. "Forschungspreis: Rolle der Ärzteschaft in der NS-Zeit." *Deutsches Ärzteblatt* 109(45): A-2219/B-1811/C-1775.

Dickman, Steven. 1989. "Scandal over Nazi Victims' Corpses Rocks Universities." *Nature* 337: 195.

Dörner, Klaus. 1967. "Nationalsozialismus und Lebensvernichtung." *Vierteljahreshefte für Zeitgeschichte* 15: 121–52.

Doneith, Thorsten. 2007. "August Mayer: Direktor der Universitätsfrauenklinik Tübingen 1917–1949." MD diss., University of Tübingen. Accessed 19 September 2013. http://tobias-lib.uni-tuebingen.de/volltexte/2007/2963/.

Ebbinghaus, Angelika, and Klaus Dörner, eds.. 2001. *Vernichten und Heilen: Der Nürnberger Ärzteprozess und seine Folgen*. Berlin: Aufbau-Verlag.

Eckart, Wolfgang Uwe. 2012. *Medizin in der NS-Diktatur: Ideologie, Praxis, Folgen*. Wien: Böhlau Verlag.

Eley, Geoff. 2003a. "Hitler's Silent Majority? Conformity and Resistance under the Third Reich, Part 1." *Michigan Quarterly Review* 42(2): 389–425

Eley, Geoff. 2003b. "Hitler's silent majority? Conformity and Resistance under the Third Reich, Part 2." *Michigan Quarterly Review* 42(3): 550–83

Esch, Peter. 1951. "Werden, Vergehen und Wiedererstehen der medizinischen Fakultät der Universität Münster (Westfalen)." In: *25 Jahre Medizinische Fakultät der Universität Münster*, ed. Universität Münster, 13–16. Münster: Verlag Aschendorff.

Evans, Richard J. 1996. *Rituals of Retribution: Capital Punishment in Germany 1600–1987*. Oxford: Oxford University Press.

Fangerau, Heiner, and Mathis Krischel. 2011. "Der Wert des Lebens und das Schweigen der Opfer: Zum Umgang mit den Opfern nationalsozialistischer Verfolgung in der Medizinhistoriographie." In: *NS-"Euthanasie" und Erinnerung: Vergangenheitsaufarbeitung— Gedenkformen—Betroffenenperspektiven*, ed. Stephanie Westermann, Richard Kühl, and Tim Ohnhäuser, 19–28. Berlin: Lit Verlag.

Faulstich, Heinz. 2000. "Die Zahl der "Euthanasie"-Opfer." In: *Die historischen Hintergründe medizinischer Ethik*, ed. Andreas Frewer and Clemens Eickhoff, 218–29. Frankfurt: Campus-Verlag.

Fornaro, Sotera. 2010. "Hochhuth, Rose Schlösinger, Sophokles: "Die Berliner Antigone." In: *Rolf Hochhuth: Theater als politische Anstalt. Tagungsband mit einer Personalbiographie*, ed. Ilse Nagelschmidt, Sven Neufert, and Gert Ueding, 197–208. Weimar: Dr. A.J. Denkena Verlag.

Friedlander, Henry. 1995. *The Origins of Nazi Genocide: From Euthanasia to the Final Solution*. Chapel Hill: University of North Carolina Press.

Gerst, Thomas. 1994. "'Nürnberger Ärzteprozess' und ärztliche Standespolitik: Der Auftrag der Ärztekammern an Alexander Mitscherlich zur Beobachtung und Dokumentation des Prozessverlaufs." *Deutsches Ärzteblatt* 91: A-1606–22.

Grundmann, Kornelia, and Gerhard Aumüller. 1996. "Anatomen in der NS-Zeit: Parteigenossen oder Karteigenossen? Das Marburger anatomische Institut im Dritten Reich." *Medizinhistorisches Journal* 31 (3–4): 322–57.

Hanauske-Abel, Hartmut. 1986. "From Nazi Holocaust to Nuclear Holocaust: a Lesson to Learn?" *Lancet* 2: 271–73.

Herrlinger, Robert. 1965. "Kurze Geschichte der Anatomischen Gesellschaft." *Anatomischer Anzeiger* 117: 1–60.

Hildebrandt, Sabine. 2009a. "Anatomy in the Third Reich: An Outline, Part 1. National Socialist Politics, Anatomical Institutions, and Anatomists." *Clinical Anatomy* 22: 883–93.

———. 2009b. "Anatomy in the Third Reich: An Outline, Part 2. Bodies for Anatomy and Related Medical Disciplines." *Clinical Anatomy* 22: 894–905.

———. 2009c. "Anatomy in the Third Reich: An Outline, Part 3. The Science and Ethics of Anatomy in National Socialist Germany and Postwar Consequences." *Clinical Anatomy* 22: 906–15.

———. 2013a. "The Women on Stieve's List: Victims of National Socialism Whose Bodies Were Used for Anatomical Research." *Clinical Anatomy* 26: 3–21.

———. 2013b. "Current Status of Identification of Victims of the National Socialist Regime Whose Bodies Were Used for Anatomical Purposes." *Clinical Anatomy* 27: 514–36.

Hochhuth, Rolf. 1967. *Der Stellvertreter*. First published in 1963. Reinbeck: Rowohlt.

———. 2011. *Die Berliner Antigone: Erzählungen und Gedichte*. First published in 1961. Stuttgart: Reclam Universal-Bibliothek.

Höpfner, Hans-Paul. 1999. *Die Universität Bonn im Dritten Reich: Akademische Biographien unter nationalsozialistischer Herrschaft.* Bonn: Bouvier Verlag.

Höss, Rudolf, Pery Broad, and Johann Paul Kremer. 1984. *KL Auschwitz as Seen by the SS.* New York: Howard Fertig.

Hofer, Hans Georg, and Karl Heinz Leven. 2003. *Die Freiburger Medizinische Fakultät im Nationalsozialismus: Katalog einer Ausstellung des Instituts für Geschichte der Medizin der Universität Freiburg.* Frankfurt: Peter Lang.

Hohendorf, Gerrit. 2010. "Nachruf: Kein Hinweis auf die Rolle im Nationalsozialismus." *Deutsches Ärzteblatt* 107(31–32): A-1520/B-1350/C-1330.

———. 2014. "The Sewering Affair." In: *Silence, Scapegoats, Self-Reflection: The Shadow of Nazi Medical Crimes on Medicine and Bioethics,* ed. Volker Roelcke, Sascha Topp and Etienne Lepicard, 131–46. Göttingen: V&R unipress.

Honolka, Bert. 1961. *Die Kreuzelschreiber: Ärzte ohne Gewissen. Euthanasie im Dritten Reich.* Hamburg: Rütten und Loening.

Hoppe, Dietrich, and Karsten Vilmar. 2010. "Gestalter im Dienste der Ärzteschaft." *Deutsches Ärzteblatt* 107(28–29): A-1409/B-1247/C-1227.

Hutton, Christopher M. 2005. *Race and the Third Reich.* Cambridge: Polity Press.

In Remembrance. 2012. "In Remembrance of the Victims of Nazi Medicine. Nuremberg, May 2012." *The Israel Medical Association Journal* 14: 529–30.

Jütte, Robert, Wolfgang U. Eckart, Hans-Walter Schmuhl, and Winfried Süss. 2011. *Medizin im Nationalsozialismus: Bilanz und Perspektiven der Forschung.* Göttingen: Wallstein Verlag.

Kater, Michael H. 1966. "Das 'Ahnenerbe': Die Forschungs- und Lehrgemeinschaft in der SS. Organisationsgeschichte von 1935–1945." PhD diss., University of Heidelberg.

———. 1974. *Das "Ahnenerbe" der SS 1935–1945: Ein Beitrag zur Kulturpolitik des Dritten Reiches.* Stuttgart: Deutsche Verlagsgesellschaft.

———. 1981. "Die nationalsozialistische Machtergreifung an den deutschen Hochschulen: zum politischen Verhalten akademischer Lehrer bis 1939." In: *Die Freiheit des Anderen: Festschrift für Martin Hirsch,* ed. Hans Jochen Vogel, Helmut Simon, and Adalbert Podlech, 49–75. Baden-Baden: Nomos.

———. 1983. *The Nazi Party: A Social Profile of Members and Leaders, 1919–1945.* Cambridge: Harvard University Press.

———. 1987a. "The Burden of the Past: Problems of a Modern Historiography of Physicians and Medicine in Nazi Germany." *German Studies Review* 10: 31–56.

———. 1987b. "Medizin und Mediziner im Dritten Reich: Eine Bestandsaufnahme." *Historische Zeitschrift* 244(2): 299–352.

———. 1989. *Doctors under Hitler.* Chapel Hill: University of North Carolina Press.

———. 2002. "The Sewering Scandal of 1993 and the German Medical Establishment." In: *Medicine and Modernity: Public Health and Medical Care in Nineteenth- and Twentieth Century Germany,* ed. Manfred Berg and Geoffrey Cocks, 213–34. Cambridge: Cambridge University Press. First published in 1997.

Kaul, Friedrich Karl. 1968. *Ärzte in Auschwitz.* Berlin: VEB Verlag Volk und Gesundheit.

Keller, Claudia. 2011. "Mörder neben Widerstandskämpfern." *Der Tagesspiegel,* 18 January 2011. Accessed March 6, 2013. http://www.tagesspiegel.de/berlin/gedenkstaetteploet zensee-moerder-neben-widerstandskaempfern/3709364.html .

Klee, Ernst. 1983. *Euthanasie im NS-Staat: Die "Vernichtung lebensunwerten Lebens."* Frankfurt am Main: Fischer Verlag GmbH.

———. 2003. *Das Personenlexikon zum Dritten Reich: Wer war was vor und nach 1945?* Frankfurt am Main: S. Fischer.

———. 2004. *Auschwitz, die NS-Medizin und ihre Opfer.* 3rd ed. Frankfurt am Main: Fischer Verlag.

Klee, Ernst, ed. 1997. *Dokumente zur Euthanasie.* Frankfurt am Main: Fischer Taschenbuch-verlag GmbH. First published in 1985.

Kogon, Eugen. 2006. *Der SS-Staat: Das System der deutschen Konzentrationslager.* 43rd ed. München: Wilhelm Heine Verlag. First published in 1946.

Kolb, Stephan, and Horst Seithe, eds. 1998. *Medizin und Gewissen. 50 Jahre nach dem Nürnberger Ärzteprozess-Kongressdokumentation.* Frankfurt am Main: Mabuse-Verlag.

Krohn, Klaus-Dieter, Patrick von zur Mühlen, Gerhard Paul, and Lutz Winckler, eds. 1998. *Handbuch der deutschsprachigen Emigration 1933–1945.* Darmstadt: Wissenschaftliche Buchgesellschaft.

Kudlien, Fridolf. 1985. *Ärzte im Nationalsozialismus.* Köln: Kiepenheuer und Witsch.

Kühne, Anja. 2015. "Umgang mit den Skelettfunden in Dahlem: Einfach eingeäschert." *Tagespiegel,* 26 January. Accessed 2 February 2015. http://www.tagesspiegel.de/wissen/umgang-mit-den-skelettfunden-in-dahlem-einfach-eingeaeschert/11278454.html.

Kühnel, Wolfgang. 1989. "100 Jahre Anatomische Gesellschaft." *Verhandlungen der Anatomischen Gesellschaft* 82: 31–75.

Kultusministerkonferenz. 1994. *Abschlussbericht "Präparate von Opfern des Nationalsozialismus in anatomischen und pathologischen Sammlungen deutscher Ausbildungs- und Forschungseinrichtungen." Bonn, den 15.01.1994.* Copy from the William Seidelman Collection of Papers.

Landgericht Münster. 1960. "Das Urteil gegen Dr. Johann Paul Kremer." In: *Justiz und NS-Verbrechen.* Band 17. Accessed 16 December 2013. http://web.archive.org/web/20081207153240/http://www1.jur.uva.nl/junsv/Excerpts/Kremer.htm.

Leibbrand, Werner. 1946. *Um die Menschenrechte der Geisteskranken.* Nürnberg: Die Egge.

Lifton, Robert Jay. 1998. *Ärzte im Dritten Reich.* German ed. Berlin: Ullstein Buchverlag GmbH & Co KG. First published in 1986.

Mayer, August. 1961. "Vom Geist der Medizin, vom Ungeist der Zeit und vom Wanken der Arztthrone." *Ärztliche Mitteilungen; nebst Anzeiger und wissenschaftlicher Beilage.* 46(10): 549–59.

———. 1966. "Arzttum im Dritten Reich." *Deutsches Ärzteblatt* 63(12): 785–87.

Mendes-Flohr, Paul, and Jehuda Reinharz. 2011. *The Jew in the Modern World: A Documentary History.* 3rd ed.. New York: Oxford University Press.

Messerschmidt, Manfred, and Fritz Wüllner. 1987. *Die Wehrmachtsjustiz im Dienste des Nationalsozialismus: Zerstörung einer Legende.* Baden-Baden: Nomos-Verlagsgesellschaft.

Mitscherlich, Alexander, and Fred Mielke. 1947. *Das Diktat der Menschenverachtung.* Heidelberg: Verlag Lambert Schneider.

———. 1997. *Medizin ohne Menschlichkeit: Dokumente des Nürnberger Ärzteprozesses.* Frankfurt: Fischer Taschenbuch Verlag. First published in 1960.

Mitscherlich, Alexander, and Margarete Mitscherlich. 1977. *Die Unfähigkeit zu trauern: Grundlagen kollektiven Verhaltens.* 2nd ed. München: Piper.

Müller-Hill, Benno. 1984. *Tödliche Wissenschaft.* Reinbek bei Hamburg: Rowohlt Taschenbuch Verlag.

Nałkowska, Zofia. 2000. *Medallions.* Evanston, IL: Northwestern University Press. First published in Polish 1946.

Nyiszli, Miklos. 1993. *Auschwitz: A Doctor's Eyewitness Account.* New York: Arcade Publishing. First published in 1946.

Oleschinski, Brigitte, ed. 2002. *Gedenkstätte Plötzensee.* Berlin: Gedenkstätte Deutscher Widerstand. Accessed 11 September 2014. http://www.gdw-berlin.de/fileadmin/bilder/publ/gedenkstaette_ploetzensee/englisch-screen.pdf .

Peter, Jürgen. 1994. *Der Nürnberger Ärzteprozess im Spiegel seiner Aufarbeitung anhand der drei Dokumentensammlungen von Alexander Mitscherlich und Fred Mielke.* Münster: Lit-Verlag.

Platen-Hallermund, Alice. 1948. *Die Tötung Geisteskranker.* Heidelberg: Verlag der Frankfurter Hefte.

Powers, Kevin. 2012. *The Yellow Birds.* New York: Little Brown and Company.

Proctor, Robert N. 1988. *Racial Hygiene: Medicine under the Nazis.* Cambridge: Harvard University Press.

Pross, Christian. 1991. "Breaking through the Postwar Coverup of Nazi Doctors in Germany." *Journal of Medical Ethics* 17(suppl.): 13–16.

Pross, Christian, and Götz Aly. 1989. *Der Wert des Menschen—Medizin in Deutschland 1918–1945.* Berlin: Edition Hentrich Berlin.

Rapoport, Samuel Mitja, and Achim Thom. 1889. *Das Schicksal der Medizin im Faschismus: Auftrag und Verpflichtung zur Bewahrung von Humanismus und Frieden.* Neckarsulm: Jungjohann

Reis, Shmuel. 2012. "Reflections on the Nuremberg Declaration of the German Medical Assembly." *The Israel Medical Association Journal* 14: 532–34.

Roelcke, Volker. 2007. "Trauma or Responsibility? Memories and Historiographies of Nazi Psychiatry in Postwar Germany." In: *Trauma and Memory: Reading, Healing and Making Law,* ed. Austin Sarat, Nadav Davidovitch, and Michael Alberstein, 225–42. Stanford: Stanford University Press.

———. 2014a. "Between Professional Honor and Self-Reflection: The German Medical Association's Reluctance to Address Medical Malpractice during the National Socialist Era, ca. 1985–2012." In: *Silence, Scapegoats, Self-Reflection: the Shadow of Nazi Medical Crimes on Medicine and Bioethics,* ed. Volker Roelcke, Sascha Topp and Etienne Lepicard, 243–78. Göttingen: V&R unipress.

———. 2014b. "Confronting Medicine during the Nazi Period: Autobiographical Reflections." In: *Human Subjects Research after the Holocaust,* ed. Sheldon Rubenfeld and Susan Benedict, 255–68. Cham: Springer.

Roelcke, Volker, Gerrit Hohendorf, and Maike Rotzoll, M. 2001. "Psychiatric Research and 'Euthanasia': The Case of the Psychiatric Department at the University of Heidelberg, 1941–1945." *Psychoanalytic Review* 88(2): 275–94.

Roelcke, Volker, Sascha Topp, and Etienne Lepicard, eds. 2014. *Silence, Scapegoats, Self-Reflection: The Shadow of Nazi Medical Crimes on Medicine and Bioethics.* Göttingen: V&R unipress.

Roth, Karl-Heinz. 1981. "Als ob nichts gewesen wäre: Zur Geschichte der Zwangssterilisierung seit der NS-Zeit." *Autonomie* 7: 43–60.

———. 1984a. "Grosshungern und Gehorchen: Das Universitätsklinikum Eppendorf." In: *Heilen und Vernichten im Mustergau Hamburg: Bevölkerungs- und Gesundheitspolitik im Dritten Reich,* ed. Angelika Ebbinghaus and Karl-Heinz Roth, 109–35. Hamburg: Konkret Literatur Verlag.

Roth, Karl-Heinz, ed. 1984b. *Erfassung zur Vernichtung: Von der Sozialhygiene zum "Gesetz über Sterbehilfe."* Berlin: Verlagsgesellschaft Gesundheit mbH.

Saller, Karl. 1961. *Die Rassenlehre des Nationalsozialismus in Wissenschaft und Propaganda.* Darmstadt: Progress-Verlag.

Schierhorn, Helmke. 1986. "Mitglieder der Anatomischen Gesellschaft im antifaschistischen Exil." *Verhandlungen der Anatomischen Gesellschaft* 80: 957–63.

Schmuhl, Hans-Walter. 1987. *Rassenhygiene, Nationalsozialismus, Euthanasie.* Göttingen: Vandenhoeck & Rupprecht.

Schönhagen, Benigna. 1987. "Das Gräberfeld X: Eine Dokumentation über NS-Opfer auf dem Tübinger Stadtfriedhof." *Kleine Tübinger Schriften.* Heft 11. Tübingen: Gulde-Druck GmbH.

Schultka, Rüdiger, and Michael Viebig. 2012. "The Fate of Bodies of Executed Persons in the Anatomical Institute in Halle between 1933 and 1945." *Annals of Anatomy* 194: 274–80.

Schwäbisches Tageblatt. 1989. "Weitere Reaktionen zu den NS-Präparaten im anatomischen Institut: Studenten präzisieren ihre Zweifel. Todenhöfer: Ethische Unerträglichkeit/ Kohl: Unerträglich und inakzeptabel." *Schwäbisches Tageblatt* 13 January 1989.

Seeger, Andreas. 1998. "'Gegen Schwerstverbrecher ist in Kriegszeiten die zugelassene Todesstrafe grundsätzlich die gebotene': Todesurteile des Sondergerichtes Altona/Kiel 1933–1945." In: *"Standgericht der inneren Front" Das Sondergericht Altona/Kiel 1932–1945*, ed. Robert Bohn, and Uwe Danker, 166–89. Hamburg: Ergebnisse Verlag.

Sehn, Jan. 1959. "Carl Claubergs verbrecherische Unfruchtbarmachungsversuche an Häftlingsfrauen in den Nazi-Konzentrationslagern." *Hefte von Auschwitz* 2: 3–32.

Seidelman, William E. 1996. "Nuremberg Doctors' Trial: Nuremberg Lamentation: For the Forgotten Victims of Medical Science." *British Medical Journal* 313: 1463–67.

———. 2012. "Dissecting the History of Anatomy in the Third Reich—1989–2010: A Personal Account." *Annals of Anatomy* 194: 228–36.

———. 2014. "Requiescat Sine Pace. Recollections and Reflections on the World Medical Association, the Case of Prof. Dr. Hans Joachim Sewering and the Murder of Babette Fröwis." In: *Silence, Scapegoats, Self-Reflection: The Shadow of Nazi Medical Crimes on Medicine and Bioethics*, ed. Volker Roelcke, Sascha Topp and Etienne Lepicard, 281–300. Göttingen: V&R unipress.

Siebenpfeiffer, Hania. 2005. *"Böse Lust": Gewaltverbrechen in Diskursen der Weimarer Republik*. Köln: Böhlau Verlag.

Stock, Ulrich. 1987. "Deutsche Ärzte und die Vergangenheit." *Die Zeit*, 12 June 1987, accessed 4 February 2015. http://www.zeit.de/1987/25/deutsche-aerzte-und-die-vergangenheit.

Stöhr, Philipp. 1947. *FIAT Review of German Science 1939–1945: Anatomy, Histology and Embryology*. Wiesbaden: Dieterich'sche Verlagsbuchhandlung.

Thom, Achim, and Gennadij I Caregorodcev, eds. 1989. *Medizin unterm Hakenkreuz*. Berlin: VEB Verlag Volk und Gesundheit.

Toellner, Richard. 1989. "Nehmen wir die Last auf - die Last ist die Lehre." *Deutsches Ärzteblatt* 86(33): A-2271–79.

Topp, Sascha. 2013. *Geschichte als Argument in der Nachkriegsmedizin: Formen der Vergangenheitsbewältigung der nationalsozialistischen Euthanasie zwischen Politisierung und Historiographie*. Göttingen: V&R unipress.

Universität Tübingen. 1990. "Ergebnisbericht: Überprüfung der Sammlungen des anatomischen Institutes auf das Vorhandensein von Präparaten von NS-Opfern. 3.4.1989." In: *Berichte der Kommission zur Überprüfung der Präparatesammlungen in den medizinischen Einrichtungen der Universität Tübingen im Hinblick auf Opfer des Nationalsozialismus*, by Universität Tübingen. Manuscript, collection Dr. William Seidelman.

USHMM. 2014. "United States Holocaust Memorial Museum: Survivors and Victims." Accessed 27 March 2014. http://www.ushmm.org/remember/the-holocaust-survivors-and-victims-resource-center/survivors-and-victims.

Wagner, Walter. 1974. "Der Volksgerichtshof im nationalsozialistischen Staat." In: *Quellen und Darstellungen zur Zeitgeschichte. Band 16/III. Die Deutsche Justiz im Nationalsozialismus*. Stuttgart: Deutsche Verlagsanstalt.

Walther, Peter T. 2008. "Entlassungen und Exodus: Personalpolitik an der Medizinischen Fakultät und in der Charité 1933." In: *Die Charité im Dritten Reich: Zur Dienstbarkeit medizinischer Wissenschaft im Nationalsozialismus*, ed. Sabine Schleiermacher and Uwe Schagen, 37–50. Paderborn: Ferdinand Schöningh.

Weinberger, Ruth Jolanda. 2009. *Fertility Experiments in Auschwitz-Birkenau: The Perpetrators and Their Victims*. Saarbrücken: Südwestdeutscher Verlag.

Weindling, Paul J. 1989. *Health, Race and German Politics between National Unification and Nazism, 1870–1945*. Cambridge: Cambridge University Press.

———. 1991. "Medicine in Nazi Germany and Its Aftermath." *Bulletin of the History of Medicine* 65(3): 416–19.

———. 2012. "'Cleansing' Anatomical Collections: The Politics of Removing Specimens from Anatomical Collections 1988–1992." *Annals of Anatomy* 194: 237–42.

———. 2013. "From Scientific Object to Commemorated Victim: The Children of the Spiegelgrund." *History and Philosophy of the Life Sciences* 35: 415–30.

———. 2015. *Victims and Survivors of Nazi Human Experiments: Science and Suffering in the Holocaust.* London: Bloomsbury Academic.

Weingart, Peter, Jürgen Kroll, and Kurt Bayertz. 1992. *Rasse, Blut und Gene: Geschichte der Eugenik und Rassenhygiene in Deutschland.* Frankfurt am Main: Suhrkamp.

Winau, Rolf, and Karl Heinz Hafner 1974. "'Die Freigabe der Vernichtung lebensunwerten Lebens': Eine Untersuchung zu der Schrift von Karl Binding und Alfred Hoche." *Medizinhistorisches Journal* 9: 227–54.

Wolf-Heidegger, Gerhard, and Anna Maria Cetto. 1967. *Die anatomische Sektion in bildlicher Darstellung.* New York: Karger.

World Archeological Congress. 2009. "Vermillion Accord." Accessed 23 February 2015. http://www.worldarchaeologicalcongress.org/about-wac/codes-of-ethics/168-vermillion.

Wuttke-Groneberg, Walter. 1980. *Medizin und Nationalsozialismus.* Tübingen: Schwäbische Verlagsgesellschaft.

Wuttke, Walter. 1989. "Die Aufarbeitung der Medizin im 'Dritten Reich' durch die deutsche Medizinhistoriographie." *Jahrbuch für kritische Medizin. Argument-Sonderband* 186: 156–75. Berlin: Argument Verlag.

Zimmermann, Volker. 1991. "Die Medizin in Göttingen während der Nationalsozialistischen Diktatur." *Würzburger Medizinhistorische Mitteilungen* Band 9, 393–416.

Chapter 2

ANATOMY AND RELATED SCIENCES
BEFORE 1933

German science has laid the tool into the politician's hand.
Otto Aichel and Otmar von Verschuer, 1934[1]

Before starting on a narrative of anatomy in National Socialism, it is necessary to place German anatomists and their work in their historical and interdisciplinary context. The connection between anatomy and its related sciences during the nineteenth and early twentieth centuries, especially the "racial" sciences, need to be outlined to better understand the interaction between anatomists and the NS state after 1933.

Legal, Illegal, and Discriminatory Practices
of Anatomical Body Procurement before 1933

The field of scientific anatomy began with its practitioners' assertion that they must dissect—that is, cut apart—dead human bodies in order to gain knowledge of the structure and function of living human bodies. This claim was and still is in itself an imposition on the rest of society, and over time, societies have answered the anatomists' demand in different ways. The first known systematic dissections of human bodies occurred in Alexandria around 300 BCE, and were made possible because enlightened Ptolomean rulers supported Erasistros' and Herophilus' scientific inquiries by making the bodies of executed criminals available for their work.[2] Later, religious and cultural prohibitions made formal human dissection impossible until the late Middle Ages. Galen's (130–200 CE) anatomical knowledge depended on animal dissections, and as his volumes of medical writings formed the basis for all formal medical education over the next millennium, his anatomical errors were transmitted over time. Mondino de Luzzi (1270–1326 C.E.) became one of the first physicians to reintroduce human dissection, made possible by the Vatican's consent.[3] Bodies of executed criminals were the first and only legal source for anatomical body procurement for many years, but it was difficult to acquire them,[4] and so anatomists and their students had to resort to grave robbing for much of the history of anatomy.[5]

Informal "dissectionlike" practices[6] also developed in private spaces such as delivery rooms, death chambers, and through religious practices. These encountered little resistance and were even supported by the relatives of the deceased, as the bodies of their loved ones were mostly left intact and any necessary dissection occurred in the family's familiar spaces.[7] In contrast, academic anatomical dissections were frequently open to the public and through their association with the dissection of criminals, seen by the audience as a "profound dishonor" for the dissected. The general public believed that the dissection "violated both [the body's] personhood and its social identity by rendering it unrecognizable and unsuiting it for a conventional funeral."[8] Historian Helen MacDonald formulated equally poignantly: "Learning anatomy was always as much about destroying as creating […]. These [dissecting rooms] were popular places in which science combined with curiosity, punishment, art, pleasure and—always—relationships of power."[9]

In the late seventeenth and throughout the eighteenth century, legislators took advantage of the power of the public's negative perception of anatomists' work by using the threat of dissection as a deterrent against the commitment of serious crimes. Anatomical dissection became part of capital punishment in the so-called "Murder Act" of 1752 in Britain.[10] Certain crimes became "punishable by dissection" and the practice of anatomy of this time presented itself as an "ill-defined […] mixture of punitive and medical purposes."[11] The addition of anatomical dissection following the execution as punishment was considered more frightening and ignoble by many offenders and their families than the execution itself.[12] This view of dissection started to change in the eighteenth century with an increasing public interest in the sciences. Anatomy flourished in Europe, and the body supply from executions did not satisfy the increasing needs of medical schools, especially as execution rates had dropped in some countries.[13] Consequently, governments on the European continent passed legislation allowing the use for dissection of the unclaimed bodies of "paupers," inmates of prisons, and patients at psychiatric and charitable hospitals; that is, bodies of persons whose relatives did not claim them for a burial were sent to the anatomical institutes.[14] By the end of the eighteenth century the availability of unclaimed bodies alleviated the shortage on the European continent,[15] but not in Great Britain and the United States. Legislation on dissection came later in these countries, and grave robbing and even murder for dissection continued to occur, eliciting public and sometimes violent outcries against anatomists.[16] Other countries of the British Commonwealth followed the example of the British Warburton Anatomy Act of 1832 with similar legislation.[17] In general, unclaimed bodies, and among them bodies of the executed, were the predominant sources of body procurement in anatomy until the introduction of functional body donation programs in the second half of the twentieth century.

While legislation solved the problem of body procurement for anatomists, it extended the predominant use of bodies from "the poor, the black, and the marginalized."[18] This economical and "racial" discrimination against the indigent, who could not afford proper burials, haunted the poor around the world—families in England and Germany as much as African American communities in the United States during the nineteenth century.[19] Blakely and Harrington identified a "postmortem racism" in the anatomical body procurement of the nineteenth-century United States in their archeological excavation of dissection rooms used at the Georgia Medical College between 1840 and 1880. They found that the skeletal remains unearthed there were mostly those of African Americans (79 percent versus 42 percent in the general population), whereas there were disproportionately fewer bones from persons of European American descent (21 percent versus 58 percent in the general population).[20] Harriet Martineau, a British travel writer, remarked in 1838, "In Baltimore the bodies of coloured people exclusively are taken for dissection, because the whites do not like it, and the coloured people cannot resist."[21] And in 1831 the South Carolina Medical College advertised its services with the note, "No place in the United States offers as great opportunities for the acquisition of anatomical knowledge. Subjects being obtained from among the colored population in sufficient numbers for every purpose, and proper dissection carried on without offending any individuals in our community!"[22] Similar "racial" discrimination in body procurement occurred in Europe, e.g., with the use of bodies of so-called "gypsies" in the eighteenth century,[23] and in Australia with the mortal remains of Tasmanian aborigines in the nineteenth century.[24] Discrimination and coercion in anatomy and other areas of medicine are still ongoing.[25] A current example is the predominant use of undocumented bodies of Chinese origin displayed in an array of shows traveling globally of plastinated human body specimens, which developed after 1995 in the wake of the anatomist Gunther von Hagens' controversial original "Body Worlds."[26]

Anatomy and the "Racial" Sciences in Germany before 1933

Anatomists apparently saw bodies of all so-called "races" as adequate for teaching and research. Their demand for bodies was always greater than the supply, and they accepted all bodies as suitable for dissection, irrespective of their "race". It is truly curious then that during a time when racial anthropology was developing, the so-called "racial" variations observed through the multitude of anthropometric measurements by physical anthropologists[27] were of no apparent interest for the anatomical educator and histological researcher. And this was despite the pervasive racism prevalent in so many societies and the "intra-professional anti-Semitism"[28] inherent to the German medical community. Those anatomists who were focused on anatomical education and histological research apparently did not care about the "race" of the person they or their students dissected. All bodies were taken as "typical examples" of anatomical structure in education and most research. A "racial" discriminatory value judgment, as claimed by many at the time, did not make these bodies unsuitable for anatomical purposes, and there is no evidence that anatomical educators or histological researchers distinguished between bodies from so-called "superior" and "inferior races" in their body procurement. This behavior is even more remarkable given the fact that many of the instructors in dissection courses were also involved in developing and teaching the discipline of racial hygiene.

The ideas of "racial" characteristics and a differential value of "races" are generally recognized as having roots in the nineteenth and early twentieth centuries. Historian Craig Steven Wilder sees the origins of racist ideology even earlier in the eighteenth century, as transatlantic travel for education led to an active exchange of ideas between Europe and the New World.[29] By the late nineteenth century the field of anatomy was closely associated with physical anthropology, eugenics, and the emerging science of genetics. The relationship between these disciplines and anatomy up to and during the Third Reich is little explored, even though there existed extensive overlap in personnel between these fields in research and medical education. Physical anthropology was a subdiscipline of anatomy, and its objective was the description of the physical and physiological characteristics of human beings and their varieties, often called "races." Some early anthropologists thought it incorrect to use the term "race" in connection with these human varieties, because Darwinian biology postulated a single human "race" that presented variants, but not multiple distinct and separate "races."[30] Rudolf von Virchow and his anthropological colleagues accepted the concept of "race" as a hereditary variation, but specifically denied the existence of fixed "racial" types or the superiority of certain "races" over others.[31] However, many considered *Rassenkunde* (literally, the "science of race") an essential part of anthropology. With the growing influence of the ideas of nationalism, racism, and Aryan supremacy as advanced by authors like Arthur de Gobineau, Houston Chamberlain, Ludwig Schemann, and Georges Vacher de Lapouge, some younger anthropologists and eugenicists embraced the concept of "races" and applied it to apparent physical varieties of humans.[32] The physician Alfred Ploetz saw the category of "race" as the foundation of the discipline of eugenics, a social-Darwinist concept proposed by Francis Galton. Eugenic ideas were based on the theories of evolution and human inheritance according to August Weismann, and were concerned with the hereditary quality of "human stock."[33] Eugenics was thought to be the only effective therapy against the perceived threat of an inevitable degeneration of civilized humankind. Eugenic thinking promoted the negative selection—i.e., elimination—of those seen as "unfit" for society, and positive selection for those deemed "fit."

The fear of degeneration was pervasive not only in Europe but also in other parts of the world. In the first decades of the twentieth century, eugenics was considered a progressive science and pursued by representatives of the entire political spectrum, including socialist, liberal, conservative and totalitarian political ideologies. Eugenic ideas of negative selection, including sterilization of certain groups of society, were applied to legislature in many countries, among them the United States, Sweden, Norway, Finland, and Denmark.[34] Positive selection was promoted through various measures including public health education.[35] Beginning in 1895, Ploetz called his concept of eugenics, which was based on ideas of the physician Wilhelm Schallmayer,[36] "racial hygiene" and founded the first eugenic society, the German Society for Racial Hygiene, in 1904.[37] The term, which replaced in Germany the internationally used word "eugenics," implied a betterment of public health not only through an observation and care for the social environment of a person, as addressed in the discipline of social hygiene, but also through an attention to "racial" aspects of health. After WWI many German racial hygienists increasingly included the ideas of supremacy of the northern and Aryan races, as well as anti-Semitism, in their thinking.[38] During the Weimar Republic, German politics became increasingly open to the introduction of eugenic ideas in legislation, based on economic and biological considerations.[39] First drafts of sterilization laws were proposed as early as 1923,[40] and measures of "negative selection" were openly discussed. In 1920 a publication by the lawyer Karl Binding and the physician Alfred Hoche examined the economical impact of care for "lives unworthy of living" and argued for the "mercy-killing" or so-called "euthanasia" of persons who were considered to be a burden on society.[41]

Eugenic policies became an essential part of the public political debate in a Germany suffering from financial recession, political instability, loss of traditional values, and the narcissistic insult of its military defeat in WWI. Within this debate, the idea of biologically defined "races" had taken hold through the research of anatomist Eugen Fischer, who had claimed to have found evidence for the Mendelian inheritance of racial physical traits such as hair shape and eye color, as well as character qualities.[42]

With the rediscovery of the Mendelian laws of genetics after the turn of the century, physical anthropology was no longer interpreted as only a morphological science, but also as a genetic one, particularly through the work of Fischer. The fields of anthropology, racial hygiene/eugenics and genetics, as well as statistics, psychology, and clinical medicine were closely intertwined.[43] By the 1920s racial hygiene and *Vererbungslehre* (the science of inheritance, i.e., human genetics) were taught by anatomists, anthropologists, hygienists, and biologists as part of a medical curriculum that newly included these subjects as mandatory in medical education. Institutes of racial hygiene were founded at German universities and as independent research institutes.[44] The scientific methods of these disciplines ranged from eugenic legislative measures to experimental genetics. Following the Great Depression, racial hygienists around the world, but also in Germany, focused increasingly on eugenic measures, which included sterilization and premarital counseling.[45] These population based approaches to eugenic work together with anthropological techniques, including family and twin research, became increasingly more the prime methodology in German racial hygiene in the 1930s. By then the Mendelian approach as used by Fischer had shown hardly any tangible results.[46] The lack of scientific progress, which may have also been due to insufficient communication between experimental geneticists and racial hygienists became even more apparent by 1939, when geneticist Otmar von Verschuer emphasized the need to include new genetic insights in racial hygienic concepts.[47]

Racial hygiene did not remain without opposition. Already in 1927 American biologist Raymond Pearl leveled a profound criticism at the prevailing eugenic theory and practice and cautioned: "Eugenics has fallen into disrepute [...] devotees have assigned such complex and heterogeneous phenomena as poverty, insanity, crime, prostitution, cancer etc., to the operation of single genes. [...] The propaganda phase has always gone along hand in hand with the purely scientific [...] the literature of eugenics has largely become a mingled mess of ill-grounded and uncritical sociology, economics, anthropology, and politics, full of emotional appeals to class and race prejudices".[48] Pearl continued by demanding that eugenicists return to the clear scientific facts of Mendelian genetics and reminded them to recognize the difference between genotype and phenotype. German anthropologist Karl Saller also criticized the lack of evidence for certain allegedly inherited traits like beauty,[49] and the fact that the commonly accepted concept of racial hygiene in Germany did not allow for a dynamic development of races.[50] However, Fischer's and his colleagues' ideas prevailed in a political climate that led to the rise of the National Socialists, for whom these static perceptions of race proved a welcome "scientific" foundation of their policies.

Anatomists' Involvement in Racial Hygiene as the Leading Science of NS Ideology

In 1924, during his incarceration in Landsberg Prison following the Beer Hall Putsch, Adolf Hitler read the "Baur-Fischer-Lenz," the first standard text on racial hygiene and inheritance coauthored by Fischer.[51] Hitler incorporated his views on racial hygiene and anti-Semitism in his political arguments in the book *Mein Kampf* (*My Struggle*). Over the following years a concept of static genetic traits and "races" unalterable by environmental influences became the purported scientific foundation for the racism pervading NS ideology and policies, and the growing NS movement promoted the concepts of racial hygiene eagerly.[52] In 1931 geneticist Fritz Lenz declared that National Socialism was "applied biology," a phrase often repeated, among others by Hitler's deputy Rudolf Hess in a public speech in 1934.[53] Eugenic measures were promoted by the NS leadership as the only possible therapy for a Nordic people threatened by degeneration. The individual human being was seen only as part of the whole body of the people (*Volkskörper*), thus necessitating the removal of "diseased" individuals as a drastic but unavoidable cure to save the health of the body of the people ("*Volksgesundheit*"). Due to this biologistic definition the term "diseased" in National Socialism included not just the physically sick, but also those deemed genetically, "racially," and socially "unfit" for the new society: the mentally ill, children born with malformations and diseases believed to be hereditary, homosexuals, social misfits, political opponents, certain religious groups and so-called "non-Aryan" groups including Jews, Sinti, and Roma (European gypsies), and those of African heritage. They all could be subjected to the various methods of negative selection, which ranged from sterilization, to so-called "euthanasia," incarceration, and ultimately mass murder.[54]

The racial hygienists, and among them many anatomists, welcomed the National Socialists' rise to power because they saw a bright future for the implementation of their eugenic policies, and indeed their hopes were fulfilled through full support of their science by the NS government.[55] In a

1934 special edition of *Zeitschrift für Morphologie und Anthropologie* in cele-
bration of Fischer's sixtieth birthday, the editors of the journal Otto Aichel,
chairman of anthropology in Kiel, and von Verschuer wrote: "We stand at
the turn of an era. For the first time in world history, the *Führer* Adolf Hitler puts into practice the insights on biological principles of the evolution
of the various peoples—race, inheritance, selection. It is no coincidence
that Germany is the place of this event: German science has laid the tool
into the politician's hand."[56] In 1943, Fischer recapitulated the preceding
ten years by pointing out the great good fortune of his theoretical science
being allowed to prosper in an atmosphere of general acceptance fostered
by NS ideology. He praised the practical application of racial hygiene's
"scientific results" in governmental procedures.[57] Scientists and politicians
supported each other in a dangerous symbiosis.[58] The racial hygienists were
not only bystanders in the "application" of their science, but also actively
involved in its implementation. Fischer, von Verschuer, and psychiatrist
Ernst Rüdin provided evaluations for the hereditary health courts,[59] draw-
ing up testimonies about an individual's "racial" and genetic heritage, and
thus contributing to the justification of "racial" discrimination, the physical mutilation of involuntary sterilization, expulsion, and murder.[60] There
is no indication that Fischer and his colleagues acknowledged the atrocities committed with the help of their science any time before the end of
the war.

Several anatomists worked and taught anthropology, human inheritance, and racial hygiene, among them Karl Saller, Franz Weidenreich,
Heinrich Poll, Walter Brandt of Cologne, Ferdinand Wagenseil, Eugen
Kurz and Johann Paul Kremer of Münster—who had made their careers in
these areas after WWI,.[61] Saller and Weidenreich ran into political trouble
because they disagreed with the official NS line of static races, and were
both dismissed. Also dismissed were Poll and Brandt—Poll because of his
Jewish descent and Brandt for having married a Jewish woman. There is a
special tragedy in the fate of these academics, as they ultimately became
victims of the NS application of their own field of scientific activity. Active
National Socialists Kurz and Kremer continued teaching and publishing
on racial hygienic topics in the 1930s, as did Andreas Pratje in Erlangen.[62]
Wilhelm Pfuhl, Otto Aichel of Kiel, and Otto Grosser of Prague also taught
racial hygiene.[63] Other anatomists started to focus on research on racial
hygienic topics when it became politically advantageous to do so, among
them August Hirt and Friedrich Heiderich, chairman of the anatomical
department of Münster University. Heiderich established a new division of
racial hygiene and human genetics within his institute by 1939, supported
Kurz and Kremer in their work, and promoted racial studies of the local
population.[64]

Konrad Lorenz, Nobel Laureate for ethology, represents another link between anatomy and racial hygienic thinking. After an early career in anatomy under Ferdinand Hochstetter in Vienna, he left this discipline for studies in ethology. He received funding by the German Research Foundation in 1938 through the support of anatomists Hochstetter, Alexander Pichler, and Eduard Pernkopf. They vouched for Lorenz's "growing interest in National Socialism" and the fact that his biological studies were in keeping with the general worldview of National Socialism.[65] Many of the principles on which Lorenz's ethology was based had the same ideological roots as the concept of racial hygiene, including his pronouncements on domestication and degeneration in human beings and animals, his social-Darwinist view of biology and society,[66] and his statement that social "morality" in humans was mostly inborn and not acquired.[67] Lorenz greatly welcomed the annexation of Austria by NS Germany and joined the NSDAP on 29 June 1938.[68] His career prospered and included work for the NSDAP Office of Racial Policy, and the chairmanship of the department of psychology at the University of Königsberg in 1940.[69]

Lorenz and his colleagues are examples of the way in which anatomists and their fellow scientists were intimately involved in the creation of the ideological basis for NS discriminatory policies through research on racial hygiene, in the propagation of NS ideology through medical and public education, and in the implementation of NS laws through evaluations in the heredity health courts. And while the relationship between the government and anatomists changed in many ways after 1933, the established structures of body procurement led to unforeseen new "opportunities" in NS Germany for anatomists eager to promote their educational and scientific endeavors.

Notes

1. Aichel and Verschuer 1934.
2. Staden 1989.
3. Jones and Whitaker 2009.
4. Hildebrandt 2008.
5. Ball 1928.
6. Park 2006.
7. Park 2006, 15–16.
8. Park 2006, 15.
9. Macdonald 2006, 39.
10. Richardson 1987.
11. Sappol 2002, 102.
12. Hunter 1931.
13. Stukenbrock 2003.
14. Pauser 1998.
15. Stukenbrock 2003.

16. Ball 1928; Hunter 1931; Richardson 1987; Sappol 2002.
17. Persaud 1997; Jones 1991; Gopichand 2002; Canadian Legal Information Institute 2006.
18. Halperin 2007, 489
19. Prüll 2000, 62–64; Richardson 1987; Hutton 2013.
20. Blakely and Harrington 1997.
21. Quoted after Halperin 2007, 492.
22. Quoted after Washington 2006, 116.
23. Enke 2005.
24. Macdonald 2006.
25. Washington 2006; Duke 2013.
26. Gutmann 2014, 287–308; Working 2005; Hildebrandt 2008.
27. MacDonald 2006, 96–135.
28. Kater 1999, 226.
29. Wilder 2013.
30. Evans 2010; Lösch 1997; Schafft 2004, 202–3; Weindling 1989a, 49.
31. Weindling 1989a, 48–49, 55.
32. Kühl 1994; Vogel 2000; Bergmann et al. 1989; Chamberlain 1901; Vacher de Lapouge 1939.
33. Weingart 1989; Fangerau and Noack 2006.
34. Kühl 1994; Black 2003; Stern 2005; Broberg and Roll-Hansen 1996.
35. See e.g., Cold Spring Harbor 2014.
36. Ritter 1992.
37. Klee 2003; Fangerau and Noack 2006.
38. Schafft 2004, 205–7.
39. Schmuhl 1987; Kröner 1998.
40. Weingart et al. 1992, 291.
41. Binding and Hoche 1920.
42. Lösch 1997; Gessler 2000.
43. Weindling 1989b.
44. Weingart et al. 1992, 438; Ritter 1992, 176–77; Fangerau and Noack 2006.
45. Weiss 2010, 59-65.
46. Weiss 2010, 97.
47. Roth 1999, 348; Kröner 1998.
48. Pearl 1927, 260.
49. Weingart et al. 1988, 317–18.
50. Beushausen et al. 1998; Schafft 2004, 227–30.
51. Baur et al. 1921.
52. Geiss 1988; Schafft 2004.
53. Lifton 1986, 54; Proctor 2000, 341.
54. Platen-Hallermund 1948; Lifton 1986; Klee 1985; Bäumer 1990; Weingart et al. 1992; Proctor 1994; Kröner 1996; Seidelman 1996.
55. Ritter 1992, 177–78; Müller-Hill 1984, 13–14.
56. See note 1.
57. Hofer and Leven 2003, 27.
58. Weingart 1989, 260; Weiss 2010.
59. Ritter 1992, 183; Proctor 1994; Kröner 1998.
60. Müller-Hill 1984; Massin 1999, 41.
61. See on Karl Saller: Beushausen 1998; on Franz Weidenreich: Mussgnug 1988 and Hertler 2008; on Heinrich Poll: Rothmaler 1990; on Walter Brandt: Blaschke 1988; on Ferdinand Wagenseil: Unger 1998; on Eugen Kurz and Johann Paul Kremer of Münster: Münster 1932–39; Vieten 1983.

62. Vieten 1983; Dicke 2004; Wendehorst 1993, 183 and 203, Wittern 1993, 389; Proctor 1988, 18
63. Greifswald 1933–1939; Kiel 1933–1938; Hlaváckovâ and Mísková 2001.
64. Dicke 2004, 37–39.
65. Föger and Taschwer 2001; Taschwer and Föger 2003; Deichmann 1996, 183; Mertens 2004, 294; Burkhardt 2005.
66. Kalikow 1983, 39–40.
67. Burkhardt 2005, 243.
68. Klee 2003.
69. Burkhardt 2005.

Bibliography

Aichel, Otto, and Otmar von Verschuer. 1934. "Festband Eugen Fischer zum 60. Geburtstag gewidmet." *Zeitschrift für Morphologie und Anthropologie* 34: v–vi.
Bäumer, A. 1990. *NS-Biologie*. Stuttgart: S. Hirzel Wissenschaftliche Verlagsgesellschaft.
Ball, James M. 1928. *The Body Snatchers: Doctors, Grave Robbers and the Law*. Reprint, New York: Dorset Press, 1989.
Baur, Erwin, Eugen Fischer, and Fritz Lenz. 1921. *Menschliche Erblichkeitslehre und Rassenhygiene*. 1st ed. München: J. F. Lehmanns Verlag.
Beushausen, Ulrich, Hans-Joachim Dahms, Thomas Koch, Almut Massing, and Konrad Obermann. 1998. "Die medizininsche Fakultät im Dritten Reich." In: *Die Universität Göttingen unter dem Nationalsozialismus,* ed. Heinrich Becker, Hand Joachim Dahms, and Cornelia Wegeler, 183–286. München: K. G. Saur.
Blakely, Robert, and Judith M. Harrington, eds. 1997. *Bones in the Basement: Postmortem Racism in Nineteenth-Century Medical Training*. Washington and London: Smithsonian Institution Press.
Bergmann, Anna, Gabriele Czarnowski, and Annegret Ehmann. 1989. "Menschen als Objekte humangenetischer Forschung und Politik im 20. Jahrhundert." In: *Der Wert des Menschen: Medizin in Deutschland 1918–1945,* ed. Christion Pross and Götz Aly, 121–42. Berlin: Edition Hentrich.
Binding, Karl, and Alfred Hoche. 1920. *Die Freigabe der Vernichtung unwerten Lebens*. Leipzig: Verlag Felix Meiner. Accessed 3 April 2014. http://www.staff.uni-marburg.de/~rohrmann/Literatur/binding.html. English excerpts, accessed 3 April 2014. http://germanhistorydocs.ghi-dc.org/sub_document.cfm?document_id=4496.
Black, Edwin. 2003. *War against the Weak: Eugenics and America's Campaign to Create a Master Race*. New York: Four Walls Eight Windows.
Blaschke, Wolfgang, Olaf Hensel, Peter Liebermann, and Wolfgang Lindweiler, eds. 1988. *Nachhilfe zur Erinnerung, 600 Jahre Universität Köln*. Bonn: Pahl-Rugenstein.
Broberg, Gunnar, and Nils Roll-Hansen. 1996. *Eugenics and the Welfare State: Sterilization Policy in Denmark, Sweden, Norway, and Finland*. East Lansing: Michigan State University Press.
Burkhardt, Richard W., Jr. 2005. *Patterns of Behavior: Konrad Lorenz, Niko Tinbergen, and the Founding of Ethology*. Chicago: University of Chicago Press. Pages 1–648.
Canadian Legal Information Institute. 2006. The Anatomy Act, C.C.S.M.v.A80. Accessed 23 February 2015. http://www.canlii.org/en/mb/laws/stat/ccsm-c-a80/latest/ccsm-c-a80.html.
Chamberlain, Houston Stuart. 1901. *Die Grundlagen des neunzehnten Jahrhunderts*. 3rd ed. München: Münchner Verlagsanstalt F. Bruckmann AG.
Cold Spring Harbor. 2014. Image archive of the American eugenics movement. Accessed 3 April 2014. http://www.eugenicsarchive.org/eugenics/.

Deichmann, Ute. 1996. *Biologists under Hitler*. Cambridge: Harvard University Press.

Dicke, Jan Nikolas. 2004. *Eugenik und Rassenhygiene in Münster*. Berlin: Weissensee Verlag.

Duke, Naomi N. 2013. "Situated Bodies in Medicine and Research: Altruism versus Compelled Sacrifice." In: *The Global Body Market: Altruism's Limits*, ed. Michele Goodwin, 107–24. Cambridge: Cambridge University Press.

Enke, Ulrike. 2005. "'… fuer uns das feinste Parfüm': Historische Anmerkungen zu Anatomie und anatomischem Unterricht an den hessischen Universitäten vom 16. bis zum 18. Jahrhundert." *Hessisches Ärzteblatt* 12: 819–24.

Evans, Andrew D. 2010. *Anthropology at War: World War I and the Science of Race in Germany*. Chicago: University of Chicago Press.

Fangerau, Heiner, and Thorsten Noack. 2006. "Rassenhygiene in Deutschland und Medizin im Nationalsozialismus." In: *Geschichte, Theorie und Ethik der Medizin: Eine Einführung*, ed. Stefan Schulz, Klaus Steigleder, Heiner Fangerau, Norbert W. Paul, 224–46. Frankfurt am Main: Suhrkamp.

Föger, Benedikt, and Klaus Taschwer. 2001. *Die andere Seite des Spiegels. Konrad Lorenz und der Nationalsozialismus*. Wien: Czernin Verlag.

Gessler, Bernhard. 2000. *Eugen Fischer (1874–1967): Leben und Wirken des Freiburger Anatomen, Anthropologen und Rassenhygienikers bis 1927*. Frankfurt: Peter Lang.

Geiss, Imanuel. 1988. *Geschichte des Rassismus*. Frankfurt am Main: Suhrkamp.

Gopichand, Patnaik VV. 2002. "Editorial." *Journal of the Anatomical Society of India* 51(2): 143–44.

Greifswald 1933–1939. *Vorlesungsverzeichnisse der Universität Greifswald 1933 bis 1939*. Greifswald: University Press.

Gutmann, Ethan. 2014. *The Slaughter: Mass Killings, Organ Harvesting, and China's Secret Solution to Its Dissident Problem*. Amherst, New York: Prometheus.

Halperin, Edward C. 2007. "The Poor, the Black, and the Marginalized as the Source of Cadavers in the United States Anatomical Education." *Clinical Anatomy* 20: 489–95.

Hertler, Christine. 2008. "Franz Weidenreich und die Anthropologie in Frankfurt: Weidenreichs Weg an die Universität." In: *Frankfurter Wissenschaftler zwischen 1933 und 1945*, ed. Jörn Kobes, and Jan-Otmar Hesse, 111–23. Göttingen: Wallstein Verlag.

Hildebrandt, Sabine. 2008. "Capital Punishment and Anatomy: History and Ethics of an Ongoing Association." *Clinical Anatomy* 21: 5–14.

Hlaváčková, L, Mísková, A. 2001. "Otto Grosser (1873–1951) Mediziner: Vom überzeugten Nationalisten zum aktiven Nationalsozialisten." In: *Prager Professoren 1938–1945. Zwischen Wissenschaft und Politik*, ed. Monkia Glettler, and Alena Mísková, 415–28. Essen: Klartext Verlag.

Hofer, Hans Georg, and Karl Heinz Leven. 2003. *Die Freiburger Medizinische Fakultät im Nationalsozialismus: Katalog einer Ausstellung des Instituts für Geschichte der Medizin der Universität Freiburg*. Frankfurt: Peter Lang.

Hunter, Richard H. 1931. *A Short History of Anatomy*. London: John Bale, Sons and Danielsson Ltd.

Hutton, Fiona. 2013. *The Study of Anatomy in Britain, 1700–1900*. London: Pickering & Chatto

Jones, David G. 1991. "Bequests, Cadavers and Dissections: Sketches from New Zealand History." *New Zealand Medical Journal* 104: 210–12.

Jones, David G., and Maya I. Whitaker. 2009. *Speaking for the Dead: The Human Body in Biology and Medicine*. Farnham: Ashgate.

Kalikow, Theodora J. 1983. "Konrad Lorenz's Ethological Theory: Explanation and Ideology, 1938–1943." *Journal of the History of Biology* 16(1): 39–73.

Kater, Michael H. 1999. "Das Böse in der Medizin: Nazi-Ärzte als Handlanger des Holocaust." In: *"Beseitigung des Jüdischen Einflusses…": Antisemitische Forschung, Eliten und*

Karrieren im Nationalsozialismus, ed. Fritz Bauer Institut, 219–39. Frankfurt: Campus Verlag.

Kiel 1933–1938: Christian-Albrechts Universität Kiel. *Personal- und Vorlesungsverzeichnisse 1933–1938.* Kiel: University Press.

Klee, Ernst. 2003. *Das Personenlexikon zum Dritten Reich: Wer war was vor und nach 1945?* Frankfurt am Main: S. Fischer.

Klee, Ernst, ed. 1985. *Dokumente zur Euthanasie.* Reprint, Frankfurt am Main: Fischer Taschenbuchverlag GmbH, 1997.

Kröner, Hans-Peter. 1998. *Von der Rassenhygiene zur Humangenetik.* Stuttgart: Gustav Fischer Verlag.

Kühl, Stefan. 1994. *The Nazi Connection: Eugenics, American Racism, and German National Socialism.* New York, Oxford: Oxford University Press.

Lifton, Robert Jay. 1986. *Ärzte im Dritten Reich.* German edition, Berlin: Ullstein Buchverlag GmbH&Co KG, 1998.

Lösch, Niels C. 1997. *Rasse als Konstrukt: Leben und Werk Eugen Fischers.* Frankfurt: Peter Lang GmbH.

MacDonald, Helen. 2006. *Human Remains: Dissection and Its Histories.* New Haven and London: Yale University Press.

Massin, Benoit. 1999. "Anthropologie und Humangenetik im Nationalsozialismus oder: Wie schreiben deutsche Wissenschaftler ihre eigene Wissenschaftsgeschichte?" In: *Wissenschaftlicher Rassismus: Analysen einer Kontinuität in den Human- und Naturwissenschaften,* ed. Heidrun Kaupen-Haase and Christian Saller, 12–64. Frankfurt: Campus Verlag, Frankfurt.

Mertens, Lothar. 2004. *"Nur politisch Würdige": Die DFG-Forschungsförderung im Dritten Reich 1933–1937.* Berlin: Akademie-Verlag.

Müller-Hill, Benno. 1984. *Tödliche Wissenschaft.* Reinbek bei Hamburg: Rowohlt Taschenbuch Verlag.

Münster. 1932–39. *Personal- und Vorlesungsverzeichnisse der Westfälischen Wilhelms-Univesität Münster 1932–1939.* Münster: University Press.

Mussgnug, Dorothee. 1988. *Die vertriebenen Heidelberger Dozenten: Zur Geschichte der Ruprecht-Karls-Universität nach 1933.* Heidelberg: Carl Winter.

Park, Katharine. 2006. *Secrets of Women: Gender, Generation, and the Origins of Human Dissection.* Brooklyn, NY: Zone Books.

Pauser, P. 1998. "Sektion als Strafe." In: *Körper ohne Leben: Begegnung und Umgang mit Toten,* ed. Norbert Stefenelli, 527–35. Wien: Böhlau.

Pearl, Raymond. 1927. "The Biology of Superiority." *The American Mercury* 7(47): 257–66.

Persaud, T Vid N. 1984. *Early History of Human Anatomy.* Springfield: Charles C. Thomas.

Platen-Hallermund, Alice. 1948. *Die Tötung Geisteskranker.* Heidelberg: Verlag der Frankfurter Hefte.

Proctor, Robert N. 1988. *Racial Hygiene: Medicine under the Nazis.* Cambridge: Harvard University Press.

———. 1994. "Racial Hygiene: The Collaboration of Medicine and Nazism." In: *Medical Ethics and the Third Reich: Historical and Contemporary Issues,* ed. John J. Michalczyk, 35–41. Kansas City, MO: Sheed & Ward.

———. 2000. "Nazi Science and Nazi Medical Ethics: Some Myths and Misconceptions." *Perspectives in Biology and Medicine* 43(3): 335–46.

Prüll, Cay-Rüdiger. 2000. "Der Umgang mit der menschlichen Leiche in der Medizin: die historische Perspektive." In: *Zum Umgang mit der menschlichen Leiche in der Medizin,* ed. Hans-Konrat Wellmer and Gisela Bockenheimer-Lucius, 59–69. Lübeck: Schmidt-Römhild.

Richardson Ruth. 1987. *Death, Dissection and the Destitute.* Second edition with a new afterword (2000). Chicago; London: The University of Chicago Press.

Ritter, Horst. 1992. Die Rolle der Anthropologie im NS-Staat. In: *Menschenverachtung und Opportunismus: Zur Medizin im Dritten Reich,* ed. Jürgen Peiffer, 172–78. Tübingen: Attempto Verlag.

Roth, Karl-Heinz. 1999. "Schöner neuer Mensch: Der Paradigmenwechsel der klassischen Genetik und seine Auswirkungen auf die Bevölkerungsbiologie des 'Dritten Reiches.'" In: *Wissenschaftlicher Rassismus. Analysen einer Kontinuitäet in den Huamn- und Naturwissenschaften,* ed. Heidrun Kaupen-Haas, and Christian Saller, 346–424. Frankfurt: Campus Verlag.

Rothmaler, Christiane. 1990. "Gutachten und Dokumentation über das Anatomische Institut des Universitätskrankenhauses Eppendorf der Universität Hamburg 1933–1945." "1999," *Zeitschrift für Sozialgeschichte des 20. u 21. Jahrhunderts* 2: 78–95.

Sappol, Michael. 2002. *A Traffic of Dead Bodies: Anatomy and Embodied Social Identity in Nineteenth-Century America.* 1st ed. Princeton, NJ: Princeton University Press.

Schafft, Gretchen E. 2004. *From Racism to Genocide: Anthropology in the Third Reich.* Urbana and Chicago: University of Illinois Press.

Schmuhl, Hans-Walter. 1987. *Rassenhygiene, Nationalsozialismus, Euthanasie.* Göttingen Vandenhoeck & Rupprecht.

Seidelman, William E. 1996. "Nuremberg Doctors' Trial: Nuremberg Lamentation: For the Forgotten Victims of Medical Science." *British Medical Journal* 313: 1463–67.

Staden, H. v. 1989. *Herophilus: The Art of Medicine in Early Alexandria.* Cambridge: Cambridge University Press.

Stern, Alexandra Minna. 2005. *Eugenic Nation: Faults and Frontiers of Better Breeding in Modern America.* Berkeley: University of California Press.

Stukenbrock, Karin. 2003. "Unter dem Primat der Ökonomie? Soziale und wirtschaftliche Randbedingungen der Leichenbeschaffung für die Anatomie." In: *Anatomie: Sektionen einer medizinischen Wissenschaft im 18. Jahrhundert,* ed. Jürgen Helm and Karin Stukenbrock, 227–39. Stuttgart: Franz Steiner Verlag.

Taschwer, Klaus, and Benedikt Föger. 2003. *Konrad Lorenz: Eine Biographie.* Wien: Paul Zsolnay Verlag.

Vacher de Lapouge, Georges. 1939. *Der Arier und seine Bedeutung für die Gesellschaft: Freier Kursus in Staatskunde gehalten an der Universität Montpellier 1889–1890.* Frankfurt: Verlag Moritz Diesterweg.

Vieten, Bernward. 1983. "Medizinstudenten und Medizinische Fakultät in Münster im 'Dritten Reich.'" In: *Münster- Spuren aus der Zeit des Faschismus. Zum 50. Jahrestag der nationalsozialistischen Machtergreifung,* ed. Hans-Günther Thien, 201-13. Münster: Edition Westfälisches Dampfboot.

Vogel, Christian. 2000. *Anthropologische Spuren: zur Natur des Menschen.* Stuttgart: Hirzel.

Washington, Harriet A. 2006. *Medical Apartheid: The Dark History of Medical Experimentation on Black Americans from Colonial Times to the Present.* New York: Doubleday.

Weindling, Paul J. 1989b. *Health, Race and German Politics between National Unification and Nazism, 1870–1945.* Cambridge: Cambridge University Press.

———. 1989b. "The 'Sonderweg' of German Eugenics: Nationalism and Scientific Internationalism." *British Journal for the History of Science* 22: 321–33.

Weingart, Peter. 1989. "German Eugenics between Science and Politics." *Osiris* 2nd series, 5: 260–82.

Weingart, Peter, Jürgen Kroll, and Kurt Bayertz. 1992. *Rasse, Blut und Gene: Geschichte der Eugenik und Rassenhygiene in Deutschland.* Frankfurt am Main: Suhrkamp.

Weiss, Sheila. 2010. *The Nazi Symbiosis: Human Genetics and Politics in the Third Reich.* Chicago and London: University of Chicago Press.

Wendehorst, Alfred. 1993. *Geschichte der Friedrich-Alexander-Universität Erlangen-Nürnberg 1743–1993*. München: Verlag C. H. Beck.

Wilder, Craig Steven. 2013. *Ebony & Ivy: Race, Slavery, and the Troubled History of America's Universities*. New York: Bloomsbury Press.

Wittern, Renate. 1993. "Aus der Geschichte der medizinischen Fakultät." In: *250 Jahre Friedrich-Alexander-Universität Erlangen-Nürnberg*, ed. Henning Kössler, 315–420. *Festschrift*. Erlangen: Verlagsdruckerei Schmidt.

Working, Russell. 2005. "Shock Value. Gunther von Hagens Has Outraged Critics with Macabre Publicity Stunts to Promote 'Body Worlds' and Lure 17 Million Visitors to the Exhibit Worldwide." *Chicago Tribune*, 31 July 2005.

Chapter 3

THE INTERACTION BETWEEN
THE NS STATE AND ANATOMISTS

> In common with most Germans in a secure position, I had never
> been a politician. We had a government with fair manners and
> we were content. Why should we be concerned? [...] One could
> mind one's own business or hobby just as I did, and one could
> read the newspapers and periodicals to know what was going on.
> But if something was not in order, then it was usually too late.
> Robert Meyer, dismissed embryologist[1]

Traditionally, German anatomical institutes were dependent on the
government for general legislation, approval of personnel decisions,
research funding, and body procurement. Professors were civil servants,
employees of the state with a special legal status and loyalty, and a close
collaboration between anatomical institutes and the current rulers was
common. These relationships changed during the NS period, most decisively in institutional and individual loss of political freedom, combined
with the restriction of funding to war-relevant projects and the provision
of new sources for body procurement. While many of the preexisting structures persisted in the contact between government and anatomy, others
changed in decisive manners, especially during the war. The close relationship between NS authorities and the discipline of anatomy intensified
during the war, and came to resemble in many ways the "cumulative radi-
calization" previously identified in other areas of NS governance.[2]

Politics and Universities in NS Germany

On 30 January 1933, Adolf Hitler, leader of the National Socialist Workers'
Party of Germany (NSDAP), was named the new chancellor of the Ger-

man Republic by its president, Paul von Hindenburg. Within the next two months Hitler and his party seized all governing power. At the time over half of Germany had been ruled by Prussian law and the rest by local and state laws. All of these were quickly integrated into a general legislation for a new so-called *Drittes Deutsches Reich,* a third German empire or Third Reich for short.[3] Whereas previously the individual states governed the universities within them, authority was now centralized in Berlin within the *Reichsministerium für Wissenschaft, Erziehung und Volksbildung* (Reich Ministry of Science, Education and Culture; short: *Reichserziehungsministerium* or REM), led by Bernhard Rust.[4] On 12 March 1938 Germany annexed Austria, followed by parts of Czechoslovakia on 29 September 1938, and after 1939 occupied the Alsace, parts of Poland, and the Baltics. From 1934 on, all preexisting and later newly founded German universities (Strasbourg, Posen, Prague, Dorpat) in these areas came under the direct governance of the REM. Over time, the NSDAP also asserted its influence on the universities via its student and faculty associations, the *Nationalsozialistischer Deutscher Studentenbund* (National Socialist German Students' League, NSDStB) and the *Nationalsozialistischer Deutscher Dozentenbund* (National Socialist German Lecturers' League; NSDDB). In addition, the SS *Sicherheitsdienst* SD (security service) recruited spies directly from the faculties of the universities.

It was the declared aim of the REM to reorganize the universities according to the authoritarian NS leadership concept.[5] Other goals included the "cleansing" of the faculty from so-called "non-Aryans" and dissidents, the alignment of all sciences with NS doctrine, and their utilization for war purposes.[6] With rare exceptions, most university faculties followed these intentions by a "self-alignment" (*Selbstgleichschaltung*) with these new policies.[7] The new legislation concerning the "restoration of the professional civil service" affected all universities immediately. The first of these laws, ratified on 7 April 1933, stipulated the dismissal of so-called "non-Aryan" civil servants as well as politically suspect persons, e.g., communists. Over the course of the next six years these regulations came to include persons married to "non-Aryans" and members of the Social Democratic Party, and were vague enough to be applicable to anybody whom the authorities wanted to eliminate from the university system, usually for political reasons.[8] With the ratification of the "law for the protection of German blood and German honor," the so-called Nuremberg laws from 5 September 1935, the previously used term "non-Aryan" was specified with a biologistic definition of who was considered to be of "Jewish descent" or of "mixed race." This led to another wave of civil servant removals. Sometimes the dismissals were disguised as "early retirement," such as in the case of Alfred Kohn, who was removed from the chair in histology at the university of

Prague because of his Jewish descent.[9] Overall 19.3 percent of all university faculty members listed in Germany in 1933 were dismissed (1,145 out of a total of 6,140).[10] In metropolitan cities the effect of the dismissals was especially severe. In Vienna 153 of the 197 members of the medical faculty were removed from office because of their "Jewish" descent.[11] All of the vacant positions were quickly filled with successors.

In 1933, twenty-four German universities had anatomical departments. The Austrian universities of Vienna, Graz, and Innsbruck, as well as the Czechoslovakian university of Prague came under German control in 1938. During the war, Strasbourg (Alsace), Posen (Poland), and Dorpat (Estonia) were added, bringing the total to thirty-one anatomical departments.[12] The universities of Strasbourg, Posen, and Prague were conceptualized as *Reichsuniversitäten* (universities of the German Reich) with a special bulwark function against "foreign cultural influences" and an emphasis on the supremacy of the German people.[13] Little information is available for Dorpat, which was the least functional of the *Reichsuniversitäten*.[14] Overall, the anatomical departments varied greatly in size, with one chair for gross anatomy and sometimes an additional chair for histology and embryology, as well as several more tenured and untenured positions and assistants. While German student numbers dropped from 25,000 in 1933 to 16,000 in 1939 due to the expulsion of Jewish and female students and a central regulation of student numbers by the REM, they rose again to 40,000 in 1944 as doctors were needed in the field of war.[15]

The Political Spectrum of Anatomists

If asked about their political leanings, many anatomists would have agreed with Otto Veit's 1946 pronouncement that their professional work had nothing to do with politics.[16] However, during the NS period no civil servant—faculty members of universities among them—remained politically and personally unexamined or unchallenged.[17] Even anatomists who were critical of the regime had to find some way of officially demonstrating their acquiescence with NS politics if they wanted to keep their positions and remain in Germany. Others were in full or partial support of the new regime, at least during the first few years. German academics tended to have national-conservative convictions to begin with, and as civil servants in the Prussian tradition they held a great loyalty to the state as their employer. Many had simply never learned to question political authorities. They were willing to give the new regime the benefit of the doubt and remained silent on political matters even if they disagreed.[18] At the same time, anti-Semitism was pervasive in European societies and at German

universities.[19] For many young academics, the dismissal of their Jewish colleagues meant the emergence of new professional opportunities in a very
tight academic job market. Statistics from 1932 show that only one out of
seven lecturers in medicine had the chance to become a full-time tenured
professor.[20] At the University of Berlin every position vacated through "racially" and politically motivated dismissals of academics in 1933 was filled
again within the following year.[21] Generally, most academics, especially
physicians, and among them the younger ones who still desired to advance
professionally, saw fit to join the NSDAP party itself, or at least some other
NS formation like the NSDDB, the NS Physicians' Union (*NS Ärztebund*),
or the NS People's Welfare Organization (*NS Volkswohlfahrt*), even if they
were not convinced by the new regime. Politically zealous academics who
had joined the SS (*Schutzstaffel*, NS elite troupe) or the SA (*Sturmabteilung*, brown shirts) could experience distinct career advantages due to their
political choices, especially if they were also competent anatomists.

So far, 233 names of anatomists working at the thirty-one German anatomical departments between 1933 and 1945 have been identified (see
appendix, table 1). They include chairmen, professors, and assistants; there
are twelve women among them. Between 1933 and 1945, sixty-four anatomists held chair positions. Political information is available for 176 persons,
and all political data on individuals mentioned here stem from the sources
quoted in table 1. A total of fifty-four of these 176 anatomists suffered an
interruption of their careers for so-called "racial" and political reasons.
These interruptions ranged from career disadvantages to forced emigration
and incarceration. Of the remaining 122 anatomists, ninety-four joined
the NSDAP, forty-two the SA, fourteen the SS, and fifty other minor NS
organizations. Multiple memberships were common. Ten anatomists held
no memberships in NS political groups at all. There was also a group of
nine who only joined minor organizations. Some of this latter group were
established anatomists who had their differences with the NS regime,
among them Hermann Stieve, Philipp Stöhr Jr., and Ferdinand Wagenseil.

German anatomists at the time were usually trained as physicians, often
with an additional degree in the biological sciences. Their high rate of
NSDAP membership lies above the 44.8 percent reported for physicians
from a sample based on members of the *Reichsärztekammer* (Reich physicians' association) investigated by Kater.[22] The same sample showed an
SA membership of 26 percent and SS membership of 7 percent.[23] Another
study shows that as early as 1936, 31 percent of all physicians had joined
the NSDAP, 21 percent the SA and 4 percent the SS.[24] These numbers
are all much higher than those of the general population or of any other
professional group. In Kater's sample of teachers the comparable membership data between 1925 and 1943 was 23.8 percent for NSDAP (with 10

percent joining before 1933), 11 percent for the SA, and 0.4 percent of all male teachers for the SS.[25] Kater noted that by conservative estimates physicians were seven times more likely to become members of the SS than the general population, while teachers had an SS membership rate comparable to the rest of the population. Only lawyers were more likely to join the SS.[26] Kater explained the high membership numbers for physicians with economic reasons, as physicians became the highest earning professionals in the Third Reich.[27] On the other hand, a certain "social elitism" and its claim to "technical and intellectual perfection" may have made the SS particularly attractive for physicians.[28] The anatomists examined here were professors and assistant professors and a few junior employees, so most were civil servants, while the physicians in Kater's study also comprised large numbers of those employed in community hospitals and private practice. The overrepresentation of academic civil servants among the anatomists examined here may explain their high NSDAP membership rate. This interpretation is supported by the high rate of NSDAP membership in a group of academics studied by Adam. He reported from the University of Tübingen that 129 of the 160 faculty members were members of the NSDAP by 1945.[29] On the other hand, the number of disrupted careers in anatomy was comparable to the data reported for all German academics in 1933 by Grüttner and Kinas, who identified 1,145 expelled scholars in Germany, reflecting 19.3 percent of a total of 6,140 academics in 1933.[30] Clearly, those anatomists who remained active in their field in NS Germany had found an arrangement with the politics of the time.

NS Governmental Influence on Personnel Decisions in Anatomy

Academic advancement in NS Germany depended to a large extent on some measure of loyalty to the regime, as the NS government pursued— with varying degrees of success—its goal of controlling the universities.[31] Apart from university officials, particularly the deans of the medical faculties, the REM and the NSDAP via the NSDDB or the NSDStB took part to varying degrees in recruitment decisions.[32] All candidates were investigated by the secret service, so that no liberal or dissident academic had any chance of professional advancement.[33] The REM elicited evaluations for the appointment of chairman positions not only in the traditional manner by asking the opinion of leading anatomists, but also by evaluations from the local NSDDB.[34] The NSDAP contributed assessments through the office of Alfred Rosenberg (*Amt Rosenberg*), Hitler's representative for ideological education. The active National Socialists Enno Freerksen and August Hirt served as expert referees on anatomy for the *Amt Rosenberg* in 1942–43.[35] The anatomists Benno Romeis, Robert Heiss, and Titus von Lanz reported that Max Clara and Robert Wetzel, who were both leaders of their local NSDDB, provided evaluations of all candidates for chairman-ships in anatomy for the NSDAP.[36] Students from the NSDStB also voiced their opinions concerning recruitments.[37] While the REM tried to abolish the traditional appointment process for chairmanships in general, this pro-cedure was never really changed in anatomy. However, whereas anatomists themselves always stressed the importance of academic excellence in a candidate and were often supported in this by the respective dean, the NS government tried and succeeded to exert its influence in several instances of candidates with less than stellar professional evaluations. Traditionally, the medical faculty produced a list of three candidates ranked according to the recommendations from nationally leading anatomists, and the uni-versity would then start negotiations with them.[38] In some cases the NS authorities intervened directly by changing the list ranking or removing suggested names and inserting names of other anatomists, often active Na-tional Socialists, thus forcing certain anatomists into positions against the will of their colleagues.

This was certainly the case with Max Clara and his recruitments to Leipzig and Munich. Documents assembled by Winkelmann and Noack[39] paint a vivid picture of Clara's appointment process to the chairmanship in anatomy at the University of Leipzig. It all started with the search for a successor of Rudolf Fick (1866–39), chairman of anatomy in Berlin (and possibly the most prestigious position in German anatomy), who retired in 1934. The NSDAP, through the office of the chief of staff of the SA, tried to promote Max Clara as a candidate. They argued that, apart from scientific expertise, the "political attitude" of the candidate was decisive.[40] Clara's case was simultaneously pushed forward at the University of Leipzig, where the faculty had drawn up the usual list of three candidates, naming Alfred Bennighoff, Hans Petersen, and Stieve. For reasons unknown, these names were not mentioned again. Instead the faculty was given a choice, most likely by one of the representatives of the NSDAP or the REM, between Clara and Wetzel.[41] The anatomists' opposition against Clara rested on the fact that he had been recruited straight out of private practice and, while an accomplished histologist, he had no experience in medical educa-tion or gross anatomy. Fick was appalled by Clara's appointment in Leipzig and voiced his and presumably many of his colleagues' concerns about the search for his successor in a letter to the dean of the medical faculty in Berlin in December 1934. He first expounded the necessity of finding an experienced department leader and recognized scholar of anatomy to succeed him in order to uphold the status of German anatomy's scientific standards in the eyes of the world. Fick felt that the impending recruitment of the politically well-connected but professionally inexperienced Clara in Leipzig was a sign of such a loss of scientific standard, and was afraid that such a decision could be made in Berlin, too.[42] Ultimately Stieve, who was not a party member, was chosen for Berlin, while Clara became chair in Leipzig and remained there until 1942. Then Clara transferred to Munich, again against the wishes of the faculty he was joining.[43] He had negotiated this move with Max de Crinis, a high-ranking party member and advisor on medical science at the REM.[44]

The recruitment of Ernst Theodor Nauck to the chair of anatomy in Freiburg was another example of direct government intervention, and followed the pattern of Clara's appointment in Leipzig.[45] On 27 April 1935 the faculty presented a list with four candidates for the succession of Wilhelm von Möllendorff, who had immigrated to Zurich. For unknown reasons, neither Benninghoff, Goerttler, Petersen, nor Stöhr were chosen. Instead, the REM inserted the name of Nauck on the list, and he was recruited. Nauck was an active NS party member since May 1933 and joined the SS on 15 November 1941.[46] As a spy for the SS security service he contributed to an atmosphere of distrust at the University of Freiburg.[47] In contrast to Clara, at the time of his recruitment to Freiburg Nauck was professionally accepted by his colleagues. Benninghoff, who had collaborated with him, considered Nauck to be scientifically competent and personable.[48] Nauck's teacher Ernst Göppert had been of a friendly disposition toward him, but noted an unfavorable change in his personality due to his membership in the NSDAP.[49]

Robert Wetzel joined the NSDAP and the SA and spied for the SS security service like Nauck.[50] He encountered little opposition from the medical faculty at his recruitment to the chair of anatomy in Tübingen in 1936. The academic senate had advertised the position for a leadership personality that was "fully grounded in National Socialism".[51] At this time Wetzel was interim chair at the University of Giessen, and the evaluations of his scientific work and even of his political efforts were not completely favorable to a promotion. However, Wetzel was a favorite of the students, and his Swabian descent was considered an asset in communicating the needs of body procurement to the general population. Wetzel was ranked as number one in the recruitment process and promptly appointed.[52]

Wilhelm Blotevogel became a member of the NSDAP in 1933 and a leader of the NS Lecturers' League in Hamburg in 1934. The local NSD-StB proposed Blotevogel's recruitment as successor of Heinrich Poll, who had been dismissed for "racial" reasons.[53] The medical faculty elicited evaluations from Benninghoff and asked specifically about the local anatomists Brodersen and Blotevogel.[54] Benninghoff clearly ranked the scientific work and personalities of Brodersen and several external anatomists higher than

Blotevogel on his list.[55] Kurt Goerttler, another early National Socialist,[56] was finally chosen for the position. In March 1935 Benninghoff was again asked for his opinion when the REM wanted to recommend Blotevogel for the chair of anatomy in Breslau, which had become vacant after Heinrich von Eggeling's retirement.[57] Benninghoff had not changed his opinion: he found Blotevogel's work epigonal, his scientific interest lacking, and his impact at scientific meetings weak. He thought him personable and be-lieved he would get along well with the students, but did not expect much more of him.[58] Despite this, Blotevogel was recruited for the position in Breslau. Von Eggeling and Benninghoff were convinced that this was due to Blotevogel's close connections with authorities in the NSDAP. In 1946 Benninghoff called him a "decided winner of the previous system".[59]

As far as is known at this point, the most autocratic behavior by the REM seems to have been employed in the recruitment of August Hirt for the chair of anatomy at the university of Greifswald. According to Schneck, the REM directly appointed Hirt to this position on 1 April 1936 without any further consultation of the medical faculty.[60] At the time Hirt had been interim chair at Frankfurt University; however, his reputation among colleagues was mixed. In a 1938 evaluation of Hirt, Benninghoff wrote that his scientific work seemed to be serious, but that it was unclear how much of his most important publications on fluorescence microscopy had been produced by his former collaborator, pharmacologist Philipp Ellinger, who had been dismissed for "racial" reasons. Benninghoff also mentioned that Hirt had made no significant contributions at scientific meetings and that the main reasons that Hirt's colleagues had not mentioned him for new positions lay in his "controversial personality".[61] It remains unclear whether Benninghoff's term "controversial personality" only referred to Hirt's often loud and abrupt nature[62] or also to his activism for National Socialism as manifested in his memberships in NSDAP and SS, and his close association with Heinrich Himmler, leader of the SS.[63] Somebody in the NS hierarchy had taken a special interest in Hirt's career, an example of which can be seen in the inquiry by the dean of the medical faculty at the University of Cologne, who asked for Benninghoff's evaluation of Hirt specifically after having been prompted by unnamed authorities to con-sider Hirt for the succession of the dismissed Otto Veit.

Another man who clearly profited from his close connections with the NSDAP in terms of career advancement was Hellmut Becher. Becher was one of the candidates for the succession of Nauck in 1936 as chair of anatomy at Marburg, after Nauck had been recruited to Freiburg and Karl Zeiger's interim position as chair had been deemed unsatisfactory. Of the three scientifically qualified candidates, Adolf Dabelow was indispensable in Munich, Benninghoff was seen as politically questionable, and Becher was ultimately chosen.[64] Becher had joined the SA in 1934, the NSDAP and SS in 1937, and became the local leader of the NSDDB after his transfer to the University of Münster. His appointment in Münster had not only been supported by the faculty but also by de Crinis. From 1941 to 1945 Becher was dean of the medical faculty.[65]

Finally, the rapid career arc of Enno Freerksen can only be understood on the background of his political connections.[66] He had joined the NSDAP in 1932, the SA in 1933, the SS in 1938, and the SS security service in 1941.[67] He also held leadership positions in the NSDDB and NSDStB. When Benninghoff was recruited to Marburg in 1941, the list for his replacement as chair of anatomy included Dabelow, Zeiger, and Hans Schreiber. Dean Hanns Löhr, an active National Socialist, insisted on the addition of Klaus Niessing and Enno Freerksen to the list.[68] Apparently Freerksen already had good connections to the REM at that time, as Buddecke reports an agreement between Freerksen and de Crinis concerning Benninghoff's replacement in spring 1941. Thus Freerksen, a thirty-year-old rather inexperienced *ausserordentlicher* professor of anatomy, became the successor of one of Germany's leading anatomists. He was promoted to full professor in February 1945.[69] According to Ratschko, Freerksen was being groomed for a leadership position at the University of Kiel.[70]

The recruitments of Clara, Wetzel, Nauck, Blotevogel, Hirt, Becher, and Freerksen demonstrate the importance of the political position of candidates, as well as the various degrees of direct interference by the REM in the decision processes at the universities. With the possible exception of Blotevogel and Freerksen, these candidates, who held NS political leadership roles, had also been considered reasonably competent at least in some areas of anatomy. However, political ambition and activism alone were not sufficient to ensure career advancement in anatomy,[71] as the example of Werner Blume shows. Blume, an anatomist at the University of Göttingen, had joined the NSDAP in 1923. As a leader of the local NSDDB he was politically very active in administrative issues, which brought him in conflict with some of his colleagues. His career in anatomy stalled, and his younger colleague Erich Blechschmidt was chosen over Blume as successor to the chair in anatomy after Hugo Fuchs's retirement in 1941.[72] Blume's scientific achievements were considered mediocre at best. Benninghoff had been prompted to evaluate Blume's professional work and reputation among his peers and ranked him as the least capable of several candidates under consideration for Fuchs's succession.[73] A similarly unsuccessful supporter of the NS regime was Johann Paul Kremer, from the University of Münster.

The NS authorities' success in inserting several of their "own" candidates into anatomical departments throughout Germany often caused resentment in the community of anatomists, who insisted on the priority of scientific and educational excellence in a candidate. Despite these "political" appointments, the most prestigious chairs of anatomy in Berlin and Munich were awarded in 1935 to men who were not party members: Hermann Stieve and Walther Vogt.[74] Both were anatomists of international renown. And while it could be argued that it might have been possible to appoint a nonparty member in the early years of the Third Reich but not later on, it has to be pointed out that even the politically suspect Benninghoff was recruited to the chair in Marburg as late as 1941, at a time when he had not yet joined the NSDAP. So, while the political activities of a candidate were important, they were not decisive in every case, and final decisions depended, apart from government intervention and the scientific standard of the candidate, on the political and scientific expectations of the local faculty. Thus the faculty in Berlin clearly insisted on the highest scientific standard in their candidate, while the faculty in Tübingen looked for a leadership personality with a firm NS background. The postwar fate of the "political" candidates was varied and often depended on their ability to forge amicable contacts with the local faculty and other anatomists during NS times.

National Socialist Research Funding in Anatomy

Apart from their efforts to influence personnel decisions, the NS governmental authorities pursued their aim of an alignment of the sciences with NS ideology through research funding. Science had to serve war purposes and further the dominance and health of the German people. In general, research funding of German science was based on grants from the *Notgemeinschaft der Deutschen Wissenschaft* (emergency association of German science), which was *"gleichgeschaltet"* (aligned with) and administered by the REM from 1934 on. In 1937 the name was changed to *Deutsche Forschungsgemeinschaft* (DFG, German research association), and a *Reichsforschungsrat* (RFR, Reich research council) was created to support certain branches of science that were deemed particularly important for Germany's future.[75] From 1939 on, research projects had to be recognized as vital for the war effort to be financially supported by the government or for the government to grant temporary exemption of scientists from the military.[76] NS party membership was not necessarily a prerequisite for DFG support; however, NSDAP party members tended to have higher and longer funding than nonmembers.[77] Another funding program was Heinrich Himmler's *Forschungs- und Lehrgemeinschaft das Ahnenerbe e.V.* (research and teaching community ancestral heritage).[78] It supported some research projects directly through its *Ahnenerbe* foundation, which was mostly financed through private contributions.[79] Regional foundations also provided financial support, e.g., the *Provinzialverband der Provinz Westfalen* (regional association of the province of Westphalia).[80] In rare cases research was backed by German and international nonpolitical private foundations, and was thus less dependent on political alignment of the recipient. One example is the research institute led by Oskar and Cécile Vogt, who had been considered politically suspect and removed from their leadership position at the KWI for Brain Research, which was privately funded by their sponsor Gustav Krupp von Bohlen und Halbach.[81] In addition, the Vogts received money from the Rockefeller Foundation,[82] which also supported other research groups in Germany, among them Eugen Fischer's work in racial anthropology.[83] Funding from the Rockefeller Foundation became potentially life-saving for those scholars of anatomy who wanted to emigrate, whether their expatriation was forced following dismissal or they chose voluntary exile for political reasons. Such grants were instrumental in helping Franz Weidenreich and Ernst and Berta Scharrer.[84]

Current research has uncovered funding by the government and other groups for some anatomists, but information is far from complete. No scientist was forced to apply for research grants, all requests were voluntary.[85] The contents of research projects ranged widely from basic scientific questions of cell biology to inquiries in racial anthropology. In 1939 Wolfgang Bargmann, then in Leipzig, applied to the *Notgemeinschaft* for support of studies in epithelial cell growth in culture[86] and was granted 1,200 *Reichsmark* to employ a research technician, but was denied the purchase of a centrifuge. In 1944 in Königsberg he still pursued cell culture studies, trying to assess the influence of vitamins on epithelial growth. Leading physicians Karl Brandt and Ferdinand Sauerbruch supported his grant application to the RFR, and Bargmann was given a *Wehrmachtsauftrag* (commission by the German armed forces) on 8 January 1945 with a budget of 1,000 *Reichsmark*.[87] It is unlikely that any such work was carried out given the impending end of the war and Bargmann's flight from Königsberg. Bargmann's work may have been of interest for the government because of possible implications for wound healing, an essential field of study in war medicine.

Another successful grant application to the DFG/RFR was made by Hilde Krantz, Benninghoff's assistant, for a project on the influence of narcosis on the cellular nucleus in vitro. Krantz was awarded 5,000 *Reichsmark* for the establishment of a cell culture laboratory in 1944, but the money was never actually paid out.[88] Again, it can be assumed that the positive funding decision was based on the potential application of Krantz's future results to the field of traumatology. The DFG funded another cell biology project, Ferdinand Wagenseil's study of cells from, among other sources, an executed man.[89] Between 1941 and 1944 Enno Freerksen received funds totaling 16,800 *Reichsmark* for various studies, including those on blood vessels and blood production.[90] The DFG/RFR also funded work in practical anatomy, which enabled von Lanz to continue work on his atlas of surgical anatomy after his dismissal.[91]

Other projects focused on racial anthropology. Friedrich Heiderich, chairman of anatomy at the University of Münster, was a seasoned histologist and anatomist. He harbored an interest in physical anthropology, which extended to paleoanthropology and racial anthropology.[92] In 1937 he helped develop a concept for research on the local paleoanthropology and geographical anthropology of Westphalia.[93] In collaboration with the *Provinzialinstitut für Westfälische Landes- und Volkskunde* (regional institute for Westphalian regional studies and folklore) and the *Rassenpolitische Amt des Gaus Westfalen* (office of racial politics of the region of Westphalia), Heiderich had one of Eugen Fischer's students investigate prehistoric skulls from the region. He intended to produce a complete anthropological survey of the Westphalian population.[94] This project was financed by the *Provinzialverband der Provinz Westfalen*[95] but ended when Heiderich died in 1940. Another project in racial anthropology was planned in Vienna and received funding approval by the DFG/RFR on 22 April 1944. Eduard Pernkopf, chair of anatomy in Vienna, had agreed to collaborate with Lothar Loeffler, chair of racial biology, on a study of "prisoners of war from alien races," with a budget set at 3,000 *Reichsmark*. The researchers wanted to explore the "opportunities" for anthropometrical studies on prisoners of different "races", which were held in camps in the vicinity of Vienna.[96]

August Hirt's research was extensive and ranged from innovative methodology in microscopy to cancer studies and abusive human experiments. All of these received funding from various agencies within the NS regime. His early studies focused on the histology of the autonomous nervous system and intravital luminescence microscopy. The REM funded the salary for an assistant to apply this new technique to carcinoma studies in 1937,[97] and in 1935 Hirt received 1,780 *Reichsmark* from the DFG/RFR.[98] Later Hirt directed his research toward new fields of inquiry, possibly because he thought them more in line with the government's and specifically Himmler's wishes. The *Ahnenerbe* supported his activities financially as well as administratively; additional funding came directly from the University of Strasbourg and the DFG/RFR,[99] the latter granting him 9,000 *Reichsmark* each in 1941 and 1942.[100] His work included exposing prisoners at the concentration camp Natzweiler-Struthof to mustard gas in 1942, possible medical experiments on the sterilization of men, and plans for a so-called "racial" skeleton collection. The poison gas experiments and the work on vitamin metabolism were also funded by the DFG/RFR, and the former studies were registered in a secret file.[101]

The NS authorities funded a wide variety of anatomical research, based on the applicability of the results to NS purposes and the political evaluation of the applicant. However, their funding decisions shaped anatomical investigations to a lesser degree than the changes in the body supply provided to the anatomists for research, particularly the increase in bodies of executed persons.

Laws, executions and anatomical body procurement in NS Germany

The REM was generally responsible for the anatomical institutions at the universities and organized the body supply together with the *Reichsjustizministerium* (RJM, Reich Ministry of Justice). The latter functioned as the central coordinating authority for distributing the bodies of executed persons to the anatomical departments.[102] The specifics of delivering unclaimed bodies were regulated by state anatomy laws, which sometimes included the right to object to anatomical dissection by the deceased person's family.[103] A Prussian law from 1877 and other local laws—e.g., Hamburg 1907 and 1919—granted the anatomical institutes the right to use the bodies of executed persons for dissection under the condition that they were unclaimed. This legislation was reinforced by Prussia on 6 October 1933, when anatomists from Halle (Stieve), Göttingen (Fuchs), and Jena (Hans Böker) criticized a perceived lack of compliance by the authorities. The regulations were again reinforced by the RJM on 22 October 1935, and the REM on 18 February 1939.[104] Anatomists protested against the need to inform the families of the executed and resented the relatives' opposition to dissection.[105] Thus some anatomists asked for access to the bodies of the executed before the bodies left the execution facility in cases where the families requested their loved ones' bodies for burial. This request was granted in some jurisdictions, and anatomists dissected bodies secretly.[106] In June 1937 the RJM decreed that the bodies of all persons executed for high treason could not be claimed by their families any longer, but were to be delivered to an anatomical department.[107] Finally on 26 November 1942, a circular by the RJM announced that families of executed persons would not be informed of the date of the execution any longer, thus making it practically impossible for them to claim the bodies in time for a burial.[108] Despite this regulation, some families succeeded in retrieving the bodies of their loved ones even after the delivery to the anatomical department, as a letter by Clara, then chairman in Munich, reveals. He wrote on 31 May 1943 to the director of the Munich prison and execution site Stadelheim, complaining about the fact that families had tried to recover bodies from the anatomical department, and suggested, in order to prevent such incidents, that the prison should not share information about the delivery of the bodies to the anatomical department.[109] This complaint was possibly connected to the fate of the young dissidents Christoph Probst, Sophie Scholl, and Hans Scholl, who had been executed on 22 February 1943 and whose bodies were likely retrieved by their families from the anatomical department. In the later war years the fate of the bodies of the executed and their use for anatomical purposes had become well known among the general population,[110] thus Clara's wish for secrecy came too late.

In 1942 Germany had 240 independent correctional facilities,[111] and between 1935 and 1937 all places of execution had been initially centralized to fourteen facilities, rising to twenty-one by the end of the war. The measure was meant to increase the efficiency of the execution procedures.[112] With a decree from 1933 and amendments in 1935 and 1939, the RJM allocated to each place of execution certain anatomical institutes, which were to receive bodies from these facilities. The universities of Greifswald and Berlin received bodies from the Berlin-Plötzensee facility, and Berlin also received bodies from the Brandenburg-Görden facility; the University of Breslau from the Breslau and Posen facilities; the Universities of Frankfurt am Main, Giessen, and Marburg from the Frankfurt-Preungesheim facility; the Universities of Hamburg, Rostock, and Kiel from the Hamburg-Stadt facility; the Universities of Bonn, Münster, and Cologne from the Cologne facility (Münster also from Dortmund); the University of Königsberg from the Königsberg and Posen facilities; the Universities of Munich, Erlangen, Würzburg, and Innsbruck from the München-Stadelheim facility; the Universities of Freiburg, Tübingen, and Heidelberg from the Stuttgart facility; the University of Leipzig from Dresden; the Universities of Halle and Jena from the Weimar and Halle facilities; the University of Göttingen from the Wolfenbüttel facility; the Universities of Vienna and Graz from the Vienna facility; the Universities of Danzig, Posen, Strasbourg, and Prague from the facilities in each of those cities.[113]

A circular by the RJM from 28 December 1936 decreed as the general method of execution decapitation by guillotine.[114] Only special circumstances led to execution by shooting, for example by the GeStapo. Hanging was reserved for the crime considered most heinous, high treason, and the execution sites were furnished with butcher hooks for hanging.[115] The GeStapo also performed hangings with mobile gallows that were transported to the place of execution and most commonly used to execute forced laborers.[116] With the escalating political situation during the war, the legal definition of high treason came to encompass, aside from political offenses, actions like black marketeering, listening to enemy radio, mail theft, looting, theft during blackout, black-market butchering, the telling of political jokes, doubting the success of the NS regime and its war (which was called "defeatism"), and so-called *Rassenschande* ("racial" defilement). In addition, forced laborers from Poland and other countries were victims of special laws that led to frequent executions (*Volksschädlingsverordnung*, law concerning persons harming the German people, 5 September 1939; *Kriegswirtschaftverordnung*, law concerning the war economy, 4 September 1939; *Polenstrafrecht*, punitive law for Poles, 4 December 1941).[117] On 31 March 1943 the RJM decreed that the bodies of executed Polish citizens or persons of Jewish descent were under no circumstance to be released to their families.[118] While the *Volksgerichtshof* (people's court) dealt with prominent cases of political "crimes" committed by Germans and foreigners, the *Reichskriegsgericht* (Reich court martial or military court) was responsible for military trials, and the *Sondergerichte* (NS special courts) were specifically put in place to deal swiftly and rigorously with all other "offenses against the German people."[119]

Prison authorities cooperated closely with anatomists in terms of announcing upcoming executions and the collection of the bodies. Many execution facilities provided special dissection rooms for the anatomists, situated next to the execution chamber to facilitate quick tissue removal.[120] During an internal hearing at the university in Würzburg in 1958, Herrlinger gave a detailed description of such a set-up for the execution site in Posen. After the local prosecutor notified the anatomical institute of an upcoming execution, Herrlinger took position in the dissection room next to the execution chamber at the assigned time. Immediately after the decapitation, the body of the prisoner was pushed on a trolley into the dissection room, where Herrlinger then collected blood from the cut carotids (neck arteries) and opened the abdomen to remove the spleen. All this happened within forty to eighty seconds after the prisoner's death.[121] Kurt Neubert, chair of anatomy in Würzburg at the time, served as expert witness in Herrlinger's hearing and reported similar experiences from his work in Tübingen and Würzburg.[122] In Tübingen the anatomical institute had been located directly adjacent to the courthouse. During the 1920s executions were performed in the open yard of the courthouse, from which a door led into the anatomical institute. The execution site in Würzburg did not offer a special dissection room; however, Neubert and his superior Hans Petersen were allowed to set up dissection tables and tools, covered by black clothes, directly in the execution chamber within a few meters of the guillotine, to be able to start their work immediately after the decapitation of the prisoner.

Removal of tissues in the execution chambers was preferable to possibly lengthy transportation times in order to avoid postmortem deterioration of particularly sensitive "material." Bodies were usually transported in special vans owned by the anatomical institutes,[123] sometimes also by train.[124] Only in Berlin were the distances so short that Stieve was able to move the bodies quickly enough to the anatomical institute, which could be done in 20 to 30 minutes. And here again the prison authorities accommodated the needs of the anatomists by agreeing to arrange the executions early enough in the day to make further processing compatible with the schedule of the anatomical institute.[125]

The exact events can be reconstructed for the execution and body transport of Maria Diecker and Ewald Funke in Berlin.[126] Diecker, a thirty-two-year-old waitress from Gelsenkirchen, had been accused of spying and sentenced to death for high treason by the *Volksgerichtshof* on 9 May 1940. Her petition for clemency from 13 May 1940 was denied on 21 May 1940. On 16 July 1940 the prosecutor at the *Volksgerichtshof* issued an order concerning the procedure for the execution, a multipaged, preprinted form that was completed by hand with the specifics of each case. The date of execution was set for 20 July 1940 at 5:45 AM. The prisoner was to be informed of the date of execution on 19 July at 8:00 PM. Detailed instructions from the prosecutor were given to the prison authorities in a multi-part document, referring, among other things, to provisions for the executioner. Interestingly, the form, which was a standard paper printed before the execution, contains a passage noting that the body had not been claimed by the relatives and thus the anatomical institute was to be notified.[127] Furthermore, the form contained a note to the director of the anatomical department in Berlin, which informed him of the name of the convicted, verdict, and dates of verdict and execution; it also declared that the body was transferred for research and teaching purposes, and that neither the body nor any information about it was to be surrendered to the relatives. Strictest secrecy was to be maintained. The document also contained an admission ticket to the prison for the day of the execution for one employee of the anatomical department.[128] The prosecutor's reminder to keep the delivery of the body to the anatomical department secret seems curious, as Maria Diecker's verdict and execution were publicly announced on thirty red posters throughout the city.[129] Apparently Maria Diecker knew that her death was to be advertised in some manner and left a last handwritten appeal to the *Volksgerichtshof* to desist from announcing her death in the newspaper so that her parents might be saved from this "disgrace".[130] Diecker was executed on 20 July 1940. The anatomical technician would have already been admitted to the prison, and following the execution, the body would have been transported six kilometers to the anatomical department at the University of Berlin and dissected immediately.[131] Maria Diecker became number twelve on Stieve's 1946 list of the persons whose bodies he had used for research.[132] She was killed nine days before her thirty-fourth birthday. The remains of her body were most likely cremated in one of the three Berlin crematories and her ashes interred anonymously in a mass grave, perhaps in the municipal cemetery of Altglienicke.[133]

Another example of the extensive administrative interaction between the NS authorities and the anatomical department in Berlin is the case of the German communist and clerical worker Ewald Funke.[134] He had been imprisoned on 17 May 1936, was sentenced to death for high treason by the *Volksgerichtshof* Berlin on 16 August 1937, and was executed on 4 March 1938.[135] The prosecutor's instructions concerning the formal proceedings and notification of the anatomical department were the same in 1938 as in 1940. Ewald Funke's name was not among the ones on Stieve's list, which contained mostly names of women.[136] As he was executed before the number of executions had escalated, it is safe to assume that his body was not only transported to the anatomical facilities but also used for teaching or research purposes. The remains of his body would have shared the fate of Maria Dieker's, in that after dissection they were cremated and buried. On 24 November 1938 prison pastor E. Knodt of Tegel addressed a letter in the name of Ewald Funke's parents to the prosecutor, asking about their son's burial site. The prosecutor answered on 9 December in letters to both the parents and the pastor informing them Ewald Funke's body had been dealt with "according to existing regulations" and that information on a burial site could not be disclosed. That these "regulations" included the delivery of the bodies of executed persons to the anatomical institute was not mentioned.[137]

The collaboration between the government authorities which were responsible for body procurement and the anatomical departments was as close during the NS period as it had been before 1933. During the war years the relationship intensified, as the ministries of education and justice often accommodated petitions by anatomists, and new NS policies provided an abundance of bodies for most anatomical department. These bodies included increasing numbers of victims of the National Socialist system, especially unprecedented amounts of bodies of executed persons.

Notes

1. Novak 1949, 126.
2. Mommsen 1999.
3. Forsbach 2006; Rothmaler 1990; Koops 2005.
4. Wechsler 2005.

5. Böhm 1995.
6. Eberle 2002.
7. Forsbach 2006; Grundmann and Aumüller 1996; Thom 1990; Ribhegge 1985; Bussche 1989a; Remy 2002.
8. Mussgnug 1988, 95; Uhlig 1991, 143; Seidler and Leven 2007, 442; Grüttner and Kinas 2007, 133.
9. Körting 1968.
10. Grüttner and Kinas 2007, 139.
11. Ernst 1995; Mühlberger 1998b.
12. Müller-Wille 1991.
13. Piotrowski 1984; Wróblewska 2000, 89; Baechler et al. 2004.
14. Piotrowski 1984, 471.
15. Aumüller et al. 2001.
16. Letter from Veit to von Eggeling, 21 July 1946, University Archive Marburg Estate Benninghoff, UAM-EB; Novak 1949, 126.
17. Bussche 1989a.
18. Franze 1972, 55, 179–82; Jansen 1994; Böhm 1995; Grundmann and Aumüller 1996; Zimmermann 2000.
19. Geiss 1988; Lichtenberger-Fenz 1989, 5–6.
20. Kater 1989.
21. Walther 2008.
22. Kater 1989, 56.
23. Kater 1989, 254.
24. Rüther 1997, 167.
25. Kater 1979, 609.
26. Kater 1989, 70.
27. Kater 1989, 55, 60.
28. Kater 1987, 315.
29. Adam 1977, 153.
30. Grüttner and Kinas 2007.
31. Kelly 1985; Remy 2002.
32. Hartshorne 1938; Scheiblechner 2001; Adam 1977, 145; Grundmann and Aumüller 1996, 335.
33. Bussche 1993.
34. E.g., UAM-EB.
35. Grundmann and Aumüller 1996, 337.
36. Winkelmann and Noack 2010, "The Clara Cell", reference #18.
37. Bussche 1993.
38. Schneck 1993, "Die Berufungs- und Personalpolitik", 53.
39. Winkelmann and Noack 2010.
40. Ibid., reference 17 (additional material available in e-version of Winkelmann and Noack 2010).
41. Ibid., reference 18.
42. Ibid., reference 13.
43. Ibid., reference 18.
44. Ibid., reference 20.
45. Seidler and Leven 2007, 505.
46. Fragebogen, September 1944, University Archive Freiburg, UAF B24 Nr.2600.
47. Hellmich 1989, 181–84; Seidler and Leven 2007, 488.
48. Letter from Benninghoff to dean of medical faculty, University of Marburg, 10 January 1934, UAM-EB.

49. Letter from Göppert to Benninghoff, 22 May 1937, UAM-EB.
50. Klee 2003.
51. Scharer 2010, 818.
52. Mörike 1988.
53. Giles 1985, 126; Bussche 1993; Hildebrandt 2012; Klee 2003.
54. Letter from Medizinische Fakultät der Hamburgischen Universität, 3 October 1933, UAM-EB.
55. Letter from Benninghoff to Medizinische Fakultät der Hamburgischen Universität, 24 October 1933, UAM-EB.
56. Bussche 1993.
57. Letter from REM to Benninghoff, 14 March 1935, UAM-EB.
58. Letter from Benninghoff to REM, 18 March 1933, UAM-EB.
59. Correspondence von Eggeling/Benninghoff, letter from Beninghoff, 10 July 1946, UAM-EB
60. Schneck 1993, 58; Uhlmann and Winkelmann 2015.
61. Letter from Benninghoff to dean of medical faculty Cologne, 24 June 1938, UAM-EB.
62. Lachman 1977, 596.
63. Wechsler 2005, 126; Lang 2007.
64. Grundmann and Aumüller 1996, 341.
65. Klee 2003; Thamer et al. 2012; Aumüller et al. 2001.
66. Ratschko 2009; Ratschko 2013.
67. Aumüller et al. 2001; Ratschko 2013, 351.
68. Raschko 2013, 347.
69. Buddecke 2011.
70. Ratschko 2013, 346.
71. Giles 1993, 87.
72. Beushausen 1998, 188, 197.
73. Letter from Benninghoff to dean of medical faculty University of Göttingen, 6 June 1941, UAM-EB.
74. Winkelmann and Schagen 2009; Schütz et al. 2013.
75. Deichmann 1996; Hammerstein 1999; Mertens 2004.
76. Deichmann 1996; Grundmann 2001.
77. Mertens 2004, 212.
78. Pringle 2006; Kater 2006.
79. Pringle 2006, 139.
80. Mamali 2011, 14.
81. Reindl 2001.
82. Black 2003, 302–3.
83. Black 2003, 294.
84. Hertler 2008; Purpura 1998.
85. Roelcke 2005.
86. Beihilfeakte der Notgemeinschaft der Deutschen Wissenschaft bzw. der Deutschen Forschungsgemeinschaft, BA, R 73/10160.
87. *Karteikarte des Reichsforschungsrates*, BA, RFR, Bargmann, Dr. habl. Bal 16/72.
88. Grundmann 2001, 618.
89. Unger 1998, 86.
90. *Karteikarte des Reichsforschungsrates*, BA, RFR, Freerksen, Dr. habl. Fr 2/06.
91. Mertens 2004, 25; Bandmann et al. 1967.
92. Voit 1940.
93. Schmuhl 2003, 168.
94. Dicke 2006, 38.
95. Mamali 2011, 14.

96. *Karteikarte des Reichsforschungsrates*, BA, RFR, Bargmann, Dr. habl. Pe 1/05. Number K So RFR. 0832-3470/10-III/46

97. *Erlass* WG Nr.84 10 February 1937, letter from *Kurator* of the university to REM 27 October 1937, UAG PA 2161 Wimmer.

98. Karteikarte des Reichsforschungsrates, BA, RFR, Hirt Hir.

99. Hammerstein 1999, 425; Kater, 2006, 245–55.

100. Karteikarte des Reichsforschungsrates, BA, RFR, Hirt 1/02.

101. Karteikarte des Reichsforschungsrates, BA, RFR, Hirt Hir 1/04.

102. Noack and Heyll 2006, 133, 137.

103. Ibid., 135.

104. Viebig 2002,; Forsbach 2006; Noack 2008; Bussche 1989b; Rothmaler 1990.

105. Bussche 1989b; Waltenbacher 2008a, 218; Noack 2012, 288.

106. Noack and Heyll 2006, 12–13.

107. Ibid., 14

108. Ibid., 15.

109. Winkelmann and Noack 2010, reference 42.

110. Noack 2012.

111. Waltenbacher 2008a.

112. Noack 2008.

113. Viebig 1998; Waltenbacher 2008a; Waltenbacher 2008b, personal communication; Noack 2008; Oberkofler and Goller 1999.

114. Viebig 2002, 143.

115. Viebig 2002; Noack 2008; Waltenbacher 2008a.

116. Schönhagen 1992; Drews 1992; Richter 2009.

117. Beushausen et al. 1998, 233; Trobitzsch et al. 2002; Form and Schiller 2005.

118. Viebig 1998, 47.

119. Wagner 1974; Seeger 1998; Noack 2008

120. Noack 2008; Waltenbacher 2008a.

121. University Archive Würzburg UWü ZV PA Herrlinger H 31, Niederschrift … 30. Juni 1958, 15 Uhr, im Amtszimmer des Rektors, 4–5.

122. UWü ZV PA Herrlinger H 31, *Gutachten* 3.II. 1958.

123. Waltenbacher 2008a

124. Huter 1969.

125. Noack 2008, 137; Winkelmann and Schagen 2009,165.

126. Documents from the Gedenkstätte Deutscher Widerstand Bestand GDW/P.

127. Copy of letter from Oberreichsanwalt,16. Juli 1940, page 3; Maria Diecker Deutscher Widerstand Bestand GDW/P.

128. Ibid., 4–5.

129. Ibid., 6.

130. Copy of letter from Maria Diecker, An den Volksgerichtshof Berlin, Maria Diecker Deutscher Widerstand Bestand GDW/P.

131. Winkelmann and Schagen 2009, 165.

132. BA Ministerium der Justiz, DP1/6490.

133. Funke and Kröger 2008.

134. Ewald Funke Deutscher Widerstand Bestand GDW/P; also, Funke and Kröger, 2008.

135. Bundesstiftung Aufarbeitung, 2008; copy Verfügung Oberreichsanwalt beim Volksgerichtshof, Berlin 1. März 1938, Ewald Funke Deutscher Widerstand Bestand GDW/P.

136. Hildebrandt 2013.

137. Copy letter from E. Knodt An den Oberreichsanwalt 24.XI, 1938; copy letter Oberreichsanwalt An die Eheleute Funke, Oberreichsanwalt Strafanstaltspfarrer Herrn E. Knodt, 9. Dezember 1938, Ewald Funke Deutscher Widerstand Bestand GDW/P.

Archival Sources

Bundesarchiv (BA), Reichsforschungsrates (RFR), Gedenkstätte Deutscher Widerstand (GDW/P)
University Archives:
– Freiburg (UAF)
– Greifswald (UAG)
– Marburg (UAM), Estate (*Nachlass*) Benninghoff (EB)
– Würzburg (UWü)

Bibliography

Adam, Uwe Dietrich. 1977. *Hochschule und Nationalsozialismus: Die Universität Tübingen im Dritten Reich.* Tübingen: JCB Mohr (Paul Siebeck).

Aumüller, Gerhard, Kornelia Grundmann, Esther Krähwinkel, Hans H. Lauer, and Helmuth Remschmidt. 2001. *Die Marburger Medizinische Fakultät im "Dritten Reich."* München: K.G. Saur.

Baechler, Christian, Francois Igersheim, Pierre Racine. 2004. *Les Reichsuniversitaeten des Strasbourg et de Poznan et Les Resistances Universitaires 1941–1944.* Strasbourg: Presse Universitaires des Strasbourg.

Bandmann, H. J., D. Hamburger, H. Holzmann, and A. Kressner. 1967. "Prof. Titus Ritter von Lanz: In Memoriam." *Münchner Medizinische Wochenschrift* 109(16): 902–3.

Beushausen, Ulrich, Hans-Joachim Dahms, Thomas Koch, Almut Massing, and Konrad Obermann. 1998. "Die medizininsche Fakultät im Dritten Reich." In: *Die Universität Göttingen unter dem Nationalsozialismus,* ed. Heinrich Becker, Hand Joachim Dahms, and Cornelia Wegeler, 183–286. München: K.G. Saur.

Black, Edwin. 2003. *War against the Weak: Eugenics and America's Campaign to Create a Master Race.* New York: Four Walls Eight Windows.

Böhm, Helmut. 1995. *Von der Selbstverwaltung zum Führerprinzip: Die Universität München in den ersten Jahren des Dritten Reiches (1933–1936).* Berlin: Duncker & Humblot.

Buddecke, Julia. 2011. *Endstation Anatomie: Die Opfer nationalsozialistischer Vernichtungsjustiz in Schleswig-Holstein.* Hildesheim: Georg-Olms Verlag.

Bundesstiftung Aufarbeitung. 2008. "Ewald Funke." Accessed 9 February 2015. http://www.bundesstiftung-aufarbeitung.de/wer-war-wer-in-der-ddr-%2363%3B-1424.html?ID=4323.

Bussche, Hendrik van den, ed. 1989a. *Medizinische Wissenschaft im "Dritten Reich": Kontinuität, Anpassung und Opposition an der Hamburger Medizinischen Fakultät.* Berlin/Hamburg: Dietrich Reimer Verlag.

———. 1989b. *Im Dienste der "Volksgemeinschaft": Studienreform im Nationalsozialismus am Beispiel der ärztlichen Ausbildung.* Berlin/Hamburg: Dietrich Reimer Verlag.

———. 1993. "Personalpolitik und akademische Karrieren an der Hamburger Medizinischen Fakultät im 'Dritten Reich.'" *Akademische Karrieren im "Dritten Reich": Beiträge zur Personal- und Berufungspolitik an Medizinischen Fakultäten,* ed. Günter Grau and Peter Schneck, 19–38. Berlin: Pegasus.

Deichmann, Ute. 1996. *Biologists under Hitler.* Cambridge: Harvard University Press.

Dicke, Jan Nikolas. 2004. *Eugenik und Rassenhygiene in Münster.* Berlin: Weissensee Verlag.

Drews, Ulrich. 1992. "Die Zeit des Nationalsozialismus am Anatomischen Institut in Tübingen: Unbeantwortete ethische Fragen damals und heute." In: *Menschenverachtung und Opportunismus, Tübingen: Zur Medizin im Dritten Reich,* ed. Jürgen Peiffer, 93–107. Tübingen: Attempto.

Eberle, Henrik. 2002. *Die Martin-Luther-Universität in der Zeit des Nationalsozialismus 1933–1945*. Halle (Saale): Mdv Mitteldeutscher Verlag.

Ernst, Edzard. 1995. "A Leading Medical School Seriously Damaged." *Annals of Internal Medicine* 122 (10): 789–92.

Form, Wolfgang, and Theo Schiller. 2005. *Politische Justiz in Hessen: Die Verfahren des Volksgerichtshofs, der politischen Senate der Oberlandesgerichte Darmstadt und Kassel 1933–1945 sowie Sondergerichtsprozesse in Darmstadt und Frankfurt/M. (1933/34)*. Marburg: Elwert Verlag, Marburg.

Forsbach, Ralf. 2006. *Die Medizinische Fakultät der Universität Bonn im "Dritten Reich."* München: R. Oldenbourg Verlag.

Franze, Manfred. 1972. *Die Erlanger Studentenschaft 1918–1945*. Würzburg: Kommisionsverlag Ferdinand Schöningh.

Funke, Rainer, and Martin Kröger. 2008. "Tote irgendwo im Nirgendwo." *Neues Deutschland* (4 March), 3.

Geiss, Imanuel. 1988. *Geschichte des Rassismus*. Frankfurt am Main: Suhrkamp.

Giles, Geoffrey J. 1985. *Students and National Socialism in Germany*. Princeton, NJ: Princeton University Press.

Grüttner, Michael, and Sven Kinas. 2007. "Die Vertreibung von Wissenschaftlern aus den deutschen Universitäten 1933–45." *Vierteljahreshefte für Zeitgeschichte* 55(1): 123–86.

Grundmann, Kornelia. 2001. "Kriegswichtige Forschung." In: *Die Marburger Medizinische Fakultät im "Dritten Reich,"* ed. Gerhard Aumüller, Kornelia Grundmann, Esther Krähwinkel, Hans H. Lauer, and Helmut Remschmidt, 615–49. München: K.G. Saur.

Grundmann, Kornelia, and Gerhard Aumüller. 1996. "Anatomen in der NS-Zeit: Parteigenossen oder Karteigenossen? Das Marburger anatomische Institut im Dritten Reich." *Medizinhistorisches Journal* 31 (3–4): 322–57.

Hammerstein, Notker. 1999. *Die deutsche Forschungsgemeinschaft in der Weimarer Republik und im Dritten Reich. Wissenschaftspolitik in Republik und Diktatur 1920–1945*. München: Verlag C. H. Beck.

Hartshorne, Edward Y. 1938. "The German Universities and the Government." *Annals of the American Academy of Political and Social Science* 200: 210–34.

Heiber, Helmut. 1991. *Universität unterm Hakenkreuz: Teil I: Der Professor im Dritten Reich. Bilder aus der akademischen Provinz*. München: KG Saur.

Hellmich, Herrman-Josef. 1989. "Die medizinische Fakultät der Universität Freiburg i. Br. 1933–1945: Eingriffe und Folgen nationalsozialistischer Personalpolitik." Dissertation. Medical Faculty University of Freiburg.

Hertler, Christine. 2008. "Franz Weidenreich und die Anthropologie in Frankfurt: Weidenreichs Weg an die Universität." In: *Frankfurter Wissenschaftler zwischen 1933 und 1945*, ed. Jörn Kobes, and Jan-Otmar Hesse, 111–23. Göttingen: Wallstein Verlag.

Hildebrandt, Sabine. 2012. "Anatomy in the Third Reich: Careers Disrupted by National Socialist Policies." *Annals of Anatomy* 184: 251–56.

———. 2013. "The Women on Stieve's List: Victims of National Socialism Whose Bodies Were Used for Anatomical Research." *Clinical Anatomy* 26: 3–21.

Huter, Franz, ed. 1969. *Hundert Jahre Medizinische Fakultät Innsbruck 1869–1969: Band 1*. Innsbruck: Kommissionsverlag der Östereichischen Kommissionsbuchhandlung.

Jansen, Christian. 1994. "Das Verhältnis der Heidelberger Hochschullehrer zum Nationalsozialismus vor und während der Machtergreifung." In: *Hochschule und Nationalsozialismus,* ed. Walter Kertz. Braunschweig: Universitätsbibliothek der Technischen Universität, 23–35.

Kater, Michael H. 1979. "Hitlerjugend und Schule im Dritten Reich." *Historische Zeitschrift* 228(3): 572–623.

———. 1987. "Medizin und Mediziner im Dritten Reich: Eine Bestandsaufnahme." *Historische Zeitschrift* 244(2): 299–352.

————. 1989. *Doctors under Hitler.* Chapel Hill: University of North Carolina Press.

————. 2006. *Das "Ahnenerbe" der SS 1935–1945. Ein Beitrag zur Kulturpolitik des Dritten Reiches.* 4th ed. Munich: R. Oldenbourg Verlag.

Kelly, Reece C. 1985. "German Professoriate under Nazism: A Failure of Totalitarian Aspirations." *History of Education Quarterly* 25 (3): 261–80.

Klee, Ernst. 2003. *Das Personenlexikon zum Dritten Reich. Wer war was vor und nach 1945?* Frankfurt am Main: S. Fischer.

Körting, Walther. 1968. "Die Deutsche Universität in Prag: Die letzten hundert Jahre ihrer medizinischen Fakultät." *Schriftenreihe der Bayerischen Landesärztekammer* Band 11. München: Richard Pflaum Verlag.

Koops, Tilman. 2005. "Auf den Spuren der Reichsuniversitäten Posen und Strassburg im Bundesarchiv." In: *Les Reichsuniversitäten des Strasbourg et de Poznan et Les Resistances Universitaires 1941–1944,* ed. Christian Bächler, and François Igersheim, and Pierre Racine, 17–34. Strasbourg: Presse Universitaires des Strasbourg.

Lachman, Ernest. 1977. "Anatomist of Infamy: August Hirt." *Bulletin of the History of Medicine* 51: 594–602.

Lang, Hans-Joachim. 2007. *Die Namen der Nummern: Wie es gelang, die 86 Opfer eines NS-Verbrechens zu identifizieren. Überarbeitete Ausgabe.* Frankfurt am Main: S. Fischer Verlag.

Lichtenberger-Fenz, Brigitte. 1989. "Österreichs Universitäten und Hochschulen: Opfer oder Wegbereiter der nationalsozialistischen Gewaltherrschft? (Am Beispiel der Universität Wien)." In: *Willfährige Wissenschaft. Die Universität Wien 1938–1945,* ed. Gernot Heiss, Siegfried Mattl, Sebastian Meissl, Edith Saurer, and Karl Stuhlpfarrer, 3–15. Wien: Verlag für Gesellschaftkritik.

Mamali, Ioanna. 2011. "Psychiatrische und Nervenklinik Münster: Anfänge der Universitätspsychiatrie in Westfalen zur Zeit des Nationalsozialismus." Dissertation. Medical faculty of the Westfälische Wilhelms-Universität, Münster.

Mertens, Lothar. 2004. *"Nur politisch Würdige": Die DFG-Forschungsförderung im Dritten Reich 1933–1937.* Berlin: Akademie-Verlag.

Mörike, Klaus D. 1988. *Geschichte der Tübinger Anatomie.* Tübingen: JCB Mohr (Paul Siebeck).

Mommsen, Hans. 1999. *Von Weimar nach Auschwitz: Zur Geschichte Deutschlands in der Weltkriegsepoche. Ausgewählte Aufsätze.* Stuttgart: Deutsche Verlagsanstalt.

Mühlberger, Kurt. 1998b. "Enthebungen an der medizinischen Fakultät 1938–1945: Professoren und Dozenten." *Wiener Klininische Wochenschrift* 110(4-5): 115–20.

Müller-Wille, Michael. 1991. "Europäische Universitäten 1665–1990." In: *325 Jahre Christian-Albrechts-Universität zu Kiel: Jubiläumsfestakt am 15. November 1990,* ed. Rektorat der Universität Kiel, 31–82. Kiel: G+D Grafik und Druck GmbH.

Mussgnug, Dorothee. 1988. *Die vertriebenen Heidelberger Dozenten: Zur Geschichte der Ruprecht-Karls-Universität nach 1933.* Heidelberg: Carl Winter.

Noack, Thorsten. 2008. "Begehrte Leichen: Der Berliner Anatom Hermann Stieve (1886–1952) und die medizinische Verwertung Hingerichteter im Natinoalsozialismus." *Medizin, Gesellschaft und Geschichte. Jahrbuch des Instituts für Geschichte der Medizin der Robert Bosch Stiftung* 26: 9–35.

————. 2012. "Anatomical Departments in Bavaria and the Corpses of Executed Victims of National Socialism." *Annals of Anatomy* 194: 286–92.

Noack, Thorsten, and Uwe Heyll. 2006. "Der Streit der Fakultäten: Die medizinische Verwertung der Leichen Hingerichteter im Nationalsozialismus." In: *Geschichte der Medizin- Geschichte in der Medizin,* ed. Jörg Vögele, Heiner Fangerau, and Thorsten Noack, 133–42. Hamburg: Literatur Verlag.

Novak, Emil, ed. 1949. *Autobiography of Dr. Robert Meyer (1864–1947).* New York: Henry Schuman.

Oberkofler, Gerhard, and Peter Goller, eds. 1999. *Die medizinische Fakultät Innsbruck: Faschistische Realität (1938) und Kontinuität unter postfaschistischen Bedingungen (1945). Eine Dokumentation.* Innsbruck: Universitätsarchiv.

Piotrowski, Bernard. 1984. "Die Rolle der 'Reichsuniversitäten' in der Politik und Wissenschaft des hitlerfaschistischen Deutschlands." In: *Universities during World War II: Materials of the 49th International Symposium Held at the Jagiellonian University on the Anniversary of "Sonderaktion Krakau," Crakow October 22–24, 1979. Zeszyty naukowe Uniwersytetu Jagiellonskiego,* ed. Józef Buszko, and Irena Pszyńska. *Prace historyczne = Universitas Jagellonica Cracoviensis Acta scientiarum litterarumque. Schedae historicae* 643 (72): 467–86.

Pringle, Heather. 2006. *The Master Plan: Hitler's Scholars and the Holocaust.* New York, Hyperion.

Purpura, Dominick P. 1998. "Berta Scharrer." In: *National Academy of Sciences US: Biographical Memoirs. Volume 74,* 289–307. Washington DC: National Academic Press.

Ratschko, Karl-Werner. 2009. "Ernst Holzlöhner, Hans Gerhard Creutzfeldt und Enno Freerksen: Drei Kieler Medizinprofessoren im 'Dritten Reich.'" In: *Wissenschaft an der Grenze. Die Universität Kiel im Nationalsozialismus,* ed. Christoph Cornelissen, and Carsten Mish, 135–50. Essen: Klartext Verlagsgesellschaft.

———. 2013. "Kieler Hochschulmediziner in der Zeit des Nationalsozialismus: Die Medizinische Fakultät der Christian-Albrechts-Universität im 'Dritten Reich.'" Dissertation Philosophische Fakultät Christian-Albrechts-Universität Kiel.

Reindl, Josef. 2001. "Believers in an Age of Heresy? Oskar Vogt, Nikolai Timofeeff-Ressovsky and Julius Hallervorden at the Kaiser Wilhelm Institute for Brain Research." In: *Science in the Third Reich,* ed. Margit Szöllösi-Janze, 211–42. Oxford and New York: Berg.

Remy, Stephen P. 2002. *The Heidelberg Myth: The Nazification and Denazification of a German University.* Boston: Harvard University Press.

Ribhegge, Wilhelm. 1985. *Geschichte der Universität Münster: Europa in Westfalen.* Münster: Verlag Regensberg.

Richter, Gunnar. 2009. *Das Arbeitserziehungslager Breitenau (1940–1945): Ein Beitrag zum nationalsozialistischen Lagersystem.* Kassel: Verlag Winfried Junior. Accessed 11 March 2013. https://kobra.bibliothek.uni-kassel.de/bitstream/urn:nbn:de:hebis:34-20111205 39885/1/RichterArbeitserziehungslagerBreitenau.pdf.

Roelcke, Volker. 2005. "Die Deutsche Forschungsgemeinschaft (DFG) und die medizinisch-biologische Forschung 1920–1970. Kommentar." *Medizinhistorisches Journal* 40: 215–22.

Rothmaler, Christiane. 1990. "Gutachten und Dokumentation über das Anatomische Institut des Universitätskrankenhauses Eppendorf der Universität Hamburg 1933–1945." "1999", *Zeitschrift für Sozialgeschichte des 20. u 21. Jahrhunderts* 2: 78–95.

Rüther, Martin. 1997. "Ärztliches Standeswesen im Nationalsozialismus 1933–1945." In: *Geschichte der deutschen Ärzteschaft,* ed. Robert Jütte, 143–93. Köln: Deutscher Ärzte-Verlag.

Scharer, Philipp. 2010. "Robert F. Wetzel (1898–1962): Anatom, Urgeschichtsforscher, Nationalsozialist: Eine biographische Skizze." In: *Die Universität Tübingen im Nationalsozialismus,* ed. Urban Wiesing, Karl-Rainer Brintzinger, Bernd Grün, Horst Junginger, and Susanne Michl, 809–31. Stuttgart: Franz Steiner Verlag.

Scheiblechner, Petra. 2001. "1200 Wissenschaftler der 'ostereichischen' medizinischen Fakultäten und deren Mitgliedschaft bei NS-Teilorganisationen." In: *Medizin und Nationalsozialismus in der Steiermark,* ed. Wolfgang Freidl, Alois Kernbauer, Richard H. Noack, and Werner Sauer, 170–90. Innsbruck: Studienverlag Innsbruck.

Schmuhl, Hans-Walter. 2003. *Rassenforschung an Kaiser-Wilhelm-Instituten vor und nach 1933.* Göttingen: Wallstein Verlag.

Schneck, Peter. 1993. "Die Berufungs- und Personalpolitik an der Greifswalder Medizinischen Fakultät zwischen 1933 und 1945." In: *Akademische Karrieren im Dritten Reich. Beiträge zur Personal- und Berufungspolitik an Medizinischen Fakultäten*, ed. Günther Grau, and Peter Schneck, 51–62. Berlin: Pegasus Druck und Verlag.

Schönhagen, Benigna. 1992. "Das Gräberfeld X auf dem Tübinger Stadtfriedhof: Die verdrängte 'Normalität' nationalsozialistischer Vernichtungspolitik." In: *Menschenverachtung und Opportunismus, Tübingen: Zur Medizin im Dritten Reich*, ed. Jürgen Peiffer, 69–92. Tübingen: Attempto.

Schütz, Mathias, Jens Waschke, Georg Marckmann, and Florian Steger. 2013. "The Munich Anatomical Institute under National Socialism: First Results and Prospective Tasks of an Ongoing Research Project." *Annals of Anatomy* 195: 296–302.

Seeger, Andreas. 1998. "'Gegen Schwerstverbrecher ist in Kriegszeiten die zugelassene Todesstrafe grundsätzlich die gebotene': Todesurteile des Sondergerichtes Altona/Kiel 1933–1945." In: *"Standgericht der inneren Front" Das Sondergericht Altona/Kiel 1932–1945*, ed. Robert Bohn, and Uwe Danker, 166–89. Hamburg: Ergebnisse Verlag.

Seidler, Eduard, and Karl-Heinz Leven. 2007. *Die Medizinische Fakultät der Alberts-Ludwigs-Universität Freiburg im Breisgau: Grundlagen und Entwicklungen*. Freiburg: Verlag Karl Alber.

Thamer, Hans-Ulrich, Daniel Droste, and Sabine Happ. 2012. *Die Universität Münster im Nationalsozialismus. Kontinuitäten und Brüche zwischen 1920 und 1960*. Vols. 1 and 2. Münster: Aschendorff Verlag.

Thom, Achim. 1990. "Von 1933–1945." In: *575 Jahre Medizinische Fakultät der Universität Leipzig*, ed. Ingrid Kästner and Achim Thom, 162–202. Leipzig: Johann Ambrosius Barth.

Trobitzsch, Renate, Arnold Jürgens, and Wilfried Knauer. 2002. *Justiz im Nationalsozialismus: Über Verbrechen im Namen des Deutschen Volkes*. Baden-Baden: Nomos-Verlagsgesellschaft.

Uhlig, Ralph, ed. 1991. *Vertriebene Wissenschaftler der Christian-Albrechts-Universität zu Kiel (CAU) nach 1933*. Frankfurt am Main: Peter Lang Verlag.

Uhlmann, Angelika, and Andreas Winkelmann. 2015. "The Science Prior to the Crime: August Hirt's Career Before 1941." *Annals of Anatomy*, published online http://dx.doi.org/10.1016/j.aanat.2014.10.001.

Unger, Michael. 1998. *Ferdinand Wagenseil (1887–1967): Integrer Forscher und Bewahrer der Medizinischen Fakultät Giessen*. Giessen: Wilhelm Schmitz Verlag.

Viebig, Michael. 1998. *Das Zuchthaus Halle/Saale als Richtstätte der Nationalsozialistischen Justiz (1942–1945)*. Halle: Druckerei Heinrich John.

Viebig, Michael. 2002. "Zu Problemen der Leichenversorgung des Anatomischen Institutes der Universität Halle vom 19. bis Mitte des 20. Jahrhunderts." In: *Beiträge zur Geschichte der Martin-Luther-Universität 1502-2002*, ed. Hermann-Josef Rupieper, 117–46. Halle: MDV, Mitteldeutscher Verlag Halle.

Voit, Max. 1940. "Friedrich Heiderich +." *Anatomischer Anzeiger* 90: 185–91.

Wagner, Walter. 1974. "Der Volksgerichtshof im nationalsozialistischen Staat." In: *Quellen und Darstellungen zur Zeitgeschichte. Band 16/III. Die Deutsche Justiz im Nationalsozialismus*. Stuttgart: Deutsche Verlagsanstalt.

Waltenbacher, Thomas. 2008a. *Zentrale Hinrichtungsstätten: Der Vollzug der Todesstrafe in Deutschland von 1937–1945. Scharfrichter im Dritten Reich*. Berlin: Zwilling.

———. 2008b; personal communication via electronic mail December 16, 2008.

Walther, Peter T. 2008. "Entlassungen und Exodus: Personalpolitik an der Medizinischen Fakultät und in der Charité 1933." In: *Die Charité im Dritten Reich Zur Dienstbarkeit medizinischer Wissenschaft im Nationalsozialismus*, ed. Sabine Schleiermacher and Uwe Schagen, 37–50. Paderborn: Ferdinand Schöningh.

Wechsler, Patrick. 2005. "La Faculté des Medicine de la 'Reichsuniversität Strassburg' (1941–1945) a L'Heure Nationale-Socialiste." "Dissertation." Faculté de Medicine de Université de Strasbourg.

Winkelmann, Andreas, and Thorsten Noack. 2010. "The Clara Cell: A 'Third Reich Eponym'?" *European Respiratory Journal* 36: 722–27.

Winkelmann, Andreas, and Udo Schagen. 2009. "Hermann Stieve's Clinical-Anatomical Research on Executed Women During the 'Third Reich.'" *Clinical Anatomy* 22(2): 163–71.

Wróblewska, Teresa. 2000. *Die Reichsuniversitäten Posen, Prag und Strassburg als Modell Nationalsozialistischer Hochschulen in den von Deutschland besetzten Gebieten.* Torun: Marszalek.

Zimmermann, Susanne. 2000. *Die medizinische Fakultät der Universität Jena während der Zeit des Nationalsozialismus.* Berlin: Verlag für Wissenschaft und Bildung.

Chapter 4

THE NS STATE AND
THE *ANATOMISCHE GESELLSCHAFT*

Als Anatomen geht uns die Politik nichts an.
(Politics are of no concern to us as anatomists.)

Otto Veit, 1946[1]

In 1886 the *Anatomische Gesellschaft* (Anatomical Society) was founded
as the main organizing body of German-speaking anatomists with the
declared goal of embracing an international membership.[2] Its official organ was the *Anatomischer Anzeiger,* renamed Annals of Anatomy in 1992.
In 1932 the society had 310 members: 144 Germans and the remainder
from Europe and overseas.[3] The annual meetings—at which presentations
in German, English, French, and Italian were made and later published
in the *Anatomischer Anzeiger*—were interrupted by the two world wars.
The *Anatomische Gesellschaft* was forced to react to the political challenges
brought on by the NS regime, as the new government imposed its influence not only on the universities but also on all professional organizations
and scientific societies.

Professional Medical Societies during the NS Period

Most medical professional organizations in Germany had been founded
in the nineteenth century and traditionally operated independently from
outside rule. However, the NS regime aimed to centralize control through
the Reich Ministry of the Interior, which sought an alignment of all professional medical associations with official policies, and by 1938 had gained
oversight of sixty-five medical scientific societies. The reorganization proceeded through several steps in 1934 and 1935, requiring the societies' co-

operation through consultations, and expecting all leaders to be of Aryan descent and to fully support the regime. The ministry also controlled the societies' contacts with international colleagues, including travel to international meetings.[4]

Societies reacted in different ways, from voluntary self-alignment with NS policies to insistence on independence within the limits of a totalitarian regime. The leading associations of physicians in Germany, *Ärztevereinsbund* and *Hartmannbund,* declared their full support for Hitler and his party as early as 21 March 1933, and all state and regional branches of these associations were aligned with the new regime over the following months.[5] An elimination of all Jewish members and society leaders followed.[6] The *Hartmannbund* was integrated into the administration of health insurance providers in 1933. It collaborated with the government in excluding "non-Aryan" physicians from the official health-care compensation systems, thus effectively eliminating any professional opportunities for 95 percent of all Jewish physicians within six years.[7]

Representatives of the *Deutsche Gesellschaft für Kinderheilkunde* (Society of Pediatricians) attended a meeting of scientific societies at the Reich Ministry of the Interior on 14 July 1933, where they discussed a merger with centralized and NS-controlled health organizations. The society was able to avoid changes of its bylaws concerning the removal of "non-Aryan" members, but it did not or could not prevent the "voluntary" withdrawal of membership of 75 percent of all Jewish colleagues over the next six years. Also, changes in the bylaws accepting the authoritarian NS leadership principle were introduced.[8] The *Deutsche Pathologische Gesellschaft* (Society of German Pathologists) lost its Jewish president Gotthold Herxheimer, whose successor declared the society's loyalty to Hitler and the new state in 1934. The bylaws were changed in favor of the authoritarian NS leadership principle. From 1934 to 1935, 15 percent of all members "voluntarily" withdrew from the society.[9]

Similar decisions were made by the *Deutsche Ophthalmologische Gesellschaft* (German Ophthalmological Society) in 1934. While this society did not actively exclude any members for "racial" or political reasons, members "voluntarily" withdrew, of which a large contingent were Jewish.[10] The renowned German research society *Leopoldina* expelled Jewish members, among them Albert Einstein and anatomists Alfred Kohn, Franz Weidenreich, Max Flesch, and Alfred Fischel.[11]

Several professional societies were quite obvious in their elimination of Jewish members. In 1933 the *Deutsche Gesellschaft für Urologie* (German Society of Urology), which had many prominent Jewish members and leaders, was replaced by the NS-loyal *Gesellschaft Reichsdeutscher Urologen* (society of Reich-German urologists).[12] In May 1933 the board of directors of the

Deutsche Orthopädische Gesellschaft (German orthopedic society) resigned and Jews were banned from the following interim leadership. The society was integrated in one of the central NS governmental working groups, and its secretary worked closely with the Reich Ministry of the Interior. The bylaws were changed according to the NS leadership principle. While the society did not exclude Jewish members actively at first, they were struck from the roster in 1939.[13] In October 1933, at the national meeting of the society for gynecology, its president Walter Stoeckel promoted the self-alignment of the society with the NS government policies[14] and convinced his Jewish colleagues on the board of directors to voluntarily resign from their positions and potential Jewish speakers at the conference to withdraw their talks. Stoeckel acknowledged the "deplorable" fate of colleagues who lost their positions due to the new legislation, but pointed out that these sacrifices were necessary for the "restoration of the health of the German people".[15] The society then sent words of loyalty via telegram to Hitler. In 1934 the *Gesellschaft für Gerichtliche und Soziale Medizin* (the German Society for Forensic Medicine) changed not only its bylaws by excluding "non-Aryans," but also decided on a new name that explicitly included its German origin, *Deutsche Gesellschaft für die gesamte gerichtliche Medizin* (German Society of All of Forensic Medicine).[16]

Currently, a systematic comparison of the motivations behind the decisions of each professional society is still missing. It is unclear why several changed their bylaws while others tried to avoid this. Were some of them under more pressure by the government than others? Were the decisions dependent on the personal politics of the leaders of the societies? Or was the fact decisive that various societies were based on an international membership and others not? Several of these points became important in the decisions of the *Anatomische Gesellschaft*. However, it has to be noted that all professional societies arranged themselves somehow with the new government, some enthusiastically and others less so. In the end, many Jewish and politically dissenting members were lost due to protest, emigration, economic reasons, or death.

The *Anatomische Gesellschaft* in the Third Reich

The official postwar historiography of the *Anatomische Gesellschaft* maintained that the society protected its persecuted members and resisted any NS efforts from within and without to convert it from an international society into a purely German one.[17] Until recently, a closer scrutiny of the society's decisions had been deemed impossible due to the loss of archival records from that period. However, new sources have been discovered and reveal that the history of the *Anatomische Gesellschaft* during the NS period was much more complex and contentious than has been assumed. Among the newly explored documents are the official statements of the society published in its proceedings and its member registries, as well as papers from the estate of Alfred Benninghoff, who was a longtime board member.[18] An investigation of the recently discovered papers of Heinrich von Eggeling has just been published.[19]

Winkelmann's analysis of the society's official publications in the *Verhandlungen der Anatomischen Gesellschaft* (proceedings of the Anatomical Society) shows that, in contrast to other professional medical societies, none of the statements by the *Anatomische Gesellschaft* supported the new NS regime at any time. Max Clara, as president of the 1939 meeting in Leipzig, did not even use his address of welcome to voice any political proclamations.[20] Also, the bylaws of the society remained unchanged, even though the discussion at the 1934 meeting in Würzburg had touched on the introduction of the authoritarian NS leadership principle. An analysis of the membership lists from 1932 to 1950 included names for 453 German and international members. A total of fifty-seven of these members suffered from a disruption of their careers due to NS policies, meaning anything from a loss of career advancement to dismissal, incarceration, and death in a concentration camp.[21] Of this group, forty remained on the membership roster until at least 1939 or until their death, and thirteen were still members in 1950.[22] Of the seventeen memberships that ended between 1933 and 1939, eleven had been noted to have "dropped out" or "resigned," while six were cancelled because of nonpayment of fees.[23] Indeed, some of these anatomists had emigrated by then and did not pay fees. While the *Anatomische Gesellschaft* did not actively exclude persecuted members, it certainly lost several of them during the NS period, just as the other professional societies. Winkelmann concluded that the society differed from many of the others in that it did not accede to anti-Jewish regulations or the NS leadership principle, and was able to maintain its international status against the calls for a purely German society. While the background to these decisions could not be explained from the official statements and membership lists, it became much clearer with an analysis of Benninghoff's estate papers. The letters he exchanged with his colleagues between 1934 and 1950 allow an intimate insight into the decision-making processes of the leaders of the society.[24]

Heinrich von Eggeling's tenure as secretary of the *Anatomische Gesellschaft* lasted from 1918 to 1949. Von Eggeling (1869–1954) was associate professor of anatomy in Jena from 1902 to 1922 and then chaired the department of anatomy at the University of Breslau from 1922 until his retirement in 1934. His work focused on comparative anatomy and physical anthropology.[25] In Stieve's 1938 address on the occasion of von Eggeling's seventieth birthday, he characterized the anatomist as a fine gentleman with great knowledge of human nature. He praised his ability to balance the needs of an active international society in the time directly after WWI and his selfless work as publisher of the *Anatomischer Anzeiger*.[26] As will be discussed, von Eggeling fulfilled his duties as secretary of the society with great diligence and did so effectively during politically turbulent times, never wavering from his nondramatic style of leadership. Indeed, the author of his obituary called von Eggeling an "old-school career diplomat."[27] His political affiliations remain unknown. After his retirement he moved to Berlin, but after the 1943 bombings there, he relocated to Neustadt am Rübenberge, a rural town close to Hannover. In his function as secretary of the *Anatomische Gesellschaft* he corresponded extensively with the members of the board of directors, Stieve and Benninghoff, who both had been elected to this position in 1934.

Stieve's personality was in many ways the direct opposite of von Eggeling's, as he was unabashedly assertive and self-assured, often autocratic in his decisions, and an outspoken German conservative nationalist who never joined the NSDAP. Benninghoff (1890–1953) was chairman of anatomy at the University of Kiel from 1927 to 1941 and from then on in Marburg until his death in 1953.[28] Apart from his work on a textbook of functional anatomy, he developed a unifying concept of the human body that integrated psychological and biological principles of human nature.[29] Benninghoff did not approve of the NS regime, but felt compelled to join the NSDAP in 1941 under pressure from the SS officer Freerksen, his former colleague and successor in Kiel. After the war Freerksen contested Benninghoff's accusation of having forced him into the NSDAP membership, and the two became estranged.[30]

Martin Heidenhain and Siegfried Mollier belonged to the group of "elder statesmen" of the society and sat on its board until 1934. Heidenhain (1864–1949) was famous for his innovative histological techniques and served as chair of anatomy in Tübingen from 1917 to 1933. He was internationally recognized as one of the leaders of German anatomy.[31] Mollier (1866–1954) served as chairman for histology and embryology in Munich from 1902 to 1935.[32]

In addition to Germans, the board of the society always included international members. Between 1931 and 1938 these included Ivar Broman of Sweden, Joseph Schaffer of Austria, Granville Harrison of the United States, and Tivadar Huzella of Hungary. However, the international members were not involved in the discussions concerning German politics, or in any society decisions in reaction to these politics. In 1938 a new board was elected with Torsten Hellman from Sweden, Eduard Pernkopf, Walther

Vogt, and Max Clara. Pernkopf (1888–1955) held the chair of anatomy in Vienna from 1933 to 1945, and Vogt (1888–1941) had the same position in Munich from 1935 until his death in 1941.[33] Anatomist Robert Wetzel (1898–1962) was also involved in society discussions at the time, and he became one of the main forces behind the push for a purely German society.

While other professional medical societies actively engaged in questions of political self-alignment as soon as possible, the *Anatomische Gesellschaft* addressed this issue somewhat later, but still before the completion of the governmental reorganization of the health administration in late 1934.[34] On 16 February 1934 von Eggeling sent a letter directly to Hitler, asking him for guidance in the process of aligning the society with the new policies.[35] This document reveals the full scale of von Eggeling's diplomatic skills. He explained in great detail the international character of the society and its important position in the world of science, especially in competition with the equally international French anatomical society. Von Eggeling attributed the continued dominance of the *Anatomische Gesellschaft* over its French counterpart to its international character, held in the face of demands from some of members for a purely German entity. Von Eggeling also addressed the question of a "Germanification" of the society and the need for decisions before its next meeting in April 1934. Specifically, he mentioned Heidenhain as a member of the board, who, von Eggeling claimed, had retired from his teaching position because of his "non-Aryan" descent. Also, von Eggeling emphasized that the important contact with the "non-Aryan" Italian members Guiseppe Levi (Turin) and Tullio Terni (Padua) would have to be severed if "Aryanization" was to be pursued. Von Eggeling ended by declaring that he and Mollier, the current leaders of the *Anatomische Gesellschaft*, did not feel entitled to make such far-reaching decisions by themselves and were thus seeking a ruling from Hitler.

Von Eggeling's letter is remarkable insofar as it addressed the pertinent questions of a "Germanification" and exclusion of "non-Aryan" members directly, seemingly asking for advice from the authorities. It used the arguments of national pride, that is the German arch-rivalry with France, and the danger for the German nation to lose international supremacy. In this manner von Eggeling attempted to manipulate the response he wished for: the command to maintain the continuity of an international society. The introduction of the NS-leadership principle for the society was not addressed. On 6 April 1934, Arthur Gütt of the Reich Ministry of the Interior answered the letter and, in collaboration with the Ministry for Foreign Affairs, followed von Eggeling's reasoning and allowed the society to remain international. Gütt did, however, hew to the official political line by insisting that, in the future, members of the board of directors of the *Anatomische Gesellschaft* had to be of "Aryan" descent.[36]

While von Eggeling achieved his goal, he felt another threat to the internationality of the society in the form of a circular from 10 October 1934, in which the Reich Ministry of Health—in collaboration with the Ministry of the Interior—informed all scientific societies of plans of merging them into a single collaborative elite working structure under the oversight of the ministry.[37] Von Eggeling wondered if the circular's specifics—the consulting for the Reich Ministry of Health by German anatomists—could be fulfilled by creating an affiliated German section within the international society, possibly even under the leadership of an NSDAP party member. Benninghoff saw no need for immediate action but suggested, in case this became necessary, to name a Berlin-based liaison for the German members of the society. He suggested anatomists Rudolf Mair and Eugen Fischer as candidates.[38] A letter from the Reich Ministry of the Interior to the *Anatomische Gesellschaft* from 1 August 1935 explicitly confirmed the following to the board of a new "German Section": Stieve as president, Benninghoff as his deputy, von Eggeling as secretary, and Vogt of Munich, Wilhelm Pfuhl of Frankfurt, and Blotevogel of Breslau as members. While Pfuhl and Blotevogel were NSDAP party members, the others were not at that time.[39] Von Eggeling saw the creation of the "German Section" as a mere formality, but Stieve took his role as president quite seriously.

In April 1934 Andreas Pratje (1892–1963), a politically active National Socialist from Erlangen,[40] asked at the society meeting in Würzburg[41] for an alignment of the society with NS policies and requested the introduction of the NS leadership principle to abolish "senseless" rounds of voting.[42] He also specified these ideas in a letter to von Eggeling dated 9 January 1935. While von Eggeling considered Pratje's suggestions to be on the whole acceptable,[43] the bylaws were never changed significantly.[44]

Meanwhile Martin Heidenhain, one of the oldest and most prestigious members of the society, was severely affected by the political changes. In July 1933 the government of the state of Württemberg had lowered the retirement age for academic teachers from seventy to sixty-eight years, thereby effectively eliminating some no-longer welcome professors like Heidenhain, who had a Jewish grandfather. He was forced to step down from his chair position in the same year, but the popular professor was able to continue teaching in a lesser role until 1939, when his health took a turn for the worse. He died in 1949.[45]

At the 1934 meeting of the *Anatomische Gesellschaft* in Würzburg an incident occurred that greatly upset Heidenhain, and caused him to write a letter of protest to the newly elected Stieve and Benninghoff on 4 May 1934.[46] He reminded them that at the preceding society meeting in Lund in 1932, it had been decided not to elect a new board but rather to deputize the old one, consisting of Broman, Schaffer, Mollier, and Heidenhain.

Heidenhain was told by Mollier that it was going to be his turn to preside over the next meeting. Meanwhile, von Eggeling had written to Hitler in early 1934 and received the answer, which essentially demanded the exclusion of Heidenhain from the board of directors of the *Anatomische Gesellschaft*. Mollier communicated this result to Heidenhain in a letter, which only reached Tübingen on 22 April, when Heidenhain had already left for the meeting in Würzburg. When Heidenhain arrived, Mollier told him that younger colleagues might object to Heidenhain's presidency of the opening ceremonies, and that he was afraid that the press might find out about the situation. Heidenhain waived his rights to leading the opening ceremony, but only as a matter of expediency, not of principle. However, Mollier also declared that Heidenhain could not deliver his introductory talk. Heidenhain thought this was an outrageous decision, and deeply hurt, he wrote: "I believe the rejection of my talk, a work carefully prepared over many weeks, a scientific work within a scientific society, to be subjectively a monstrosity, objectively a severe injury to my person."[47] He was so outraged by Mollier's behavior that he only desisted from immediate departure in consideration of his wife, who had accompanied him.

Heidenhain explained in his letter to Stieve and Benninghoff that the Law for the Restoration of the Professional Civil Service did not apply in his case due to certain exclusion clauses that he fulfilled. He considered it an irreparable failure and illegal action by the leadership of the *Anatomische Gesellschaft* to have deprived him of his rights as president of the meeting and member of the board without attending to the details of this law. He emphasized that up to this point he had not been discriminated against regarding his descent by any authority or societal group, but found himself now denounced by his own colleagues, and he would never accept an apology from Mollier. He expected the new board to solve this conflict by sending a declaration on Heidenhain's legal status and an official apology to all society members. Stieve and Benninghoff answered with placating letters assuring Heidenhain of their great respect and admiration, but they did not respond to this request. Indeed, it seems that Benninghoff could not quite fathom the depth of Heidenhain's feeling of humiliation and betrayal when he wrote, "At this first meeting after the coup d'état, it was not pertinent to follow the niceties of §3 of the Law for Civil Servants but to ensure a smooth run of the meeting," and that Mollier's intention had been to save Heidenhain any unpleasantness.[48] It is not known if this matter, which according to Stieve had been noted as far away as the Netherlands, was ever resolved. There is no indication that von Eggeling's skills as negotiator were employed in this conflict.

Von Eggeling's diplomatic qualities are shown again to full effect in the proceedings of the society. He had a unique way of hiding potentially controversial issues in plain sight. For example, he reported in the official minutes of the meeting in Leipzig in 1938, "Discussion of a call for a meeting of directors of German anatomical institutes about the acquisition of material for courses of microscopy."[49] However, in a letter to Benninghoff, von Eggeling referred to a request by Eduard Nauck in the following manner: "Also, Mr. Nauck asks for a meeting of the directors of anatomical institutes about the acquisition of material from the bodies of executed persons".[50]

Von Eggeling proceeded in similar "fact-obscuring" fashion with his bland reference to Pratje's petition for new bylaws for the *Anatomische Gesellschaft* at the 1934 meeting.[51] Wolfgang Bargmann remembered this incident as "some members asking for the introduction of the '*Arierparagraph*'" (shorthand for the Law for the Restoration of the Professional Civil Service in 1933).[52] However, Pratje's letter to von Eggeling in January 1935 revealed that he had asked for a full alignment of the society with NS principles of leadership and membership. The question of an exclusion of "non-Aryan" members he deferred to the leadership.[53]

There is also no indication from the official proceedings that the meeting in Würzburg had been a crucial one, as it was the first one to occur under the new NS regime. No conflicts were reported in the proceedings.[54] Apart from the Pratje incident, the meeting may indeed have been very calm and quiet on its surface, as Mollier and von Eggeling had tried their utmost to avoid all confrontation with so-called "younger" members holding NS sympathies, and the decision about Heidenhain remained secret. Bargmann remembered Friedrich Wassermann—another so-called "non-Aryan"—frowning at Pratje's political proposal. Mollier's course of action concerning Heidenhain looked very much like "preemptive obedience." Given the fact that the government ruling came only shortly before the meeting, Mollier could have ignored it for a few days longer to allow Heidenhain a dignified farewell from his role as leader of the society. Even if Mollier was afraid of younger colleagues and the press, he could have trusted men like Stieve to help him keep order. At the very least, Stieve and Benninghoff could have shown more empathy for the plight of their older colleague. Heidenhain had clearly been wronged, and even if all leaders of the society thought that the political calculations of the day demanded this iniquity, they could at least have shown some outspoken regret for their behavior and sympathy for their renowned colleague beyond their reassurance of respect for him.

The society functioned in its usual manner until late 1942, when Robert Wetzel again suggested a purely German anatomical society. While national and international meetings of the *Anatomische Gesellschaft* had ceased in 1939, Wetzel arranged a conference of anatomists in Tübingen on 5–7 November 1942.[55] Among the speakers were the SS members Hirt and Freerksen and other government loyalists.[56] Wetzel followed up the conference with a petition for the founding of a German anatomical society.[57] Forty-three colleagues attended the "retreat for anatomists [original: *Anatomenlager*] of the committee on human biology within the office for sciences/Reich lecturers' leadership organization." A list of attendants of this meeting is not known. Wetzel reported that on 5 November 1942, all colleagues unanimously voted for the creation of an association of German anatomists. The preliminary board of directors of this new organization was to include Robert Wetzel as president, Pernkopf as deputy, and Klaus-Dieter Niessing from Leipzig as secretary. They decided on the name "*Deutsche Anatomenschaft*" (German association of anatomists) and its location in Munich. Wetzel had also drafted bylaws for the new association,[58] which provided that the president was to be the main authority—thus following the NS-leadership principle—and the members German professors at German universities. Pernkopf felt this new group could harm the existence of the original international society and disagreed with Wetzel's draft, which had already been signed by Wetzel, Clara, Petersen and Hirt.

Wetzel thought Pernkopf's attitude toward this issue as overly timid. On 8 October 1943 he wrote to Benninghoff in his function as an official of the NSDDB, apparently after a previous phone conversation with him that addressed this future German anatomical society.[59] He reported that only two days before, the Reich Ministry of the Interior had informed him that the creation of a German society was expected, and that the ministry had questions for Benninghoff. He wanted to send the draft first to some "closer allies" like Kurt Neubert of Rostock, Arno Nagel of Halle, Niessing, Freerksen, and Kurt Goerttler of Heidelberg, and later on as a circular to all anatomists. Wetzel did not expect a potential competition between the new German organization and the old international one, but envisioned every German anatomist to be a member of both, and saw von Eggeling as the main obstacle to his own idea of a purely German society.[60]

For von Eggeling 1934 and 1943 were not the first years in which he had to counter demands for a purely German *Anatomische Gesellschaft* and the threat of ending its international character. In 1920, such a proposal had been brought forward by Carl Hasse, emeritus of anatomy in Breslau, at the society meeting in Bonn, but it was not discussed further.[61] The problem of "Germanification" seemed to have been solved until this new initiative by Wetzel. Contrary to Kühnel's report,[62] it was not Pernkopf, who was the most active protagonist for this German group, but Robert Wetzel. Pernkopf's procrastination contributed to delays in the matter, and the petition seems to have been ultimately lost or forgotten in times of escalating war action.

This "*Anatomenlager*" was also remarkable for another, deeply insidious event. According to Lang, it was here that Hirt proposed an entirely new kind of anatomical body procurement for anatomy: making a person's death part of the research design, and possibly even actively kill for anatomical purposes. This method already conformed to the lines of Hirt's own human experiments, which were supported by the SS organization *Ahnenerbe*. Lang quotes Hirt reporting in a letter to Wolfram Sievers, the managing director of the *Ahnenerbe*: "Others are gradually becoming aware that something could happen here" and "I have been commissioned to compose guidelines for the collection of materials for all German anatomists."[63] Just as with the "*Deutsche Anatomenschaft*," there is no documentation following up on these plans. However, it can be surmised that the atmosphere at the November 1942 meeting must have been a radicalizing one that fostered political extremism, including a willingness to transgress traditionally existing ethical boundaries.

The *Anatomische Gesellschaft* after the War

In the summer of 1945 the Allied Control Council prohibited all professional societies, including the *Anatomische Gesellschaft*.[64] The *Anatomische Anzeiger* and the proceedings of the society, the society's main means of communication, were no longer published. A first circular on the postwar status of the society was mailed by Stieve to all colleagues on 28 January 1946, who called himself director of the German section of the society.[65] In it he explained that the last board of directors was now defunct, as Hellman and Vogt had died, and Clara and Pernkopf had lost their positions. He wanted to reconnect the remaining members of the society as soon as possible so they could share information on the status of anatomical institutes and their personnel. He also thought it would be particularly important to gather as many facts as possible in light of media reports accusing anatomical departments of atrocities—e.g., soap production from human bodies, referring to the scandal surrounding Rudolf Spanner in Danzig. Stieve concluded by listing all information he had on the individual university institutes and anatomists, and congratulated Titus von Lanz in Munich, Hermann Hoepke in Heidelberg, and Otto Veit in Cologne on their reinstatement to their old positions after having been "undeservedly removed" in previous years.[66]

Benninghoff responded by sending his own circular to all colleagues on 7 February 1946, adding to and correcting Stieve's information.[67] He sent separate versions of the circular to von Eggeling and Stieve in which he advised caution concerning the distribution of details on imprisonments and dismissals, as some colleagues might not want these facts widely known.[68] He also added in his letter to Stieve that he hoped the society would be revived and wished him personally all the best in these times, clearly referring to Stieve's position in the Soviet occupation zone.

Veit was highly incensed by Stieve's message and dismissed his congratulations outright, as he felt Stieve had insulted him in previous professional encounters. He called Stieve's letter "typical Stieve pomposity and tactlessness."[69] Von Eggeling was also somewhat critical of Stieve's intervention, as he disliked Stieve's mention of the German section of the society, which von Eggeling had hoped could be disregarded in the postwar situation. At the same time he doubted the practicability of Benninghoff's idea of leaving out information on imprisonments and dismissals in times when people were searching for information about each other. He felt that previous political stances could not be ignored any longer, especially in "severe cases," among which he counted Hirt, Wetzel, Clara, and Blotevogel, who he believed had achieved their positions because of political connections.[70]

It was difficult for von Eggeling and Stieve to collaborate, as the former found himself increasingly isolated from his colleagues due to unreliable mail delivery, lack of telephone service, severe travel limitations across borders of the different occupation zones, and difficulties with such basic needs as paper for circulars.[71] He worried whether the publishers of the *Anatomische Anzeiger* would (or could) resume its production. He was also concerned with "getting rid of members that don't belong to us any longer," but did not specify whom he was referring to, other than talking about persons who had "not done much for the society." While he called NS career anatomists later the *"Firma WETZEL-HIRTH [sic] und Genossen"* (company of Wetzel-Hirth and Co.) and praised Stieve's battle against them,[72] he objected to the removal of Pernkopf and Clara from the board of directors solely for political reasons.[73]

To von Eggeling's doubts that the German section of the society should try to reconnect internationally at this time, Benninghoff answered that he had already heard from Switzerland through inquiries about his textbook, which signified for him the renewal of international contacts after the war and hopes for a future meeting of anatomists.[74] However, he believed that such a meeting should be postponed until the end of the denazification process and not yet include members from abroad.[75] Concerning the removal of members from the society, Benninghoff reminded von Eggeling in January 1947 that Benninghoff wanted to exclude those who had committed iniquities and harmed others as well as those who had achieved their positions only through political connections.[76]

Stieve's sixtieth birthday prompted another exchange of letters between von Eggeling, Benninghoff, and Stieve in May and June of 1946.[77] Benninghoff and von Eggeling did not share Stieve's opinion that the board of the society now consisted of the three of them, and considered Stieve's one-sided determination as undemocratic. They considered only von Eggeling, the secretary, as a legitimate representative of the *Anatomische Gesellschaft*. Stieve also intended to begin negotiations for the society with the occupying forces, which had started to recognize other professional organizations. At the same time von Eggeling had handed in a formal application for approval of the *Anatomische Gesellschaft* in the British zone. He had doubts about Stieve's freedom of action within Soviet occupation.

Parallel to the exchange between Benninghoff, Stieve, and von Eggeling, Veit, Stöhr, Bargmann, and Blechschmidt had a similar discussion.[78] Veit (1884–1972) was chairman of anatomy at the University of Cologne from 1927 to 1937, when he was removed from this position because of a Jewish grandfather. He was reinstated by the British authorities in 1945 and finally retired in 1957.[79] Stöhr (1891–1979) was chairman of anatomy at the University of Bonn from 1935 to 1962.[80] Wolfgang Bargmann (1906–78), chairman of anatomy at the University of Kiel from 1946 to 1974,[81] and Erich Blechschmidt (1904–92), chairman of anatomy at the University of Göttingen from 1941–73,[82] were anatomists of a younger generation and also involved in this effort to revive the society. Bargmann and Blechschmidt had contacted Veit and Stöhr in 1946 to discuss a meeting of anatomists, possibly in Göttingen. While Veit and Stöhr agreed about the meeting, they preferred Bonn as the host city.

The British military authorities approved the September 1946 conference in Bonn for all "members of the society," as Veit had specifically put it. He wanted to avoid inviting only those anatomists currently working who had been approved by the authorities, because this would have excluded colleagues who were still suspended for political reasons. Only the anatomists in the Soviet zone would be barred from attendance due to travel restrictions. Veit did not want to wait until all denazification processes were finished, as he correctly anticipated that these would last years.[83]

At least fourteen anatomists attended the meeting, called "*Anatomen-Versammlung*" (Meeting of Anatomists), and this was considered a success.[84] Among them was Benninghoff along with three colleagues from Marburg, and von Eggeling presided over the meeting. No formal matters of the society were discussed; however, a publication of the results in the FIAT (Field Information Agency, Technical) reviews, edited by Stöhr, was presented.[85] As Benninghoff had anticipated, Stieve was upset about his inability to attend[86] and called the conference a "Nazi-meeting," presumably because of the presence of anatomists who had been dismissed by the occupying forces for political reasons, and Stieve claimed that this harmed German anatomy in the eyes of other countries.[87]

The next conference was again held in the British occupied zone in Bonn in September 1947, under the presidency of von Eggeling. Heidelberg and Marburg were located in the American zone, where members who had not yet undergone denazification were forbidden to participate.[88] Although nearly seventy participated in the event, colleagues from the Soviet zone were still missing. Bluntschli from Switzerland and Glees from England contributed talks, thus indicating signs of newly budding international relationships. In another sign of international connection during the summer of 1947, Benninghoff exchanged letters with Häggquist in Sweden, telling him about the current situation in Germany and describing the mood of many Germans as "depressed."[89]

The publication of the *Anatomischer Anzeiger* was granted by the Soviet authorities in March 1947. However, the publisher, Gustav Fischer-Jena, was located in the Soviet occupation zone and beholden to the legislation of the Soviet Military Administration. The publication was only granted under the conditions that Stieve was designated as coeditor with von Eggeling, who had previously held this position alone since 1918, and that there were only 4 issues of the journal per year. Von Eggeling held the opinion that creating a new journal might be a better choice than continuing with the *Anatomischer Anzeiger* under these restrictions,[90] and Benninghoff agreed with him. Against Benninghoff's and von Eggeling's wishes, Stieve continued to extend his power within the society not only by assuming the leadership for the journal, but also for the society in general by negotiating its recertification in Berlin.[91] All of these circumstances may have contributed to von Eggeling's growing disenchantment with his colleagues and his decision to retire as editor of the *Anatomische Anzeiger* in 1948, which left Stieve in an acerbic mood against him.[92] Von Eggeling continued to pursue the idea of a new anatomical journal, the short-lived *Anatomische Nachrichten* (*Anatomical News*) and found a publisher by February 1949.[93]

A third informal meeting of anatomists took place in Bonn from 25 to 28 April 1949, at which the *Anatomische Gesellschaft* was officially refounded.[94] Von Eggeling had by then retired from his position as secretary and did not attend, giving his limited financial and living circumstances as well as the lack of a connection with a university in his rural location as reasons.[95] Bargmann was elected as the new secretary, and Veit, Stöhr, Curt Elze of Würzburg, and Dietrich Starck of Frankfurt were elected as the new board of directors. Sixty anatomists attended the conference, which was presided over by Albert Hasselwander of Erlangen.

According to von Eggeling, the new leadership of the society expected him to share the editorship of the new anatomical journal, *Anatomische Nachrichten* with the society's new secretary, Bargmann. Von Eggeling refused this for formal reasons of tradition, and because he did not trust Bargmann's motivation.[96] However, a year later von Eggeling had to acknowledge that, while the first issue of his new journal *Anatomische Nachrichten* had been launched successfully, the new leadership of the society did not share his doubts about the quality and freedom of the *Anatomischer Anzeiger*, but continued to use it as the main organ of communication with the members. While he was not willing to support this decision, he had to admit defeat for his plans with the new journal, as no further funding was granted. He saw his role in the rebuilding of the society at an end.[97] The ongoing conflicts with colleagues like Stieve as well as von Eggeling's lack of appreciation of the younger colleagues' efforts, an increasing sense of isolation, and possibly his age (he turned 80 in 1948) may have contributed to his resignation from the post of secretary of the society and as editor of the *Anatomischer Anzeiger*. Herrlinger called this manner of ending a long, successful, and dutiful career a "*Missklang*" (jarring note).[98]

Meanwhile Bargmann's negotiations with the British leadership of anatomists resulted in the first official postwar invitation for German anatomists to an international meeting, the Federative International Congress of Anatomy in Oxford in July 1950.[99] Benninghoff and two of his colleagues attended the meeting. In a letter to von Eggeling, Benninghoff compared the Oxford meeting with the meeting of the *Anatomische Gesellschaft* in Kiel which followed in August, and concluded that, while Oxford was satisfactory, the scientific standard of presentations in Kiel was higher. He was supported in this evaluation by Swedish colleagues. He also reported that Bargmann had resigned from his post as secretary of the society, purportedly because of a planned sabbatical abroad.[100] The reasons for Bargmann's resignation are not yet documented, but Benninghoff's hope, expressed in a letter to von Eggeling, that the newly elected secretary Maximilian Watzka would be more conciliatory than Bargmann, points to the strong tensions that had developed between the older and younger members of the society.[101] Oxford allowed Benninghoff to reconnect with Walter Brandt, a colleague from the anatomical institute in Bonn, who had emigrated from NS-Germany for political reasons.[102] Benninghoff told Brandt that the German anatomists' reception in England had been much more cordial than he had expected and that he had met old friends and made new ones.

Although Bargmann's name is often first mentioned in regards to the postwar revival of the *Anatomische Gesellschaft*, certainly justified in terms of reconnecting international colleagues,[103] the Benninghoff papers testify that von Eggeling, Stieve, and Benninghoff, and later Veit and Stöhr, were also actively involved in consolidating the remaining membership and rebuilding the society. Conflicts certainly still erupted, as opinions varied about the exclusion of formerly active National Socialists. The outcome of these discussions is not documented and a closer analysis of the lives of anatomists in Germany shows that there was a strong continuity of personnel during and after the war in German anatomy, including some very active National Socialists.

With Heinrich von Eggeling's resignation, an era of the *Anatomische Gesellschaft* came to an end. During his tenure as secretary from 1918 to 1948, he served as the keeper of the international character of the society, thereby avoiding nationalistic influences after WWI and the alignment of the society with NS ideology later on. His allies in this endeavor were Mollier until 1934, and then the autocratic Stieve, who with his aggressive and strong personality had great impact on other members. At the same time Benninghoff's quiet influence and wide connections within the society may have helped stabilize the politically undecided part of the membership. In the end, the demands of such vocal NS anatomists as Wetzel and Pratje were never fulfilled. However, a closer look at the society leaders and their decisions reveals that they were not infallible, as incidents like the Heidenhain controversy illustrate. As Winkelmann's investigation has shown and the Benninghoff papers confirm, the society did not defend its vulnerable members as decisively as Kühnel and Schierhorn have suggested.[104] The papers show personal conflicts intermingled with political pressures reaching far into the postwar refounding period. They also reflect the political realities for German professionals trying to reestablish their science in a country divided by the occupation forces. The leaders of the society in these difficult times died within three years of each other: Stieve in 1952, Benninghoff in 1953, von Eggeling and Mollier—both a generation older than the other two—in 1954.

In his address of welcome at the first official postwar meeting of the *Anatomische Gesellschaft* in 1950, Veit only hinted at the "difficult fate that came over the world" since the last meeting in Budapest in 1939, not mentioning any rifts within the society itself, but remembering the members deceased since then.[105] Two years later Stöhr spoke more clearly in his speech celebrating the fiftieth meeting of the society. He stated that anatomical departments were institutions of the state and thus dependent on the goodwill of politicians. Further he acknowledged that the "destructive forces of the Second World War" and the years leading up to it had "left deep marks" in the previously productive *Anatomische Gesellschaft*, essentially leading to a loss of ten years' worth of mental labor and a distinct lack of suitable young candidates for leading positions in German anatomy.[106] Stöhr recognized that "some members, carried by the governing party and disoriented about reality" had caused a dangerous conflict in the society, and characterized the general situation at the end of the war as a "mental, political and material field of ruins."[107] Stöhr was apparently much more ready to acknowledge the past and future difficulties of the society than was his colleague Veit, who was very conciliatory despite the fact that he himself had suffered so much. The reality of the history of the *Anatomische Gesellschaft* confirms once again science's close association with and dependence on the governing politics, thus refuting Veit's opinion that "politics are of no concern to us as anatomists."[108]

Notes

1. Letter from Veit to von Eggeling, 21 July 1946, University Archive Marburg Estate Benninghoff, UAM-EB.
2. Schierhorn 1980.
3. Winkelmann 2012.
4. Thom 1991.
5. Rüther 1997, 143.
6. Rüther 1997, 147; Jachertz 2008.
7. Rüther 1997, 143.
8. Seidler 2000, 31; Jahnke-Nückles 1992, 54.
9. Lampert 1991.
10. Rohrbach 2007.
11. Gerstengarbe et al. 1995; Initiative Stolpersteine Frankfurt am Main 2009.
12. Krischel et al. 2010.
13. Thomann and Rauschmann 2001.
14. Schagen 2010.
15. Ibid., 209.
16. Herber 2002, 208.
17. Herrlinger 1965; Schierhorn 1980; Schierhorn 1986; Kühnel 1989.
18. Winkelmann 2012; Hildebrandt 2013a.
19. Winkelmann 2015a and b.
20. Winkelmann 2012, 248.
21. Hildebrandt 2012.
22. Winkelmann 2012, 248.
23. Ibid., 249.
24. Hildebrandt 2013a.
25. Herrlinger 1956.
26. Schierhorn 1980; Stieve 1939.
27. Herrlinger 1956, 374.
28. Dabelow 1954; Grundmann and Aumüller 1996.
29. Dabelow 1954; Niessing 1953.
30. Aumüller et al. 2001.
31. Jacobj 1952.
32. Lanz 1955.
33. Spemann 1942.
34. Thom 1991.
35. Copy of letter from von Eggeling to *Reichsführer*, 16 February 1934, University Archive Marburg, Estate Benninghoff, UAM-EB.
36. Letter from Gütt to von Eggeling, 6 April 1934, UAM-EB.
37. Letter from president of the Reich Ministry of Health to Anatomical Society, 10 Oc-

tober 1934; letter from von Eggeling to board of directors of the *Anatomische Gesellschaft*, 26 October 1934; UAM-EB; see also Thom 1991.

38. Letters from Benninghoff to von Eggeling, 1 and 28 November 1934, UAM-EB.
39. Letter from Reich Ministry of the Interior to the AG, 1 August 1935, UAM-EB.
40. Wittern 1993; Isabel Braun, personal communication.
41. Winkelmann 2012.
42. Copy of letter from Pratje to von Eggeling, 9 January 1935, UAM-EB.
43. Letter from von Eggeling to colleagues, 12 January 1935, UAM-EB.
44. Winkelmann 2012.
45. Mörike 1988, 70.
46. Letter from Heidenhain to members of board, Stieve, and Benninghoff, 4 May 1934, UAM-EB.
47. Ibid., 1, reverse side.
48. Letter from Benninghoff to Heidenhain, 19 May 1934, UAM-EB.
49. Von Eggeling 1939.
50. Letter von Eggeling to Benninghoff, 5.August 1938, UAM-EB.
51. Von Eggeling 1934.
52. Bargmann 1971.
53. Copy of letter from Pratje to von Eggeling, 9 January 1935, UAM-EB.
54. Von Eggeling 1934.
55. Lang 2007; Lang 2013.
56. Winkelmann 2015.
57. Copy draft of letter from Wetzel to Reich Ministry of the Interior, 31 May 1943, UAM-EB.
58. Copy, bylaws, Deutsche Anatomische Gesellschaft e.V., undated, UAM-EB.
59. Letter from Wetzel to Benninghoff, 8 October 1943, UAM-EB; Hildebrandt 2013a.
60. Hildebrandt 2013a.
61. Kühnel 1989.
62. Ibid., 54.
63. Lang 2013, 373.
64. Winkelmann 2012.
65. Letter from Stieve to Benninghoff and other members, 28 January 1946, UAM-EB.
66. Hildebrandt 2012.
67. Letter from Benninghoff to members of AG, 7 February 1946, UAM-EB.
68. Letter from Benninghoff to Stieve, 7 February 1946, and von Eggeling, 19 February 1946, UAM-EB.
69. Letter from Veit to Benninghoff, 19 February 1946, UAM-EB.
70. Letter from von Eggeling to Benninghoff, 3 March 1946, UAM-EB.
71. Letter from von Eggeling to Benninghoff, 16 April 1946, UAM-EB.
72. Letter from von Eggeling to Benninghoff, 29 April 1946, UAM-EB.
73. Letter from von Eggeling to Benninghoff, 21 June 1946, UAM-EB.
74. Letter from Benninghoff to von Eggeling, 9 May 1946, UAM-EB.
75. Letter from Benninghoff to von Eggeling, 28 June 1946, UAM-EB.
76. Letter from Benninghoff to von Eggeling, 12 January 1947, UAM-EB.
77. Correspondence Stieve, von Eggeling, Benninghoff, May and June 1946, UAM-EB.
78. Letter from Veit to von Eggeling, 21 July 1946; further correspondence Veit, von Eggeling, Benninghoff, July and August 1946, UAM-EB.
79. Hildebrandt 2012.
80. Fleischhauer 1981.
81. Hildebrandt 2013b.
82. Hinrichsen 1992.
83. Letter from Veit to von Eggeling, 21 July 1946, UAM-EB.

84. Von Eggeling 1946.
85. Stöhr 1947.
86. Letter from Benninghoff to von Eggeling, 15 October 1946, UAM-EB.
87. Letter from Stöhr to Benninghoff, 16 April 1947, UAM-EB.
88. Hildebrandt 2013a; von Eggeling 1948.
89. Letter from Benninghoff to Häggquist, summer 1947, UAM-EB.
90. Letter from Benninghoff to von Eggeling, 3 April 1947, UAM-EB.
91. Correspondence Stieve, Benninghoff, von Eggeling, March and April 1947, UAM-EB.
92. Herrlinger 1956; letter from von Eggeling to Benninghoff, 19 November 1948, UAM-EB.
93. Letter from von Eggeling to Benninghoff, 5 February 1949, UAM-EB.
94. Herrlinger 1965.
95. Letter from von Eggeling to Benninghoff, 8 May 1949, UAM-EB.
96. Letters from von Eggeling to Benninghoff, 8 May 1949 and 23 December 1949, UAM-EB.
97. Letter from von Eggeling to Benninghoff, 6 December 1950, UAM-EB.
98. Herrlinger 1956, 382.
99. Letter from Bargmann to Benninghoff, 5 January 1950, UAM-EB.
100. Bargmann 1951.
101. Letter from Benninghoff to von Eggeling, 26 September 1950, UAM-EB.
102. Correspondence Brandt, Benninghoff, December 1950, UAM-EB; Hildebrandt 2012.
103. See Herrlinger 1965; Kühnel 1989; Schierhorn 1980.
104. Kühnel, 1989, 53; Schierhorn 1980, 184, 185; Schierhorn 1986, 959.
105. Veit 1951, 3.
106. Stöhr 1952, 3.
107. Ibid., 4.
108. See note 1.

Archival Sources

University Archive Marburg, Estate Benninghoff, UAM-EB

Bibliography

Aumüller, Gerhard, Kornelia Grundmann, Esther Krähwinkel, Hans H. Lauer, and Helmuth Remschmidt. 2001. *Die Marburger Medizinische Fakultät im "Dritten Reich."* München: K.G. Saur.

Bargmann, Wolfgang. 1951. "Geschäftliches." *Verhandlungen der Anatomischen Gesellschaft* 97: 243–45.

———. 1971. "Friedrich Wassermann 13. August 1884–16. Juni 1969." *Anatomischer Anzeiger* 128: 1–15.

Dabelow, Adolf. 1954. "Alfred Benninghoff +." *Anatomischer Anzeiger* 100: 157–65.

Eggeling, Heinrich von. 1946. "Bericht über die Anatomentagung in Bonn am 2. und 3. September 1946." *Ärztliche Wochenschrift* 1: 250–53.

———. 1948. "Anatomentagung in Bonn am 3., 4. und 5. September 1947." *Ärztliche Wochenschrift* 3: 372–82.

Eggeling, Heinrich von, ed. 1934. "Stand der anatomischen Gesellschaft nach Schluss der zweiundvierzigsten Versammlung (Würzburg 1934)." *Verhandlungen der Anatomischen Gesellschaft* 78: 247–60.

Eggeling, Heinrich von, ed. 1939b. "Stand der anatomischen Gesellschaft nach Schluss der sechsundvierzigsten Versammlung (Budapest 1939)." *Verhandlungen der Anatomischen Gesellschaft* 88: 313–28.

Fleischhauer, Kurt. 1981. "In Memoriam Philipp Stöhr Jr." *Anatomischer Anzeiger* 150: 239–47.

Gerstengarbe, Sybille, Heidrun Hallmann, and Wieland Berg. 1995. "Die Leopoldina im Dritten Reich." *Die Elite der Nation im Dritten Reich—Das Verhältnis von Akademien und ihrem wissenschaftlichen Umfeld zum Nationalsozialismus.* Acta Historica Leopoldina 22: 167–212.

Grundmann, Kornelia, and Gerhard Aumüller. 1996. "Anatomen in der NS-Zeit: Parteigenossen oder Karteigenossen? Das Marburger anatomische Institut im Dritten Reich." *Medizinhistorisches Journal* 31(3–4): 322–57.

Herber, Friedrich. 2002. *Gerichtsmedizin unterm Hakenkreuz.* Leipzig: Militzke Verlag.

Herrlinger, Robert. 1956. "Heinrich von Eggeling +." *Anatomischer Anzeiger* 102: 373–82.

———. 1965. "Kurze Geschichte der Anatomischen Gesellschaft." *Anatomischer Anzeiger* 117: 1–60.

Hildebrandt, Sabine. 2012. "Anatomy in the Third Reich: Careers Disrupted by National Socialist Policies." *Annals of Anatomy* 184: 251–56.

———. 2013a. "Anatomische Gesellschaft from 1933 to 1950: A Professional Society under Political Strain—The Benninghoff Papers." *Annals of Anatomy* 195: 381–92.

———. 2013b. "Wolfgang Bargmann (1906–1978) and Heinrich von Hayek (1900–1969): Careers in Anatomy Continuing through German National Socialism to Postwar Leadership." *Annals of Anatomy* 195: 283–95.

Hinrichsen, Klaus V. 1992. "In memoriam des Anatomen und Embryologen Erich Blechschmidt (1904–1992)." *Annals of Anatomy* 174: 479–84.

Initiative Stolpersteine Frankfurt am Main. 2009. *7. Dokumentation 2009,* 29–32. Accessed 31 July 2014. http://www.stolpersteine-frankfurt.de/downloads/doku2009.pdf.

Jachertz, Norbert. 2008. "NS-Machtergreifung: Freudigst fügte sich die Ärzteschaft." *Deutsches Ärzteblatt* 105(12): A-622/B-549/C-537.

Jacobj, Walther. 1952. "Martin Heidenhain." *Anatomischer Anzeiger* 99: 80–94.

Jahnke-Nückles, Ute, 1992. *Die Deutsche Gesellschaft für Kinderheilkunde in der Zeit der Weimarer Republik und des Nationalsozialismus.* Dissertation, Medizinische Fakultät der Universität Freiburg.

Kühnel, Wolfgang. 1989. "100 Jahre Anatomische Gesellschaft." *Verhandlungen der Anatomischen Gesellschaft* 82: 31–75.

Lampert, Udo. 1991. "Zur Situation der Pathologischen Anatomie an den deutschen Hochschulen während des zweiten Weltkrieges." In: *Der Arzt als "Gesundheitsführer": Ärztliches Wirken zwischen Resourcenerschliessung und humanitärer Hilfe im Zweiten Weltkrieg,* ed. Sabine Fahrenbach, Achim Thom, Gerhard Baader, N. Decker, and Wolfgang Uwe Eckart, 143–50. Frankfurt: Mabuse-Verlag.

Lang, Hans-Joachim. 2007. *Die Namen der Nummern: Wie es gelang, die 86 Opfer eines NS-Verbrechens zu identifizieren.* Überarbeitete Ausgabe. Frankfurt am Main: S. Fischer Verlag.

———. 2013. "August Hirt and 'Extraordinary Opportunities for Cadaver Delivery' to Anatomical Institutes in National Socialism: A Murderous Change in Paradigm." *Annals of Anatomy* 195: 373–80.

Lanz, Titus von. 1955. "Siegfried Mollier zum Gedenken." *Münchner Medizinische Wochenschrift* 19: 642–43.

Mörike, Klaus D. 1988. *Geschichte der Tübinger Anatomie*. Tübingen: JCB Mohr (Paul Siebeck).

Niessing, Klaus. 1953. "Alfred Benninghoff +." *Gegenbaurs Morphologisches Jahrbuch* 93: 113–28.

Rohrbach, Jens Martin. 2007. *Augenheilkunde im Nationalsozialismus*. Stuttgart: Schattauer.

Rüther, Martin. 1997. "Ärztliches Standeswesen im Nationalsozialismus 1933–1945." In: *Geschichte der deutschen Ärzteschaft*, ed. Robert Jütte, 143–93. Köln: Deutscher Ärzte-Verlag.

Schagen, Udo. 2010. "Walter Stoeckel (1871–1961) als (un)politischer Lehrer: Kaiser der deutschen Gynäkologen." In: *Geschichte der Berliner Universitäts-Frauenklinik: Strukturen, Personen und Ereignisse in und ausserhalb der Charité*, ed. Matthias David and Andreas Ebert, 200–18. Berlin: de Gruyter.

Schierhorn, Helmke. 1980. "Die Multinationalität der anatomischen Gesellschaft und die Mehrsprachigkeit ihrer Versammlungen." *Anatomischer Anzeiger* 148: 168–206.

———. 1986. "Mitglieder der Anatomischen Gesellschaft im antifaschistischen Exil." *Verhandlungen der Anatomischen Gesellschaft* 80: 957–63.

Seidler, Eduard. 2000. *Kinderärzte 1933–1945: entrechtet—geflohen—ermordet*. Bonn: Bouvier.

Spemann, Hans. 1942. "Walther Vogt zum Gedächtnis." *Wilhelm Roux' Archiv der Entwicklungsmechanik* 141: 1–14.

Stieve, Hermann. 1939. "Heinrich von Eggeling zum siebzigsten Geburtstag." *Anatomischer Anzeiger* 88: i–vi.

Stöhr, Philipp. 1947. *FIAT Review of German Science 1939–1945: Anatomy, Histology and Embryology*. Wiesbaden: Dieterich'sche Verlagsbuchhandlung.

———. 1952. "Eröffnungsrede." *Verhandlungen der Anatomischen Gesellschaft* 50: 3–8.

Thom, Achim. 1991. "Wandlungen der Wirkungsformen und Funktionen medizinisch-wissenschaftlicher Gesellschaften unter den Bedingungen der faschistischen Diktatur in Deutschland." *Zeitschrift für die gesamte Hygiene* 37(3): 124–26.

Thomann, Klaus Dieter, and Michael Rauschmann. 2001. "Orthopäden und Patienten unter der nationalsozialistischen Diktatur." *Der Orthopäde* 30: 696–711.

Veit, Otto. 1951. "Begrüssungsworte." *Verhandlungen der anatomischen Gesellschaft* 48: 3–5.

Winkelmann, Andreas. 2012a. "The Anatomische Gesellschaft and National Socialism: A Preliminary Analysis Based on Society Proceedings." *Annals of Anatomy* 194: 243–50.

———. 2015a. "The Anatomische Gesellschaft and National Socialism: An Analysis Based on Newly Available Archival Material." *Annals of Anatomy* 201: 17–30.

———. 2015b. "The *Nachlass* (estate) of Heinrich von Eggeling (1869-1953), long-time secretary of the *Anatomische Gesellschaft*." *Annals of Anatomy* 201: 31–37.

Wittern, Renate. 1993. "Aus der Geschichte der medizinischen Fakultät." In: *250 Jahre Friedrich-Alexander-Universität Erlangen-Nürnberg*, ed. Henning Kössler, 315–420. *Festschrift*. Erlangen: Verlagsdruckerei Schmidt.

Chapter 5

ANATOMISTS WHO BECAME VICTIMS OF NS POLICIES

> Every German, whether Jew or Christian, has to
> account for his actions between 1933 and 1945.
> Hans Elias, 1979[1]

Hans Elias, a passionate and multitalented scientist, formulated this demand at the end of an eventful life, far away from his mother country from which he had to flee, and for which he was still yearning. As a young anatomist, Elias was highly respected by Hermann Stieve, a leader in the field, and became his good friend. Why was it, then, that the one had to leave Germany and the other could stay and prosper? Why did Elias come to call his friend Stieve a murderer after the war? The following chapters will approach these questions by looking at the political spectrum of anatomists in Germany between 1933 and 1945. There are two groups to discuss: those who kept their jobs and stayed, and those whose careers were disrupted. The ones who stayed found a variety of ways to live with the new regime, from wholehearted support to cautious opposition. The fates of the others ranged from loss of academic positions to emigration and death in concentration camps. An account of individual biographies illustrates the many ways in which NS policies affected the lives of these scientists.

Scholars of Anatomy as NS Victims

Much of the research on the history of medicine in the Third Reich has dealt with the influence of NS policies on academia and the documentation of medical crimes and their perpetrators. Only recently has the identification of the victims of NS policies come into focus.[2] However, the reconstruction of biographies of those affected by the Third Reich presents specific problems, as obituaries written by colleagues as well as recent editions of standard encyclopedias often fail as sources of information. Many neglect to include newer information or tend to skim over the years of the NS regime in the biographies of both the victims and the perpetrators of crimes, thereby contributing to the postwar silence on all issues related to this time. Peter Voswinckel specifically criticized the *Neue Deutsche Biographie* (New German Biography) and the *Deutsche Biographische Enzyklopädie* (German Biographical Encyclopedia).[3] Also, the more comprehensive and informative new edition of Isidor Fischer's encyclopedia edited by Voswinckel unfortunately remains incomplete, as the second of 2 volumes was never published.[4] Thus researchers must draw from a wide variety of sources, including newly accessed archival materials, to find information on the fate of scholars of anatomy. It is sometimes quite frustrating and saddening to realize how some lives have left hardly any detectable traces.

For a long time it was uncertain how many scholars' careers were disrupted following NS policies. In a first study by Helmke Schierhorn,[5] which focused on members of the *Anatomische Gesellschaft*, the author suspected that his list of twenty-nine anatomists in what he called "antifascist exile" was incomplete. However, the true number of anatomists affected by NS persecution is much higher, including additional German and international scholars of anatomy and scientists employed in anatomical institutes, who were not members of the *Anatomische Gesellschaft*.[6] Among them were those who held positions as research students, assistants, professors, or emeritus at a university department of anatomy in Germany from 1933–45; members of the *Anatomische Gesellschaft* in 1932 and the following years including the postwar period; scholars of anatomy listed by the Society for the Protection of Science and Learning, the *Notgemeinschaft Deutscher Wissenschaftler im Ausland* in 1936 (Help Organization of German Scholars in Foreign Countries),[7] and listed in the Report of the Committee for the Study of Recent Immigration from Europe;[8] and scholars of other medical disciplines who taught anatomy or pursued anatomical research for at least part of their career. The disruption of a career could mean an outright dismissal of the scholar from academic position because of legislation concerning Jewish descent or political dissent; imprisonment for political reasons or Jewish descent; serious interruption and delay of a career for political reasons; "voluntary" emigration due to a lack of professional prospects because of the scholar's Jewish descent or political opinions; termination of the membership in the German scientific elite organization Leopoldina; or the disruption could be due to direct war action like occupation and the closure of a university by German authorities.[9]

Altogether, there were 527 scholars of anatomy whose careers were dis-rupted by NS policies. A majority of 453 were German and international members of the *Anatomische Gesellschaft,* of which the Germans made up nearly half.[10] The remaining seventy-four nonmembers and postwar mem-bers were mostly younger professors and assistants at German anatomical departments. Biographical information is available for 462 of them, and there were six women and eighty men whose careers were disrupted by NS policies (see appendix, table 2). Fifty-one careers were disrupted because of the scholar's so-called Jewish descent, thirty-two cited political reasons for dismissal and emigration, and in three cases the situation was unclear. Fourteen scholars emigrated "voluntarily," as they either anticipated their dismissal or saw no professional future for themselves, or felt they could no longer live in NS Germany. Sixty-two of the eighty-seven scholars were members of the *Anatomische Gesellschaft* and fifty-four worked in German anatomical departments. Overall, forty-three emigrated to the United Kingdom, Canada, Sweden, Belgium, Switzerland, Netherlands, France, Spain, Czechoslovakia, India, Bolivia, Turkey, the United States, and the USSR. Of those, only few returned to academic positions or retirement in Germany and Austria after the war: Paul Glees, Harry Marcus, Stephanie Martin-Oppenheim, Georg Politzer, and Carla Zawisch-Ossenitz. Twenty-one scholars remained close to their original workplace, sometimes in a different position or in private practice and research. Sixteen of those who remained were reinstated to their old or similar positions either soon after their dismissal or after the war. Twelve scholars were imprisoned, and five of them died in concentration camps. Four scholars died of natural causes between 1933 and 1945. In addition, Heinrich Joseph committed suicide, Piotr Slonimski was killed in war action, and Paul Roethig was institution-alized after a nervous breakdown and died in a psychiatric institution. The fates of four scholars, Karl-Ludwig Gieschen, Carl-Theodor Kempermann, Harry Weissberg, and Paul Quast, remain undetermined.

Factors Influencing the Fate of Scholars

There are several distinct factors that contributed to the path that each of these scholars' lives took. Fates depended on the stage of career, field of expertise, reason for dismissal, nationality, and the country of residence and exile.

Most of those in the early stages of their career decided to emigrate, altogether twenty-five of thirty-three of these younger scholars. Five of them survived the Third Reich in their home countries, and the fate of the others is unknown. Of the twenty-five emigrants, seventeen moved on to successful academic careers, the majority in the United States. Scholars with fully established careers were much more likely to remain in their countries of origin in whatever feasible professional capacity. Of the forty-four established individuals, twenty-one stayed on, fifteen emigrated, three died of natural causes between 1933 and 1945, three were killed by the NS regime, and Heinrich Joseph committed suicide together with his wife. The group of ten retired scholars, who were all of so-called Jewish descent, included two scholars who perished in concentration camps, four who died a natural death, three emigrants, and one whose fate remains unclear.

The emigration of some scholars was facilitated by grants, e.g., a Rockefeller foundation scholarship for Ernst Albert Scharrer,[11] but others had to leave the country without help or concrete plans for the future.[12] The emigration opportunities of scholars with established careers were very much dependent on their areas of expertise. It was particularly difficult to find an emigration sponsor for anatomists who had focused on racial anthropology, as the German concept of it and racial hygiene had become increasingly controversial in international circles. Potential emigrants had to convince their future colleagues of the quality of their education and research,[13] which may be the reason why the otherwise well-established Heinrich Poll was unable to find sponsors in the United States.[14] Franz Weidenreich received a grant from the Rockefeller Foundation on the basis of his paleontological expertise, rather than his work in racial anthropology.[15] Other successful emigrations of established scholars were possible with the help of former pupils,[16] family,[17] or rescue organizations.

The reason for career disruption was also an important determining factor in the future fate of the scientists. Scholars of so-called Jewish descent (or whose spouses were of Jewish descent) were more likely to emigrate. Thirty of the fifty-one scholars of Jewish descent emigrated, seven stayed in their country of origin, seven died of natural causes, one committed suicide, two perished in concentration camps, and the fate of four is unclear. Survival in the country of origin was dependent on help from colleagues and family. Of the thirty-three who were dismissed from their positions or quit for political reasons, eleven emigrated, while eighteen remained in their home country, one died of natural causes, two died in concentration camps, and one fate remains unclear.

The scholar's nationality and country of residence or exile also played a decisive role in determining one's fate. Here the history of individual European nations during the NS period, especially those with their own anti-Semitic legislation or under German occupation, was important for the fate of the scholars, as were the different countries of exile and their often restrictive policies toward emigrants.[18] Many emigrants had to move on from their first country of exile to other countries that seemed safer.

In 1938 fascist Italy adopted "racial laws" similar to German legislation from 1933 and 1935. Scholars of Jewish descent were banned from teaching in Italian schools and universities, and foreigners of Jewish descent were expelled from the country.[19] As in Germany, this law affected established and internationally well-known scholars like anatomists Guiseppe Levi of Torino and Tullio Terni of Padua, both members of the *Anatomische Gesellschaft*.[20] It also touched those who had emigrated from Germany to Italy earlier, like Hans Elias.[21] After a short exile in Belgium, Levi returned to Italy and survived in hiding until 1945. Terni remained in Italy. Both were reinstated in their old positions after the war, but Terni committed suicide in 1947 after again being removed from his position because of his history of fascist leanings. Other comparative anatomists and histologists removed by anti-Semitic laws, but not members of the *Anatomische Gesellschaft*, were Ettore Ravenna (Modena), Michelangelo Ottolenghi (Sassari), and Enrico Emilio Franco (Pisa).[22] Ravenna was reinstated to his old position after the war.[23] Ottolenghi emigrated via Ecuador to Canada in 1939.[24] Franco was appointed to the staff of Mt. Scopus medical center in Jerusalem in 1939.[25]

The situation was particularly appalling in the occupied countries. On 15 March 1939 NS Germany occupied parts of Czechoslovakia and effectively divided it into an NS-controlled "Protectorate of Bohemia and Moravia" and an independent Slovakia.[26] On 17 November 1939 all Czech universities were closed,[27] including Charles University in Prague and Czech University in Brno/Brünn. The German University in Prague and German Technical University in Brünn remained open, as did the University of Bratislava/Pressburg in the independent state of Slovakia. However, academics of Jewish descent were dismissed from the German universities of Prague and Brünn and persecuted if they did not choose emigration.

Zdenek Frankenberger, who was not of Jewish descent, had just started in his new position as head of the department of histology and embryology at the Czech university of Prague when the school was closed. He had to accept a nonacademic position as *Prosektor* at the Vinorahdy hospital, where he survived the NS occupation. He was reinstated to his old position after the war.[28] Robert Altschul, a research fellow at the histological institute of the German university of Prague, was of Jewish descent and decided to emigrate in order to escape German persecution in 1939. He had a successful academic career in anatomy in Canada.[29] Alfred Kohn, emeritus chair of histology at the German university in Prague, was also of Jewish ancestry and was deported to the concentration camp of Terezin/Theresienstadt in 1943 at the age of seventy-six years. His membership in the German scientific elite organization Leopoldina had been revoked in 1938. Kohn survived his imprisonment and lived in Prague after the war.[30]

Franz Theodor Münzer, a member of the *Anatomische Gesellschaft* and neurologist at the German University in Prague, was dismissed because of his Jewish descent in 1939. He worked in private practice in Prague until at least 1940 and was imprisoned in Terezin on 20 November 1942. From there he was transported to Auschwitz on 23 October 1944, where he presumably perished shortly thereafter.[31] Jan Florian was chair of the department of embryology and histology at the Czech university at Brünn and pursued private research after the closing of the university. An active role in the Czech national movement led to his imprisonment by the GeStapo in 1941. He was transferred from the prison in Brünn to the Austrian concentration camp Mauthausen, where he suffered solitary confinement and torture. Jan Florian was shot to death on 7 May 1942.[32]

Conditions were similar in Poland. Directly after the invasion the German occupational forces decreed the closing of all Polish higher schools and research institutions as well as the dissolution of all scientific societies. Polish academics were persecuted relentlessly, and 315 professors and lecturers perished during the war, that is 38 percent of nearly a thousand academics.[33] Three members of the *Anatomische Gesellschaft* were affected by these events. Among those killed was Kazimierz Telesfor von Kostanecki, the highly respected emeritus chair of descriptive anatomy at the university of Cracow and chairman of the Polish Academy of Sciences and Letters.[34] He and his colleague Henryk Fryderyck Hoyer Jr., emeritus chair of comparative anatomy, became victims of the *Sonderaktion Krakau* (special action Cracow).[35] On 15 November 1939, German occupying forces in Cracow rounded up 173 academics at the Jagellonian university and sent them to the concentration camp Sachsenhausen. Several of the older and sick professors died within a short time due to the brutal camp conditions, among them von Kostanecki. Hoyer was released through intervention by his German brother and worked in private practice during the war, and was reinstated to his academic position after the war.[36] Piotr Wacław Slonimski Sr. was a professor of histology at the university of Warsaw and lost his position with the closing of this institution. However, Polish academics founded a secret underground university in Warsaw and other Polish cities,[37] and Slonimski became a member of its faculty. This secret Polish university was distinct from the medical school created in the Jewish ghetto.[38] Slonimski also worked as a physician for the Polish resistance home army and was killed during the Warsaw uprising in 1944.[39]

NS German forces occupied the Netherlands in 1940, and while schools of higher education remained open, the occupation had dire consequences for many academics. Gerard Carel Heringa, chair of histology at the University of Amsterdam, was a known pacifist and correspondent of Albert Einstein. He discontinued his membership in the *Anatomische Gesellschaft* in 1934.[40] On 29 January 1942, Heringa was removed from his position, arrested with other citizens of Amsterdam, banned to the town of Assen, and his was house looted. He was later sent to a concentration camp with his wife, who died while imprisoned there. After the war, Heringa was reinstated to his former position.[41] Martinus Willem Woerdeman was a professor of anatomy at the University of Amsterdam and member of the *Anatomische Gesellschaft*. In 1941 he was arrested after students denounced him, held in prison for a month, and was then reinstated to his position.[42] Among the young scholars who emigrated was Peter DeBruyn. He had worked in the fields of preventative medicine and bacteriology at the University of Leyden and left for Chicago in 1941, were he had a successful academic career in anatomy.[43] Very little is known about the fate of Stephanie Martin-Oppenheim, an anthropologist who had immigrated to the Netherlands after an imprisonment in Memmingen. She was seized by the NS forces and incarcerated in Terezin. After her liberation she spent the remaining years of her life in Germany.[44]

Of the nine members of the *Anatomische Gesellschaft* working in the Soviet Union before the war, the fate of only one is known. Paul Vonwiller, anatomist in Moscow, returned to his native Switzerland and accepted a position at the University of Zürich in 1939.[45] There is no information for Wera Dantschakoff (Moscow), Alexandra Paulinovna Hartmann-Weinberg (St. Petersburg), Michael Ivanizky (Moscow), Alexi Lawrentjew (Odessa), B. J. Lawrentjew (Moscow), Moissey Muehlmann (Baku), or W. Tonkoff (St. Petersburg). Given the impact of the war on these regions, a disruption of some of these scholars' careers has to be assumed. In Moissey Muehlmann's case it is known that the Soviet authorities discriminated against him, probably for political reasons, as early as 1930.[46] He was listed as a member of the *Anatomische Gesellschaft* until 1939[47] and may have become a victim of the Stalinist anti-German purges of the Western regions of the USSR, which began in 1936 and culminated in 1941.[48]

Turkey became a haven of refuge for several scholars of Jewish descent. In the wake of Kemal Atatürk's sweeping political and educational reforms, the modern University of Istanbul was founded on 1 August 1933,[49] and academics for this new institution were recruited from Western countries. Philipp Schwartz, a pathologist who had been dismissed by the NS regime from his position at Frankfurt University in 1933, seized this opportunity to facilitate the emigration of other persecuted academics, and founded the *Notgemeinschaft deutscher Wissenschaftler im Ausland* (emergency committee of German scholars in exile).[50] Two scholars of anatomy, Karl Loewenthal and Tibor Peterfi, were able to immigrate to Turkey with the help of Schwartz's organization. The two held the position of professor for histology and embryology consecutively, Loewenthal from 1933 to 1938 and

Peterfi from 1939 to 1946.[51] Loewenthal, a Berlin pathologist dismissed in 1933 because of Jewish descent,[52] came to Turkey through France, and in 1938 moved on to the United States where he became chief of the laboratory at Union Hospital and Newport Rhode Island Hospital.[53] Peterfi, visiting researcher and departmental head at the KWI for Biology in Berlin, was dismissed because of Jewish descent and political reasons. After research visits in the United Kingdom and Denmark, he moved on to Turkey, which he left in 1946 to return to his native Hungary.[54] At this time he was already suffering from severe depression and was unable to work for most of his remaining years.[55]

A small number of émigrés ended up in British India, among them Georg Politzer, an anatomist and radiologist from the University of Vienna. In 1937 Politzer had accepted an invitation from the Maharaja of Patalia to establish a radiography facility in this state. After the annexation of Austria by NS Germany in 1938, Politzer and his wife decided to stay in India.[56] On 3 September 1939 the government of British India interned all male enemy subjects over sixteen years of age, including those of Jewish descent.[57] Politzer was among the interned and only set free through the intervention of a British government official.[58] He returned to the University of Vienna and its department of anatomy in 1951. Politzer had to fight for the official recognition of the professional credentials that he had achieved before leaving Austria, e.g., the reinstatement of his *Venia legendi*, which had been revoked in 1938. He died shortly after gaining the position of assistant professor.[59]

The United Kingdom (UK) was the first port of call for many emigrants, but due to the restrictive immigration policies of the home office and the British medical establishment, and despite the help of several aid organizations, many medical emigrants had to move on to other countries.[60] The situation was more difficult in England than in Scotland,[61] and easier for those who worked in research than for their clinical colleagues.[62] Several scholars of anatomy found new employment at least for a period of time in the UK through the help of the Society for the Protection of Science and Learning.[63] Some were able to stay and build new careers, such as Paul Glees who, after establishing himself as an anatomist in the UK, returned to Germany and the anatomy department of Göttingen in 1961.[64] Others moved on to countries of the British Commonwealth, some of which actively recruited new physicians.[65] Among the latter was Eric Wermuth, who after two years in the UK immigrated to Newfoundland in 1940. While he worked there as an intern in various hospitals, he was repeatedly wrongfully accused of spying for NS Germany. In 1948 he moved to London, Ontario, where he had a private practice until 1980.[66] The seventy-six-year-old neuroanatomist Adolf Wallenberg was of Jewish descent and had to leave his position as director of internal and psychiatric medicine at the municipal hospital in Danzig after the semiautonomous city-state came under German occupation in 1938. He immigrated via Holland to the UK, where he worked following an invitation by Sir Wilfrid Le Gros Clark in his anatomical laboratory in Oxford. In 1943 Wallenberg received a visa for the United States and continued to teach as a respected lecturer in the vicinity of Chicago until his death in 1949.[67]

In general, immigration of scholars to the United States depended either on sponsors in the country or direct help through personal friends and family, and could be achieved through migration via intermediate countries. Various aid organizations were involved in the rescue of these academics, among them the Rockefeller Foundation and the Emergency Committee in Aid of Displaced Foreign Scholars.[68] The latter helped with placing foreign scholars in US academic institutions, and among those Middlesex University (today, Brandeis University) in Waltham, Massachusetts, was of particular help for anatomists. Three of the seventeen immigrants who stayed permanently in the United States were given an opportunity by this institution to start their professions anew. Middlesex "increasingly became a refuge for Jews who were discriminated against elsewhere and for whom Middlesex was almost a last, desperate measure."[69] Here Richard Weissenberg, his former pupil Hans Elias, and Louis Bergmann worked as professors of anatomy and histology during the war and in the first years thereafter.[70]

The analysis of all these determinants of the scholars' fate shows that, as much as most of them tried to establish new careers and many succeeded, some could not escape the murderous intent of the NS regime.

Groups of Scholars and Their Fates

The scholars of anatomy who suffered due to NS policies were a diverse group. The youngest, Eric Wermuth, was twenty-four when he emigrated, and the oldest, Max Flesch, was eighty-seven when he was struck off the member list of the Leopoldina society.[71] Their professional fields ranged from anatomy, histology, embryology, racial anthropology, zoology and neuroanatomy, to radiology and philosophy; and their nationalities and ultimate fates were very different. In addition to Florian, Münzer, and von Kostanecki, Alexander Spitzer and Max Flesch and his wife died in Terezin.

The younger scholars generally fared better than their older established colleagues. As far as their fates are known, none of them were killed as an outcome of NS policies. The majority were of Jewish descent or had spouses of Jewish descent, had nothing much to lose in fascist Germany or Austria, and emigrated early. After hard struggles and emigration, often by several stages, most of them established thriving academic careers in their new countries.

Hartwig Kuhlenbeck (1897–1984) was a scientist who left Germany voluntarily for political reasons. He studied philosophy and medicine in Jena, and after a year in clinical practice in Mexico and three years as visiting professor of neuroanatomy in Tokyo he settled in Breslau in 1927, receiving his *Venia Legendi* in anatomy in 1928. His work focused on the relationship between brain, mind, and soul.[72] When the National Socialists came to power he refused to deal with the NS student organization or to give the obligatory Hitler salute, and immigrated to the United States in 1933. Kuhlenbeck first found work through a visiting fellowship in anatomy at Mount Sinai Hospital in New York and moved to Philadelphia as director of the anatomical institute at the Women's Medical College (today, Drexel University) in 1935. He remained there until his retirement in 1971. After the war he served as a consultant to the US army. His studies on the morphologic evolution of the brain and the origin of the cerebral cortex brought him worldwide renown.[73]

One of the most involved stories belongs to Carla Zawisch-Ossenitz (1888–1961), who emigrated from Austria and then returned there later in her career. She is counted among the younger scholars because of the late start of her career. After an early training in the fine arts, languages, and music, she studied medicine in Vienna during the 1920s with the intention of becoming a missionary. However, bad health thwarted her plans, and she accepted the position of an assistant in histology at the University of Vienna where, in 1934 at the age of forty-six years, she received her *Venia legendi* as one of the first women at the institution.[74] She was a devout Catholic, cofounder of the Catholic physicians' society *St. Lukas-Gilde*, and public educator in broadcasting. After the annexation of Austria in March 1938, her political and religious activities were deemed threatening by the new regime, and she was imprisoned from 23 March to 3 May 1938. Despite her willingness to work under the new regime after her release, her *Venia legendi* was revoked, and she decided to flee from further persecution. Her first country of emigration was France. When Paris came under German occupation, she was deported to a concentration camp in the Pyrenees, from which she was released when she fell ill. She traveled on to Barcelona, where she taught German lessons to earn a living, and finally received a US visa in 1943. She taught anatomy at a girls' college near Boston until 1946 and was instrumental in founding the Austrian University League, an aid organization for emigrant Austrian physicians in the United States. In 1946 she was recruited as director of the institute for histology and embryology in Graz as the first female professor at the university. She guided the modernization of the institute and retired in 1959.[75]

An exception from this pattern of emigration in younger scholars was Walter Krause (1910–2007). In 1934 he became an assistant at the first department of anatomy in Vienna and was dismissed in 1938 after refusing to offer the Hitler salute. He continued to work as a clinical physician in Vienna until he was accused of so-called "racial defilement" because of his relationship with Rita Smrčka, who was of Jewish descent and later became his wife. He was sentenced to a year in prison and was drafted thereafter into the army, fighting in North Africa in 1942,where he was taken prisoner by the Americans. In 1945 Krause was reinstated to his old position at the University of Vienna, and he remained there throughout his career. Rita Smrčka survived Auschwitz, where she worked as a prisoner physician.[76]

About half of the established scholars were persecuted for political reasons; the others and the retirees because of their Jewish descent or marriage to Jewish spouses. A third of this group emigrated, and the rest stayed behind in greater or lesser peril. It was nearly impossible for so-called *Volljuden* (i.e., persons whose parents were both of Jewish descent) to remain in Germany or German-occupied territory without being persecuted and threatened with destruction. Some so-called *Mischlinge* (i.e., persons who had only one parent or grandparent of Jewish descent) decided to stay with their families and endure the severe discrimination they were subjected to. The same is true for those with spouses of Jewish descent. Some of the politically persecuted scholars, such as Karl Saller and Gustav Sauser, were able to stay in their home countries and find work outside their original employment while keeping a low profile, but these were exceptional cases.[77] All others had to emigrate sooner or later, including some of the retired elderly scholars of anatomy, or face death.

Few of the established scholars had early opportunities to emigrate. Wilhelm von Möllendorff and Hans Bluntschli were offered chair positions at Swiss universities and emigrated "voluntarily," Bluntschli to Bern in 1933[78] and von Möllendorff to Zurich in 1935.[79] In von Möllendorff's case his emigration was strongly supported by the NSDAP commission for higher education in Munich, which welcomed the recruitment of a German professor to the University of Zurich, a move that simultaneously solved their problem of getting rid of an unwanted Social Democrat.[80] Heinrich Poll was not as lucky. He was dismissed from his chair position at the department of anatomy at the University of Hamburg in 1933 and tried to find a US sponsor for his emigration. Despite good international contacts, including his friendship with influential American medical educator Abraham Flexner, Poll could not find help: his work in racial anthropology did not fit into the US research landscape. He finally accepted an offer from the University of Lund in Sweden in 1939, were he died of a heart attack shortly after his arrival. His wife Clara Poll-Cords was a gynecologist. She could not accompany her husband because she was still waiting for her visa, so she was not with him at the time of his death. She travelled to Lund for his burial and committed suicide soon thereafter.[81]

Many established and retired scholars tried to postpone emigration for as long as possible, because, like Robert Meyer, they still trusted the new regime, especially during the first few years. Meyer, a venerated Berlin gynecological pathologist and embryologist, was dismissed from his official positions at the Charité in 1935, but continued his work with his colleagues' support. However, his situation as a man of Jewish descent in Germany became too dangerous by 1939, and the seventy-five-year-old accepted his US pupils' invitations and help to emigrate to the United States.[82] He continued his work at the University of Minnesota, financed by grants that his pupils sought for him.[83]

The emigration story of dermatologist and histologist Felix Pinkus was similar. He was dismissed from his position as director of the Reinickendorf Women's Hospital in 1933 and worked as a *Krankenbehandler* (caretaker of the sick) in private practice, which was the only legal position available to physicians of Jewish descent after 1933.[84] In 1939 the seventy-one-year-old Pinkus immigrated first to Oslo, then escaped to Copenhagen after Norway came under German occupation in 1940. Pinkus wanted to join his son Hermann, who had immigrated to the United States, but the only escape route open any longer led him first to Moscow, then on the Trans-Siberian railway to Vladivostok, and from there via Tokyo to Monroe, Michigan, where he worked in private practice with his son until his death in 1947.[85]

Otto Veit was one of those who stayed behind. As a *Mischling* and lifelong member of the *Bekennende Kirche* (protestant oppositional church in NS Germany), he was supposed to be dismissed as early as 1934, but the support of his colleagues kept him in place until NS pressure finally became overbearing in 1937. However, even after relinquishing his official position, he was able to pursue some of his work due to the help of pathologist Ernst Leupold, who secretly provided him with two rooms for his studies. This did not alleviate the humiliations and privations that Veit and his family were subjected to during the later years of the Third Reich, including the destruction of their home and injuries received in a bombing raid in 1944. Veit was reinstated as chair of the anatomy department in 1945 and retired in 1957 at the age of seventy-three.[86]

Like Veit, Hermann Hoepke of Heidelberg had colleagues who intervened for him with the NS authorities to keep him in his position until his final dismissal in 1937.[87] Colleagues also helped anthropologist Saller in Göttingen[88] and von Lanz in Munich. Von Lanz had been removed from his position as professor of anatomy because of his wife's status as a *Mischling*. He and his immediate family were able to survive in Munich because of the support of renowned surgeon Ferdinand Sauerbruch, who secured grant money for von Lanz to complete a book on surgical anatomy.[89] Von Lanz's teacher and father-in-law Harry Marcus, also professor of anatomy in Munich, was of Jewish descent and left Germany with several family members in 1939. After being incarcerated as an enemy alien in France, he eventually reached Bolivia where he performed research and taught anatomy, and returned to Germany after the war.[90]

Franz Weidenreich's fate was one of the most remarkable among his colleagues, as it directly reflected the political changes in his various countries of residence.[91] Born in 1873, he studied medicine with famous anatomist and anthropologist Gustav Schwalbe of Strasbourg as his mentor. Weidenreich became Schwalbe's assistant in 1900 and was promoted to associate professor of anatomy in 1904. At that time he married Mathilde Neuberger and they had three daughters. Weidenreich's early work focused on hematology and on anthropological studies on the origin of man. He also found time for politics as a member of the democratic party of the Alsace.

When the French authorities took over the University of Strasbourg after the defeat of Germany in late 1918, Weidenreich and his colleagues were dismissed, forcing him to leave his country of residence for the first time. He had trouble finding a new position, which he later attributed to anti-Semitism in German academia. Weidenreich moved to Heidelberg, where in 1922 he found temporary employment at the cancer research institute, but was pensioned off in 1924 at the age of fifty-one.[92] He obtained another position at the anthropological institute of the Von Portheim Foundation and was recruited as founding chair of physical anthropology at the University of Frankfurt in 1928.

His work over the next six years focused on paleoanthropology, with a special interest in the hominid *Sinanthropus* as described by Davison Black. His scientific theories on human "races" and the evolution of man brought him quickly into conflict with other anthropologists, especially the ones who would later collaborate with the NS regime. For him, "races" were geographical variants and not the results of heredity and evolution. He believed that "race" was not inherited but developed through the influence of the environment at a given location and was thus dynamic. This neo-Lamarckian approach was very different from that proposed by Fischer. In 1934 Weidenreich was forced by the authorities to take an unpaid leave of absence, because it was thought to be untenable that a Jew was responsible for the teaching of anthropology and racial hygiene. He accepted the invitation to work as visiting professor of anatomy at the University of Chicago.

The year 1935 became a decisive one for Weidenreich: he was dismissed from the University of Frankfurt (and from the University of Heidelberg, where he still held a formal appointment) for so-called "racial" reasons. Later his membership in the Leopoldina was annulled for the same reason.[93] The Frankfurt Institute for Physical Anthropology was subsequently disbanded and newly founded as the Institute for Hereditary Biology and Racial Hygiene with Otmar von Verschuer as the director.[94] One of von Verschuer's doctoral students in medicine was Josef Mengele, who also held a PhD in anthropology from Munich.[95] At that time Weidenreich received an offer to succeed the recently deceased Davison Black in China, and with the help of the Rockefeller Foundation Weidenreich was appointed as visiting professor of anatomy at the Peking Union Medical College. Thus he left his country of residence for the second time.

During the next few years he wrote his most important papers on *Sinanthropus pekinensis*, Peking Man, developing new theories on the evolution of humans. After years of productive studies, Weidenreich was forced to leave his country of residence for the third time as the Japanese military approached Beijing in 1941. He and his wife settled in New York, were he was a guest at the American Museum of Natural History and worked on an overview of his scientific findings and paleoanthropology.

The war years were full of personal tragedy for Weidenreich's family. His mother-in-law was killed in the Holocaust and one of his daughters, Dr. Ruth Piccagli-Weidenreich, was imprisoned in Auschwitz. Her husband, Italian navy officer and resistance fighter Italo Piccagli, was executed by fascists on 12 June 1944.[96] Another daughter, Elisabeth von Scheven, was also incarcerated in Auschwitz. Until September 1943 she had been protected by her "Aryan" husband Walter von Scheven and hidden by friends from the resistance alliance "European Union." She was one of the few of that group to survive.[97] Franz Weidenreich and his wife succeeded in bringing their two daughters, Walter von Scheven, and the von Schevens' two children to the United States, where the new immigrants were also reunited with a third Weidenreich daughter, Marion, who had left Germany at an earlier time.

In 1946 Weidenreich formulated a "Memorandum on scientists suspected as war criminals" in which he laid a major part of the guilt for NS atrocities at the feet of several of his former colleagues, among them the German anthropologists and racial hygienists Eugen Fischer, Fritz Lenz, and Ernst Rüdin. Weidenreich, who had fought for an anthropology without static "races," saw friends and family perish through crimes justified by the teachings of these men.[98] Weidenreich had another three productive years until he died on 11 July 1948. A 1946 volume containing a series of lectures was prefaced by words that describe Weidenreich's scientific and personal legacy in simple terms: "The preamble to the Declaration of Independence of the United States of America holds the truth self-evident that all men are created equal."[99]

The vacancies left by all of these scholars were quickly filled, and there is no documentation of any anatomist refusing to fill the position of a former colleague. The *Anatomische Anzeiger* did not comment on their deaths, nor did it investigate their fates until much later.[100] A wide variety of persons were affected by the career disruptions, from chairmen of anatomy to young researchers just starting out in their careers, and their fates were diverse. The families of many of these scholars suffered with them. Clara Poll-Cords and Edith Joseph committed suicide after their husbands' deaths, and Hella Flesch and Mrs. Heringa died in concentration camps.[101] Whether the scholars stayed behind in their home countries or emigrated, whether they were dismissed following anti-Semitic legislation, through direct actions of war, or whether they were political dissidents, all of them had in common that a criminal regime forced them, innocent of any crime, out of their positions and work, and changed their lives forever.

A Young Emigrant: Hans Elias

The life of Hans Elias perfectly reflects the difficulties and opportunities encountered by the younger generation of anatomists who left Germany after the forcible disruption of their careers by the NS regime. He was also a formidable diarist and correspondent and left several unpublished volumes of autobiographical writings behind, which allow a unique and intimate insight into the world of anatomy in the middle of the twentieth century.[102] Elias left his papers to libraries in the United States and Germany and, together with his family, contributed audiotapes to an oral history project of interviews of German-speaking academics who immigrated to the United States in the 1930s.[103]

Hans Elias was born in Darmstadt, Germany, on 28 June 1907 to a Jewish family of teachers, tradesmen, rabbinical scholars, and cantors.[104] He grew up in a liberal and democratic atmosphere and had Jewish and Christian friends with whom he stayed in contact throughout his life. Elias later believed that his early life among German democrats who perceived Adolf Hitler and the rising National Socialism of the 1920s as "un-German" prevented him from hating his fatherland in later life.[105]

After first focusing on the study of fine arts,[106] Elias pursued graduate studies in the natural sciences and education at the *Technische Hochschule* Darmstadt and the Universities of Berlin and Giessen.[107] Throughout his life he saw himself foremost as a teacher, but also as a scientist and artist.[108]

His 1931 doctoral thesis focused on the development of the coloring in frog skin,[109] and he created the most exquisite illustrations for his scientific papers and encouraged his students to document their findings with their own drawings.[110] After his father's death in 1931, Elias's family ran into financial trouble,[111] and he accepted a paying position at the Jewish School for the Deaf in Berlin-Weissensee, where he taught and pursued his research until 1933.[112] Felix Reich, the headmaster, was of Jewish descent and a strong nationalist. After Hitler's ascent to power, Reich demanded from his staff that they greet their students with a raised right arm and the word "*Heil*" without the "Hitler." Elias was the only teacher who refused to obey Reich and was fired. This was the first, but not the last time, that Elias did not comply with directions from superiors that ran counter to his principles, situations that led repeatedly to the loss of occupation. However, he adopted Reich's advice: "Herr Doktor Elias, one does not give up. Never make [*sic*] your enemy the pleasure of admitting defeat. Keep on working! Behave as if nothing had happened!"[113]

Before moving on to his next teaching position at a private Jewish school in Herrlingen, Swabia, Elias recognized the NS boycott against Jewish businesses on 1 April 1933 as a sign of worse things to come and organized the emigration of his mother and sister Magda from Berlin to the Netherlands. Shortly thereafter, the women moved on to Milan, Italy, were they founded a child-care facility. Elias stayed behind in Germany because he felt morally compelled to support German-Jewish youths. During the next year he worked in Herrlingen, but disagreed with his superior's pedagogic methods and resigned from his job. He came to the conclusion that he would not be able to pursue his dream of educating Jewish students in Germany due to his "unemployability," and this, coupled with a lack of finances to start his own private educational enterprise, caused him to leave Germany.[114]

In early 1934 Elias joined his family in Milan, but soon moved on to Turin, were he had found employment as tutor for the son of a wealthy family. His mornings were free for research at Giuseppe Levi's anatomical institute at the University of Turin. However, Elias's strict pedagogic ideas for his pupil did not agree with his employers' expectations, and so he lost that job.

Elias returned to Milan in the summer of 1934 where he worked on a film about the embryonic development of the European tree frog. This idea stemmed from an embryology course he had taken during his studies in Berlin, where Richard Weissenberg had encouraged Elias' plans. His first successful cinematographic project had been a collaboration with Walter Schwarz (later known as Michael Evenari) at the *Technische Hochschule* Darmstadt on the development of flowers.[115] In 1935 Elias was able to perfect his filming technique when he was granted a scholarship by the Swiss Aid Organization for German Scholars. This allowed him to work with Ernst Rüst at the Institute for Scientific Photography at the *Eidgenössische Technische Hochschule* in Zurich. The result was an educational film on the development of amphibian ova.[116] During his time in Zurich Elias also worked at the anatomical department of the university, where he became acquainted with von Möllendorf and Bargmann.[117]

By 1936 Elias had moved to Venice, where his fiancée Anneliese Buchthal worked as a physiotherapist at the Municipal Hospital of Venice.[118] He had first met her on a visit from Germany in 1933 (she had immigrated there earlier in 1933), and they were married on 11 October 1936. Because of an agreement between the two countries, the German "racial" laws of 1935 applied to German expatriates in Italy. The couple had to prove that they were both of "pure Jewish descent" according to the Nuremberg laws to be granted a wedding license. Meanwhile Elias worked as an unpaid researcher at Tullio Terni's anatomical institute at the University of Padua.[119]

Throughout his time in Italy Elias supported his family by painting portraits. His art was very successful: an image of the young Contessa Ifigenia Marini di Villafranca was accepted for exhibition at the Biennale di Venezia in 1936. Unfortunately the whereabouts of this painting are unknown. In 1937 the couple moved to Rome, where Elias had found a paid position as director of the Laboratory for Scientific Cinematography and Histology at the *Consiglio Nazionale delle Ricerche* (Italian Research Counsel), and as consultant to the International Institute for Educational Cinematography of the League of Nations.[120]

By the summer of 1938 Elias's family had convinced him to obtain US visas for all of them, as newly emigrated friends had told them about the increasingly aggressive persecution of Jews in Germany. Elias was not yet ready to move, as he dreaded another emigration and consequent unemployment. However, on 14 July 1938 Italian fascist university professors had issued an anti-Semitic manifesto that ultimately led to legislation in Italy.[121] On 1 September 1938, the Italian government issued a law that ordered all expatriate Jews to leave the country within six months, followed by another law on 2 September 1938 that commanded the dismissal of all Jewish teachers and academics. Elias received his notice on 6 September 1938. Over the next few months he finished his ongoing projects and rejected an offer to teach at the University of Istanbul. The Elias family left Italy for the United States on 1 April 1939. At the beginning of the year Elias had written in his diary: "*Per aspera ad astra.* [...] Emigration is disagreeable to me. I have seen enough already. And Darmstadt, the Odenwald and the Black Forest are entirely enough for me."[122] Throughout his years in Italy and even after his second immigration to the United States,

Elias's scientific articles continued to be published by German anatomical journals.[123] Most of Elias's early mentors and collaborators shared his fate following the political developments in their respective countries.[124]

After the family arrived in New York on 13 April 1939, Elias immediately started his search for a job. He found this difficult, as previous waves of academics had already been absorbed into the US educational system, and few positions were still available. After several trips along the East Coast, funded by aid committees, he finally found employment as a professor of biology and veterinary histology at Middlesex University in Waltham, Massachusetts.[125] Once established in the United States, Elias felt free to share his original but highly controversial ideas on education and politics with the public. He published papers on education[126] and corresponded with politicians and public figures like Chaim Weizmann, Heinrich Brüning, psychologist Erich Fromm, and German emigrant authors Klaus Mann and Thomas Mann.[127] He also continued his research and teaching, and his young family grew with the arrival of two sons.

Hans and Anneliese became US citizens on 27 November 1944.[128] In 1945 Elias found a more profitable and secure position at the Center for Communicable Diseases (CDC) in Atlanta, Georgia, where he produced medical educational films.[129] While working on a project about the human liver, Elias noticed clear discrepancies between the accepted view of liver histology as a set of "cords" and his own observations, which he described as a system of continuous "plates." He pursued his hypothesis in private research and intended to publish the results in 1948, when he was told by his supervisor that his position at the CDC did not allow the pursuit or publication of scientific work. However, Elias continued with his research and its publication,[130] and his colleague John E. Pauly summarized the results of Elias's actions aptly: "The papers made him famous, but they got him fired."[131] During their four years in Atlanta, Hans and Anneliese were appalled by their white neighbors' treatment of African American neighbors.[132] In his further career Hans went out of his way to assist minority and otherwise disadvantaged students.[133]

Hepatologist Hans Popper of the medical school at the University of Chicago had become aware of Elias's work and facilitated his recruitment as assistant professor of microscopic anatomy to this institution in 1949. Many of the faculty members, like Popper, were Jewish refugees. Elias was promoted to associate professor of anatomy in 1953 and full professor in 1960. By this time he had long been fascinated with the challenge of creating scientific images that could convey the plasticity of three-dimensional space. During his studies on liver architecture he had worked with serial cuts of tissues, a highly time-consuming process. He assumed that it should be possible to evaluate two-dimensional images mathematically in a way that would elucidate the three-dimensional structure of the tissue. In collaboration with mathematicians he developed geometric formulae that enabled him to predict structures from the two- to the three-dimensional space, a science he called "stereology." His first study objects were pasta spirals embedded in colored gelatin, but he soon advanced to complex structures like cell organelles.[134] Elias first publication on stereology in 1951 led a decade later to the foundation of the International Society of Stereology.[135]

Elias continued to pursue multiple activities: he wrote a book, *Human Microanatomy*, together with Pauly,[136] continued his collaboration with Popper, volunteered as a teacher at an elementary school, painted, developed a new distinguished career as a sculptor,[137] and supplemented his income as an illustrator for various pharmaceutical companies with a new technique of layered color transparencies.[138] Perhaps not surprisingly, his various innovative projects did not always meet with the approval of his scientific colleagues, e.g., his initiative to reintroduce Latin as a common scientific language. He published two of his research papers in Latin,[139] but had to realize that they were not read very widely—nor was he always able to understand letters of response written in classic Latin.[140]

He studied the subject of carcinogenesis throughout the rest of his life as well. He developed a controversial theory of multicentric carcinogenesis and cancer cell recruitment, which he defended avidly. He was repeatedly refused funding to study this theory, and he suspected that this was a conspiracy by his opponents.[141] This led to a lingering bitterness—this man, who at the time had already received many official recognitions of his work, wrote in a letter to his friends in December 1963, "You may say that I am ungrateful. Yes, I should be content by not having been tortured and killed by the Nazis. I should be grateful for the many blessings I have received personally. Yet, in my opinion, what one gets is nothing. Only what one accomplishes in terms of permanent values given to mankind counts. And my accomplishments are about one hundredth of what I have intended to do. [...] In summary, at the age of 56 I see my efforts in life (except in my immediate family) to have been practically useless."[142]

Elias' harsh, self-critical attitude shows how high his personal expectations were in terms of his scientific and political effectiveness. However, he still adhered to Felix Reich's advice and persisted in making his own liberal political convictions known. Elias' work in Chicago ended in 1972, when he became professor emeritus at the age of sixty-five. He and Anneliese moved to San Francisco to be closer to their sons. In 1973 they accepted an invitation for a sabbatical at Heidelberg University. Here Elias found support for his work on carcinogenesis and began another rewarding period of research. On his return to California in 1975 he became research stereologist at the University of California Medical Center at San Francisco, and at the same time fulfilled his dream of teaching younger students anatomy at the City College of San Francisco. He enjoyed these positions until his death on 11 April 1985, at the age of seventy-seven.[143]

Elias was energetic, passionate, and multitalented, and he held strong opinions. This combination of characteristics explains the great successes and joys in his life as well as his moments of defeat and despondency. Without his ability to shrewdly and fearlessly assess political situations he might not have been able to rescue his family. His accurate judgment of the danger awaiting them was confirmed by the terrible fate of his favorite uncle, Dr. Siegfried Oppenheimer, a physician at the *Staatstheater* Darmstadt, who stayed behind in Germany and ultimately perished in Terezin in February 1943.[144] Looking back on his life Elias wrote, "I had the incredible luck of not having been persecuted personally. [...] I owe my early and unencumbered emigration to my poverty and unemployment, as well as the utter hopelessness of my professional situation".[145]

Elias' life exemplifies the loss that German and Austrian anatomy sustained by the emigration of such young academics. His ability to completely follow his own insights, and his fearlessness in rejecting accepted wisdom led him to the entirely new science of stereology. Elias' fate also illustrates how hard life in exile was. He had to emigrate not only once, but twice, and endured years of unemployment and poverty. His early experience of endangerment in Germany might have left him with a tendency to suspect conspiracies later in his life. His strong will and principles also cost him several jobs and made him a difficult person to live with at times, as he wrote in 1979: "Even at the age of 72 I am still moving ahead. Who knows what the future holds? I am certainly not an easy partner for my dear wife Anneliese. But I do hope that life with me is not boring."[146]

According to Elias' younger son Thomas, religion and his Jewish descent were never of vital importance to his father's thinking, although they were certainly part of his identity and his family's history. He had had a religious education as a child, could chant the central Hebrew prayers "in a pronounced Ashkenazy accent,"[147] was proud of signing the certificate of marriage with his Hebrew name,[148] and insisted on his sons' religious education. However, Thomas felt it was the National Socialists who imposed this Jewish identity on his father as a central element of his being. Again, this was a fate Elias shared with many of his fellow German emigrants. Elias' lifelong affection for Germany was a contentious issue between him and his wife.[149] Elias saw the roots of this attitude in his liberal democratic upbringing and the fact that he had never personally experienced any anti-Semitism. Anneliese's experience in her hometown of Witten had been very different, as she had been bullied during her school years and encountered imminent danger through threats of violence from anti-Semitic protesters while walking in the street on 1 April 1933.[150] Hans Elias longed to be back in Germany and made several efforts after the war to find a position as guest researcher at a German anatomical department and finally succeeded after retirement. Elias reported that the temporary move to Heidelberg was at first very hard for his wife, but through her contacts with the new German youth and Elias' old friends she learned that the Germany of the 1970s was different from the one she had left in 1933.[151]

Elias considered himself a German and never lost his love for German literature, art and ideas, as well as for his old friends and the countryside he grew up in. In 1979, even after finally achieving financial security, living in a beautiful house on a hilltop in San Francisco, he still longed for Germany: "I feel drawn towards the home country all the time. I have interesting work here, dear colleagues, co-workers and students. Even so, this here is exile and unfamiliar. Only the Odenwald and Schwarzwald are home."[152] He never lost his Hessian accent in his German and it even carried into his English.[153] Like so many other emigrants he felt *Heimweh* (homesickness), and more than that—he experienced what Hilde Spiel, herself an emigrant, called "exile as a disease," a state that included "agonizing experiences: of homesickness, feelings of being excluded and misunderstood, insurmountable language barriers, barriers of tradition, education, habit and familial connections."[154] Yes, he had survived, but his life as a German had been disrupted forever. The *Staatsbibliothek* Berlin holds not only Elias' scientific papers and correspondence but also one of his few surviving sculptures, a depiction of Prometheus as the bringer of light.[155] Hans Elias' ambition in life was to enlighten the world, and he succeeded greatly.

Hans Elias and His German Colleagues

Elias had his first professional encounters with German anatomists when he was already living in exile. During his 1935 visit to Zurich he was impressed by von Möllendorff, in his eyes "one of the greatest contemporary histologists," who "despite his noble heritage" was affable and helpful. He also met Bargmann there, and considered him the most likeable among the German anatomists as well as the intellectual leader of anatomy in the Federal Republic of Germany after the war.[156] In September 1936 Elias attended the International Congress of Anatomy in Milan, were he encountered many of his German and international colleagues for the first time. He was greatly captivated by the feeling of unity and friendship among his colleagues, and his diary entry and memoirs sound positively jubilant:

This congress was one of the greatest feasts ever held. Everyone sensed the imminent war. This subconscious knowledge that this was one of the last occasions for this big, world-wide family of anatomists to meet intensified the glamour of this festival of companionship. [...] During this festival of international friendship and scientific brotherhood I became strangely attracted to H Stieve, a professor of anatomy at Berlin and he to me. We spent as much time together as we could.[157]

Indeed, Stieve, the leader of the German contingent at this international congress, and Elias became great friends and remained in contact throughout Stieve's life. All German anatomists, even Clara, were very friendly toward their Jewish-German colleagues, as Elias recalled.[158] Stieve and Elias met again two years later in Zurich, and after the war Stieve sent Elias a handwritten, four-page letter. In the only part of the letter that remains, Stieve asks Elias for contributions to the journal he edited. The rest of the letter so enraged Anneliese Elias that she tore it to pieces, destroying the evidence that angered her. Elias later thought that the letter might have become an important document for the Nuremberg trials. There are two sources among Elias papers in which traces of this document can be found. It should be noted that the first passage was written about three years after the letter was received and the second thirty-three years later. Elias recalled some of the contents of this letter and integrated them in his 1948–49 memoir, in which he wrote, starting with musings on the meeting in Milan 1936:

A strange pair: Stieve, the noted embryologist and little Hans Elias, a green horn [sic] in science. Stieve who wore a swastika and little Hans with the crooked nose. Stieve who later was to become a frequent user, for scientific purposes, of the bodies of massacred Jews and who is alleged (but denied it), to have ordered the removal of the womb from numerous healthy women in concentration camps, in order to study microscopically, the menstrual cycle.- Of course, at that time I took his swastika only as a piece of enameled brass forcefully imposed upon his lapel. Well, I do not know why. But the fact remains the [sic] Stieve and I liked each other tremendously and stuck together throughout the meeting, and we renewed our friendship two years later in Zürich. The astonishing thing in this matter is that I have no valid reason to like Stieve particularly. As a scientist, I do not think that he amounts to much. He is one of those opinionated pseudoscientists who believe [sic] in a theory and then try to see only those facts which seem to support their theory. *Also, his use of cadavers of the massacred which he freely admitted in a letter to me after the war, stating that these massacres, themselves deplorable, have contributed tremendously to the advancement of anatomy, was not a very commendable action.*[159] I will not form a judgement [sic] about his alleged use of uteri of healthy women, since he rigorously denies it.[160]

Elias also recalled some of the content of the letter in his 1979 German autobiography, and wrote in Stieve's voice:

Now that National Socialism, under which I have suffered greatly, has finally come to an end, I would like to again take up our friendship. *Yes, we have lived through hard times under Hitler; however, for us anatomists this was a delightful period. We were able to receive completely fresh bodies of healthy persons, as many as we wanted. Such bodies make a glorious dissection material.*[161] Now we have to again contend with the use of dried up bodies of persons who died of disfiguring diseases in hospitals.[162]

Elias continued by stating that Stieve, who had been working on reproductive organs of women, sent his assistants and students out to "spy on" sexual activity of women, and then had these women executed by the Gestapo. As proof of this he referred to a publication by Stieve from 1942 in the journal *Zentralblatt für Gynäkologie* (*Central Paper for Gynecology*), in which Stieve mentions the use of "material" from bodies of executed women.[163] However, the page number quoted by Elias (1456) does not refer to this paper but to the report on a discussion held between Stieve and his colleagues.[164] The correct page numbers are 1698–1708.

It appears that Stieve's revelations in the letter did not deter Elias from contemplating the offer of a position as lecturer of histology and embryology in Stieve's department of anatomy in Berlin after the war. How was it possible that Elias considered Stieve to be a murderer but still communicated and worked with him? The explanation for this seeming contradiction probably lies in their first meeting in Milan and their common German background. German academic medicine was tightly ruled by a rigid authoritarian style, in which Stieve impersonated the perfect professor and Elias the perfect student, even if their encounter happened in exile. Elias respected the true authority of a knowledgeable and dedicated leader, and Stieve, however questionable his character may have been otherwise, was exactly that. On the other hand, Elias must have impressed Stieve as the ideal inquisitive and hard-working younger colleague. While Stieve had been known to possess racist views,[165] his anti-Semitism seems to have been of the then "culturally accepted" kind pervasive during Germany's imperial period and thereafter.[166] He believed in "racial" differences and held collective prejudices, but had no personal aversion to individual Jewish friends. The fact that Elias was so thrilled to have such a prominent friend and was proud of this friendship may explain his inability to completely break with Stieve after the war.

If Elias was convinced that Stieve had caused the execution of human beings for use in his studies, he believed that Heinrich von Hayek, chairman of anatomy in Vienna from 1952 on and during WWII anatomist in Würzburg, had instigated the NS authorities to murder prisoners whose bodies von Hayek wanted to use for research. Elias claimed that von Hayek had been forced to leave the Federal Republic of Germany after the war because of his active political stance for National Socialism and because of what Elias called "anatomical murders."[167] At the International Congress of Anatomy in New York in 1960, von Hayek's invitation to hold the community's next meeting in Vienna in 1965 was accepted. On hearing this, Elias mobilized his Swedish colleague Carl-Hermann Hörtsjö and Gerhard Wolf-Heidegger from Basel, Switzerland,[168] to intervene against this decision with Bargmann, who was then on the board of directors of the *Anatomische Gesellschaft*. There is no documentation of Bargmann's response, but apparently he accepted their account of past wrongdoing by von Hayek and revised the decision, as the next International Congress was held in Wiesbaden, Germany. When Elias gave a presentation at the Viennese Medical Academy in 1963, von Hayek was friendly, but Elias did not talk to him. During an informal meeting with other Viennese physicians after the talk, Elias asked how it could have happened that "such a murderer could occupy the most important chair of the medical faculty." He remembered the answer as: "Well, back then it was legally sanctioned to have any number of persons imprisoned, even if they had committed the smallest crime. Thus von Hayek's imprisonments were not crimes but happened in accordance with the law."[169] Elias also counted Clara among the "anatomical murderers," as he remembered Clara's fate in his autobiography. He stated that after Clara "had made a fool of himself in Germany with his Nazi activities," he accepted a professorship in Istanbul, Turkey, where former NS activists who had to leave Germany because of their "anatomical murders" were employed after the war.[170]

It is curious then that Elias, who held no particular grudge against Germany and became a member of the *Anatomische Gesellschaft* in 1956,[171] was at the same time ready to believe the worst of some of his German colleagues. He was convinced that Stieve, von Hayek, and Clara had committed murder during the Third Reich; that they had handpicked persons who might be interesting for their respective studies, had caused their executions, and then used their bodies for anatomical studies. How did he come by these convictions and what is the truth behind these allegations? Stieve, von Hayek, and Clara published articles that left no doubt about the sources of their "material." As these papers appeared in internationally distributed journals, any attentive reader would have noted the unusually high numbers of bodies of the executed mentioned in these German studies. Elias also had firsthand information shortly after the war through Stieve's correspondence, even if the exact wording of the letter is lost. However, there is no evidence to date that Stieve, von Hayek, or Clara directly had a hand in the murder of NS victims.[172]

In the case of Stieve, there seem to have been many rumors floating around postwar Berlin and the international scientific community.[173] Seidelman remembered an informal conversation with a Holocaust survivor at a meeting in Oxford in 1988.[174] This man, an American physician and personal acquaintance of Elias, alleged that Stieve was involved in arranging the killing of a young couple who were sexually engaged in order to obtain a specimen of the woman's vagina at the height of sexual passion. The man also said that a microphotograph of such a specimen was published in one of Elias' books, and that Elias' wife, when she came across correspondence about this matter, was so upset that she destroyed the correspondence.

Seidelman's immediate response to this account was one of disbelief. While the story as a whole indeed sounds unlikely and does not align with the historical facts currently known,[175] there are several kernels of truth in it. Stieve did indeed mention the topic of orgasm in the case of a woman who died shortly after being raped by three men. He discussed the pelvic vascular engorgement and the fact that it was unlikely that this woman experienced an orgasm during her ordeal, as well as the relationship between orgasm and fertilization of ova. Stieve did not explain under which circumstances this woman had been raped but mentioned that she had been shot to death. There is no image with this case study.[176]

Elias did use one of Stieve's microphotographs in his book *Human Microanatomy*,[177] but it is of a cervix (figure 18-29), and the legend reads, "Cervix uteri, showing mucous glands," and gives the source as "Stieve: Arch. Gynaek.183:178, 1952." The image originates from a posthumous publication by Stieve from *Archiv für Gynäkologie*, but the correct year is 1953.[178] Elias used image 16, and the original legend reads, "Median-sagittal section through the cervix of a 17-year-old virgin, magnification 4 times." Stieve explained in the text that the girl had died during a bombing raid, so the specimen had nothing to do with the rape case.

The same coexistence of fact and unproven allegation exists in the quotes from Elias' memoir and autobiography concerning Stieve. While Stieve's use of the bodies delivered from the Berlin execution chambers is clearly documented, there is no such evidence concerning bodies of Jewish citizens and others from concentration camps, or the mutilation of healthy women in camps by this anatomist. Stieve had a large collection of uteri, many of them in various stages of pregnancy, and placentae, but he had received them mostly from gynecological surgeons throughout his career. There is no indication that Stieve sent personnel out to "spy on" the sexual behavior of women, as he received his clinical information on the prisoners from the prison wardens and doctor's records.[179] It may have been the true facts at the center of these rumors that led Elias and others to firmly believe in them for many years after the war. Further research may show whether the rumors were indeed only that.

Notes

1. Elias 1979, 2-2.
2. Roelcke 2007.
3. Voswinkel 2004.
4. Voswinkel 2002.
5. Schierhorn 1986.
6. Hildebrandt 2012a.
7. Anonymous 1993.
8. Davie 1947.
9. Gerstengarbe et al. 1995; Ter Meulen 2010.
10. Winkelmann 2012.
11. Purpura 1998.
12. E.g., Elias 1979.
13. Lipphardt 2009.
14. Braund and Sutton 2007.
15. USHMM 2010a, RG-19.061.01, page 9/20; USHMM 2010b, RG-19.061.02, page 9/37; Lipphardt 2009.
16. E.g., Novak 1949.
17. E.g., Mehregan 2003.
18. Mendes-Flohr and Reinharz 2011, 745.
19. Wiskemann 1970; Bosworth 2006.
20. Notice 1938; Israel 2007; Bargmann 1967; Amprino 1967; Schierhorn 1986.
21. Elias 1979.
22. Notice 1938; *Matematica* newsletter 2011.
23. Mottura 1963.
24. Contu 2008.
25. *American Journal of Public Health* 1939.
26. Czechoslovak Ministry of Foreign Affairs 1941.
27. MacDonald and Kaplan 1995.
28. Cech 2006.
29. Röder and Strauss 1983; Peiffer 1998; The Encyclopedia of Saskatchewan 2010, *Rudolph Altschul.*
30. Schierhorn 1986; Gerstengarbe et al. 1995; Ter Meulen 2010; Watzka 1959; International Tracing Service Bad Arolsen 2010a, card records #27898368.
31. Hlaváčková and Svobodny 1998; Makarová 1998.
32. Studnicka 1946; Kapeller and Tichy 1985.
33. Walczak 1978.
34. International Tracing Service Bad Arolsen 2010b, card records # 28460602_1; von Eggeling 1934; Lisowski 1986 and Nowak 1964; Lagerliste Sachsenhausen 1939.
35. August 1997.
36. Ohm 2007; Voswinckel 2002; Fischer 1932; Langenfeld 1993.
37. Redzik 2011; Bielanska-Osuchowska 2008, 148.
38. Roland 1989.
39. Feliksiak 1987; Walczak 1978.
40. Winkelmann 2012.
41. Voogd 2002; Rotblat 1979; Voswinckel 2002; Fischer 1932; Nathan et al. 1960, 235; Frieswijk 1972; Ariëns Kappers 2001, 253.
42. Fischer 1933; Ariëns Kappers 2001; Drukker 1990.
43. American Men and Women of Science 1972; Singer 1993, 29.
44. Saller 1961.
45. Bruman 1963.

46. Aschoff 1966; Weindling 1992.
47. Winkelmann 2012.
48. Ro'i 2009, 155.
49. Tachau 2002.
50. Schwartz 1995.
51. Widmann 1973.
52. Laschke 2003.
53. Anonymous 1993; *New York Times* 1944.
54. Rürup and Schüring 2008.
55. Maskar 1953.
56. Franz 2007.
57. Bhatti and Voigt 1999.
58. Wakefield 1966.
59. Archiv der Universität Wien, MED PA 417 (Sch. 49).
60. Kapp and Mynatt 1997.
61. Collins 2009.
62. Decker 2003.
63. Bentwich 1953; Baldwin 1988.
64. Röder and Strauss 1983.
65. Weindling 2009.
66. Macleod 1997; Canadian Family Physician 1987; Wermuth 1939; Bassler 2006; Baldwin 1988.
67. Wallenberg-Chermak 1963; Zeidman and Mohan 2014.
68. Duggan and Drury 1948.
69. Sachar 1995, 12.
70. Sprague 1975; Haug 1986; *NEJM* 1939.
71. Gerstengarbe et al. 1995; Ter Meulen 2010.
72. Peiffer 1998.
73. Röder and Strauss 1983; Gerlach 1986.
74. Horn and Dorffner 2000; Burkl 1963.
75. Burkl 1963; Kernbauer 1996; Medizinische Universität Wien 2011.
76. Wilde and Ellmauthaler 2007; Freidenreich 2002; International Tracing Service Bad Arolsen 2010c and d, # 28886650_1 and # 28886649_1; Krause 1998; Arias 2004; Universitätsarchiv Wien MED PA 417 (Sch. 49).
77. Spiel 1977.
78. Hammerstein 1989.
79. Seidler and Leven 2007.
80. Hellmich, 1989, 176–81.
81. Braund and Sutton 2007; Bussche et al. 1991; Villiez 2009.
82. Rudloff 2005; Ebert 2010.
83. Novak 1949.
84. Schwoch 2009.
85. Mehregan 2003.
86. Ortmann 1976; Ortmann 1986; Golczewski 1988.
87. Mussgnug 1988; Eckart et al. 2006.
88. Saller 1961.
89. Lippert 1967, 1973; Bandmann et al. 1967; Scheibe-Jaeger 2008.
90. Lanz 1950; Tonutti 1977; Anonymous 1993; Schierhorn 1986; Scheibe-Jaeger 2008.
91. Heuer and Wolf 1997; von Eggeling 1950; Gregory 1949; McCort 1957; Howells 1981.
92. Lipphardt 2008.
93. Gerstengarbe et al. 1995.

94. Hertler 2008.
95. Benzenhöfer and Weiske 2010.
96. In Ricordo 2011.
97. Gedenkstätte Stille Helden 2015.
98. Lipphardt 2008, 277–78.
99. Quoted after Hertler 2008.
100. Winkelmann 2012.
101. Pross 1955.
102. Hildebrandt 2012b.
103. *Staatsbibliothek Berlin-Stiftung Preussischer Kulturbesitz: Nachlass* Hans Elias; M.E. Grenander Department of Special Collections and Archives at the State University of New York Library at Albany, estate Hans Elias.
104. Pauly 1987; Elias 1979, chapter 1.
105. Elias 1979, 2-2.
106. Baron 2011.
107. Pauly 1987.
108. Elias 1979, 2-1.
109. Elias 1931; Haug 1986.
110. Pauly 1987.
111. Elias 1979, 5-2.
112. Elias 1979, 3-1.
113. Elias, Memoirs II, SB-EE, box 2, original in English.
114. Elias 1979, 3-1 to 3-7.
115. Elias 1933.
116. Elias 1979, 3-8 to 3-17.
117. Elias 1979, 2-9.
118. Elias audiotape, Grenander collection, 2. Interview, 1979.
119. Elias 1979, 3-20, 3-24, 4-1.
120. Pauly 1987.
121. Bayor 1972.
122. SB-EE, box 6, diary of 1939.
123. Elias 1934, 1936, 1937a, b, c, d, anatomical publications.
124. Hildebrandt 2012b.
125. Elias 1979, 4-20 to 4-26.
126. E.g., Elias 1940; Elias 1942.
127. SB-EE, box 3.
128. SB-EE, diary 1944, box 7.
129. Pauly 1987.
130. Elias 1948, 1949a, b, c, d, anatomical publications on the liver.
131. Pauly 1987.
132. Elias 1979, 7-5 to 7-7.
133. Pauly 1987.
134. Elias 1979, 12-10 and 12-11; Pauly 1987.
135. Elias 1951; Elias 1979, 8-8.
136. Elias and Pauly 1960.
137. Baron 2011.
138. Thomas Elias, son of Hans Elias, personal communication 2011.
139. Elias 1957a; Elias 1957b.
140. Elias 1979, 9-14, 9-15.
141. Elias 1979, chapter 9.
142. SB-EE box 8, diary 1964, original in English.
143. Pauly 1987.

144. Thomas Elias, personal communication 2011; Heer 2011.
145. Elias 1979, 2-6, 2-7.
146. Elias 1979, 13-9.
147. See note 136.
148. Elias 1979, 3-26.
149. See note 136.
150. Grenander collection, Elias audiotape, 1979, 1. Interview.
151. Elias 1979, 2-3/2-4.
152. Elias 1979, 4-12.
153. Grenander collection, Elias audiotape 1979, 1. Interview.
154. Spiel 1977, xxii, xxv.
155. Baron 2011.
156. Elias 1979, 2-9.
157. SB-EE box 2, memoirs III, original in English.
158. Elias 1979, 2-10.
159. Emphasis added by author.
160. SB-EE, box 2, memoirs III, original in English.
161. Emphasis added by author.
162. Elias 1979, 2-10, 2-11.
163. Stieve 1942a.
164. Stieve 1942b.
165. Stieve 1926.
166. Röhl 2002.
167. Elias 1979, 2-11.
168. Another Jewish refugee, see Hildebrandt 2012a.
169. Elias 1979, 2-12.
170. Elias 1979, 2-12, 4-11.
171. Hildebrandt 2012a.
172. Winkelmann and Schagen 2009; Winkelmann and Noack 2010; Hildebrandt 2013.
173. Winkelmann and Schagen 2009; Hansson and Hildebrandt 2014.
174. Bill Seidelman, personal communication 2010.
175. Winkelmann and Schagen 2010; Noack 2008.
176. Stieve 1952, 176.
177. Elias and Pauly 1966.
178. Stieve 1953, 183: 178–203.
179. Winkelmann und Schagen 2009.

Archival Sources

Archiv der Universität Wien, MED PA 417 (Sch. 49)
Staatsbibliothek Berlin-Stiftung Preussischer Kulturbesitz: Estate (*Nachlass*) Hans Elias, 11 boxes, short: SB-EE
M.E. Grenander Department of Special Collections and Archives at the State University of New York library at Albany, estate Hans Elias. Manuscripts and audiotapes.
United States Holocaust Memorial Museum Washington:
   – USHMM, 2010a. Records of the Emergency Committee in Aid of Displaced Foreign Scholars. United States Holocaust Memorial Museum Archive, RG-19.061.01, page 9/20 (accessed 15 August 2010)
   – USHMM, 2010b. Records of the Emergency Committee in Aid of Displaced Foreign Scholars. USHMM Archive, RG-19.061.02, page 9/37 (accessed 15 August 2010)

- International Tracing Service Bad Arolsen, 2010a. United States Holocaust Memorial Museum card records #27898368 (accessed 15 August 2010)
- International Tracing Service Bad Arolsen, 2010b. United States Holocaust Memorial Museum, card records # 28460602_1 (accessed 15 August 2010)
- International Tracing Service Bad Arolsen, 2010c. United States Holocaust Memorial Museum card records # 28886650_1 (accessed 17 August 2010)
- International Tracing Service Bad Arolsen, 2010d. United States Holocaust Memorial Museum card records # 28886649_1 (accessed August 17, 2010)

Bibliography

American Journal of Public Health. 1939. "News from the Field: Dr. Emilio Franco." American Journal of Public Health 29: 579.

American Men and Women of Science. 1972. Vol. 2, D-G. New York: Jaques Cattell Press.

Amprino, Rodolfo. 1967. "Giuseppe Levi (1872–1965)." Acta Anatomica 66: 1–44.

Anonymous. 1993. Displaced German Scholars: A Guide to Academics in Peril in Nazi Germany During the 1930s. Rockville: Borgo Press/wildside press. First published in 1936.

Arias, Ingrid. 2004. "Entnazifizierung an der Wiener Medizinischen Fakultät: Bruch oder Kontinuität? Das Beispiel des Anatomischen Institutes." Zeitgeschichte 6(31): 339–69.

Ariëns Kappers, Cornelius U. 2001. Reiziger in Breinen. Herinneringen van en Hersenonderzoeker. Amsterdam: Uitgeverij L. J. Veen.

Aschoff, Ludwig. 1966. Ein Gelehrtenleben in Briefen an die Familie. Freiburg: Hans Ferdinand Schulz Verlag.

August, Jochen. 1997. "Sonderaktion Krakau": Die Verhaftung der Krakauer Wissenschaftler am 6. November 1939. Hamburg: Hamburger Edition.

Baldwin, Nicholas. 1988. Catalogue of the Archive of the Society for the Protection of Science and Learning, 1933–87. Bodleian Library, University of Oxford. EAD version 2008. Accessed 26 January 2011. http://www.bodley.ox.ac.uk/dept/scwmss/wmss/online/modern/spsl/spsl.html.

Bandmann, H. J., D. Hamburger, H. Holzmann, and A. Kressner. 1967. "Prof. Titus Ritter von Lanz. In Memoriam." Münchner Medizinische Wochenschrift 109(16): 902–3.

Bargmann, Wolfgang. 1967. "Guiseppe Levi (1872–1965)." Anatomischer Anzeiger 121: 444–48.

Baron, Günter. 2011. "Die Prometheus-Skulptur von Hans Elias in der Staatsbibliothek zu Berlin." Bibliotheksmagazin 1/2011: 35–38. Accessed 25 February 2015. staatsbibliothek-berlin.de/fileadmin/user_upload/zentrale_Seiten/ueber_uns/pdf/Bibliotheksmagazin/Magazin_lowRes_1_2011.pdf.

Bassler, Gerhard P. Vikings to U-Boats. The German Experience in Newfoundland and Labrador. Montral: McGill-Queen's University Press.

Bayor, Ronald H. 1972. "Italians, Jews and Ethnic Conflict." International Migration Review 6(4): 377–91.

Bentwich, Norman. 1953. The Rescue and Achievement of Refugee Scholars: The Story of Displaced Scholars and Scientists 1933–1945. The Hague: Martinus Nijhoff.

Benzenhöfer, Udo, and Katja Weiske. 2010. Bemerkungen zur Frankfurter Dissertation von Josef Mengele über Sippenuntersuchungen bei Lippen-Kiefer-Gaumenspalte (1938). In: Mengele, Hirt, Holfelder, Berner, von Verschuer, Kranz: Frankfurter Universitätsmediziner der NS-Zeit, ed. Udo Benzenhöfer, 9–20. Münster: Klemm und Oelschläger.

Bhatti, Anil, and Johannes H. Voigt. 1999. Jewish Exile in India 1933–1945. New Delhi: Manohar.

Bielańska-Osuchowska, Zofia. 2008. "Bronislawa and Mieczyslaw Konopacki: Pioneers of the Application of Histochemistry to Embryology in Poland." *International Journal of Developmental Biology* 52: 147–50.

Bosworth, R. J. B. 2006. *Mussolini's Italy: Life under the Dictatorship 1915–1945.* New York: Penguin Press.

Braund, James, and Douglas G. Sutton. 2007. "The Case of Heinrich Wilhelm Poll (1877–1939): A German-Jewish Geneticist, Eugenicist, Twin Researcher and Victim of the Nazis." *Journal of the History of Biology* 41(1): 1–35.

Bruman, Franz. 1963. "Paul Vonwiller: 1885–1962." *Acta anatomica* 52: 163–74.

Burkl, Wilhelm. 1963. "Carla v. Zawisch-Ossenitz." *Anatomischer Anzeiger* 111: 79–86.

Bussche, Hendrik van den, F. Pfäfflin, and Christoph Mai. 1991. "Die Medizinische Fakultät." In: *Hochschulalltag im "Dritten Reich" Die Hamburger Universität 1933–45*, ed. Eckart Krause, Ludiwg Huber, and Holger Fischer, 1259–1387. Berlin/Hamburg: Dietrich Reimer Verlag.

Canadian Family Physician. 1987. "In Memoriam. Eric G. Wermuth." *Canadian Family Physician* 33: 1136.

Cech, Pavel. 2006. "History notes." *Vita Nostra Service 21-22.* Accessed 1 February 2011. http://old.lf3.cuni.cz/vns/vns06/vns_en_21-22.pdf.

Collins, Kenneth. 2009. "European Refugee Physicians in Scotland: 1933–1945." *Social History of Medicine* 22(3): 513–30.

Contu, M. 2008. "A Gennaio la Giornata della Memoria." *Parrocchia Santa Barbara.* Accessed 27 January 2011. http://www.parrocchiasantabarbara.it/giornali/anno-2008/gennaio-08/12-01-08.html .

Czechoslowak Ministry of Foreign Affairs. 1941. *Two Years of German Oppression in Czechoslowakia.* Woking: Unwin Brothers Limited.

Davie, Maurice R. 1947. *Refugees in America: Report of the Committee for the Study of Recent Immigration from Europe.* New York: Harper & Brothers Publishers.

Decker, Karola. 2003. "Divisions and Diversity: The Complexities of Medical Refuge in Britain: 1933–1948." *Bulletin of the History Medicine* 77(4): 850–73.

Drukker, Jan. 1990. "In Memoriam Prof. Dr. M.W. Woerdeman, anatoom-embryoloog (1892–1990)." *Nederlands Tijdschrift vor Geneeskunde.* 134(40): 1969–70.

Duggan, Stephen, and Betty Drury. 1948. *The Rescue of Science and Learning: The Story of the Emergency Committee in Aid of Displaced Foreign Scholars.* New York: Macmillan Company.

Ebert, Andreas D. 2010. "Es kommt nicht darauf an, wer Recht hat, sondern *was richtig* ist: Robert Meyers Wirken an den Frauenkliniken der königlichen Charité (1908–1912) und der Berliner Friedrich-Wilhelms-Universität (1912–1939)". In: *Geschichte der Berliner Universitäts-Frauenkliniken*, ed. Matthias David und Andreas D. Ebert, 219–37. Berlin: de Gruyter.

Eckart, Wolfgang Uwe, V. Sellin, E. Wolgast. 2006. *Die Universität Heidelberg im Nationalsozialismus.* Heidelberg: Springer.

Elias, Hans. 1931. "Die Entwicklung des Farbkleides des Wasserfrosches (Rana esculenta)." *Zeitschrift für Zellforschung* 14:55–72.

———. 1933. "Entwicklungsgeschichtliche Trickfilme nebst Bemerkungen über die Verwendung des Semper-Präparates in der Botanik." *Berichte der Freien Vereinigung für Pflanzengeographie und Systememeatische Botanik* 41 (pages unavailable).

———. 1934. "Über die Entwicklung der Chromatophoren and anderer Zellen in der Haut von Bufo viridis." *Zeitschrift für Zellforschung* 21: 529–44.

———. 1936. "Hautchromatophoren von Bombinator pachypus und ihre Entwicklung." *Zeitschrift für Zellforschung* 24: 622–40.

———. 1937a. "Zur vergleichenden Histologie und Entwicklungsgeschichte der Haut der Anuren." *Zeitschrift für Mikroskopisch-Anatische Forschung* 41: 359–416.

————. 1937b. "Plastisch-kinematographische Rekonstruktion embryonaler Vorgänge." *Archiv für experimentelle Zellforschung* 19: 507–10.

————. 1937c. "Zur vergleichenden Histologie der Haut der Anuren." *Verhandlungen der Anatomischen Gesellschaft* 83: 188–96.

————. 1937d. "Plastisch-kinematographische Rekonstruktion." *Verhandlungen der Anatomischen Gesellschaft* 83: 200–205.

————. 1940. "Liberalistic Education as the Cause of Fascism." *School and Society* 51: 593–98.

————. 1942. "The Education of the Post-War Generation." *The Scholl Review* 50(7): 504–11.

————. 1948. "Do Liver Cords Exist?" *Anatomical Record* 100(4): 23–24.

————. 1949a. "Beobachtungen über den Bau der Säugerleber." *Anatomische Nachrichten* 1: 8–19.

————. 1949b. "A Re-examination of the Structure of the Mammalian Liver: I. Parenchymal Architecture." *American Journal of Anatomy* 84: 311–33.

————. 1949c. "A Re-examination of the Structure of the Mammalian Liver: II. The Hepatic Lobule and Its Relation to the Vascular and Biliary Systems." *American Journal of Anatomy* 85: 379–456.

————. 1949d. "The Liver Cord Concept after One Hundred Years." *Science* 110: 470–72.

————. 1951. "A Mathematical Approach to Microscopic Anatomy." *Chicago Medical School Quarterly* 12: 98–103.

————. 1957a. "De Structura Glomeruli Renalis." *Anatomischer Anzeiger* 104: 26–36.

————. 1957b. "De morphologia Carcinomatis Primarii Hepatis Humani et De Eius Contextu cum Evolutione Phylogenetica et Ontogenetica." *Acta Hepatologica* 5: 1–18.

————. 1979. *Abenteuer in Emigration und Wissenschaft: Ein Beitrag zur Aufklärung des Krebsproblems.* Unpublished manuscript. M. E. Grenander Department of Special Collections and Archives, University Libraries, University of Albany, New York.

Elias, Hans, and John E. Pauly. 1966. *Human Microanatomy.* Philadelphia: F.A. Davis Company. First published in 1960.

Eggeling, Heinrich von. 1950. "Franz Weidenreich, 1973–1948." *Anatomische Nachrichten* 1: 149–59.

Eggeling, Heinrich von, ed. 1934. "Stand der anatomischen Gesellschaft nach Schluss der zweiundvierzigsten Versammlung (Würzburg 1934)." *Verhandlungen der Anatomischen Gesellschaft* 78: 247–60.

Feliksiak, Stanisław. 1987. *Słownik Biologów Polskich.* Warszawa: Państwowe Wydaw Naukowe.

Fischer, Isidor. 1932. *Biographisches Lexikon der hervorragenden Ärzte der letzten fünfzig Jahre. Band I.* Berlin: Urban und Schwarzenberg.

————. 1933. *Biographisches Lexikon der hervorragenden Ärzte der letzten fünfzig Jahre. Band II.* Berlin: Urban und Schwarzenberg.

Franz, Margit. 2007. "'Passage to India': Österreichisches Exil in Britisch-Indien 1938-1945." *Jahrbuch 2007: Namentliche Erfassung von NS-Opfern,* ed. Dokumentationsarchiv Österreichischer Widerstand, 196–223. Wien: Lit-Verlag.

Freidenreich, Harriet P. 2002. *Female, Jewish, and Educated: The Lives of Central European University Women.* Bloomington: Indiana University Press.

Frieswijk, Jouke A. 1972. "In Memoriam Prof. Dr. G.C. Heringa." *Mededelingenblad: Organ van den Nederlandse Vereniging tot Beoefening van de Sociale Geschiedenis* 41: 28–32.

Gedenkstätte Stille Helden. 2015. "Europäische Union." Accessed 25 February 2015. http://www.gedenkstaette-stille-helden.de/rettungsversuche/thema/th/europaeische-union/.

Gerlach, Joachim. 1986. "In Memoriam Hartwig Kuhlenbeck." *Anatomischer Anzeiger* 161: 89–98.

Gerstengarbe, Sybille, Heidrun Hallmann, and Wieland Berg. 1995. "Die Leopoldina im Dritten Reich." *Die Elite der Nation im Dritten Reich- Das Verhältnis von Akademien und ihrem wissenschaftlichen Umfeld zum Nationalsozialismus.* Acta Historica Leopoldina 22: 167–212.

Golczewski, Frank. 1988. *Kölner Universitätslehrer und der Nationalsozialismus. Personenge-schichtliche Ansätze.* Köln: Böhlau Verlag.

Gregory, William K. 1949. "Franz Weidenreich, 1873–1948." *American Anthropologist* 51: 85–90.

Hammerstein, Notker. 1989. *Die Johann Wolfgang Goethe-Universität Frankfurt am Main. Band I: 1914–1950.* Neuwied/Frankfurt: Alfred Metzner Verlag.

Hansson, Nils, and Sabine Hildebrandt. 2014. "Swedish-German Contacts in Anatomy 1930–1950: The Example of Gösta Häggqvist and Hermann Stieve." *Annals of Anatomy* 196: 259–67.

Haug, Herbert. 1986. "In Memoriam Hans Elias." *Anatomischer Anzeiger* 161: 185–95.

Heer, Hannes. 2011. "'Darmstädter Expressionismus.' Gustav Hartung oder die Verteidigung der Republik mit den Mitteln des Theaters." *Tribüne* 197(1): 150–60. Accessed 15 January 2014. http://www.tribuene-verlag.de/T197_Heer.pdf .

Hellmich, Herrman-Josef. 1989. "Die medizinische Fakultät der Universität Freiburg i. Br. 1933–1945. Eingriffe und Folgen nationalsozialistischer Personalpolitik." Dissertation. Medical Faculty, University of Freiburg.

Hertler, Christine. 2008. "Franz Weidenreich und die Anthropologie in Frankfurt: Weidenreichs Weg an die Universität." In: *Frankfurter Wissenschaftler zwischen 1933 und 1945*, ed. Jörn Kobes, and Jan-Otmar Hesse, 111–23. Göttingen: Wallstein Verlag.

Heuer, Renate, and Siegbert Wolf. 1997. *Die Juden der Frankfurter Universität.* Frankfurt and New York: Campus Verlag.

Hildebrandt, Sabine. 2012a. "Anatomy in the Third Reich: Careers Disrupted by National Socialist Policies." *Annals of Anatomy* 184: 251–56.

———. 2012b. "The Anatomist Hans Elias: A Jewish German in Exile." *Clinical Anatomy* 25: 284–94.

———. 2013. "The Women on Stieve's List: Victims of National Socialism Whose Bodies Were Used for Anatomical Research." *Clinical Anatomy* 26: 3–21.

———. 2014. "Stages of Transgression: Anatomical Research in National Socialism." In: *Human Subjects Research after the Holocaust,* ed. Sheldon Rubenfeld and Susan Benedict, 68–85. Cham: Springer.

Hlaváčková, Ludmila, and Petr Svobodny. 1998. *Biographisches Lexikon der Deutschen Medizinischen Fakultät in Prag 1883–1945.* Praha: Karolinum.

Howells, W. W. 1981. "Franz Weidenreich, 1873–1948." *American Journal of Physical Anthropology* 56: 407–10.

Horn, Sonia and Gabriele Dorffner. 2000. "'…männliches Geschlecht ist für die Zulassung zur Habilitation nicht vorgesehen': Die ersten an der medizinischen Universität Wien habilitierten Frauen." In: *Töchter des Hippokrates; 100 Jahre akademische Ärztinnen in Österreich,* ed. Birgit Bolognese-Leuchtenmüller and Sonia Horn, 117–38.Wien: ÖÄK Verlag.

In Ricordo. 2011. "Piccagli, Italo, 1909." Accessed 10 March 2014. http://www.inrico rdo.eu/index.php?option=com_sobi2&sobi2Task=sobi2Details&catid=0&sobi 2Id=22362&Itemid=37.

Israel, Giorgio, 2007. "Redeemed Intellectuals and Italian Jews." *Telos* 139: 85–108.

Kapeller, Karol, and Milos Tichy. 1985. "Professor Jan Florian: A Great Personality." *Folia Morphologica* 33: 305–9.

Kapp, Yvonne, and Margaret Mynatt. 1997. *British Policy and the Refugees 1933–1941.* London: Frank Kass. First published in 1941.

Kernbauer, Alois. 1996. "Carla Zawisch-Ossenitz. Eine biographische Skizze der ersten Professorin der Grazer Universität." In: *Frauenstudium und Frauenkarrieren an der Universität Graz*, ed. Alois Kernbauer and Karin Schmidlechner-Lienhart, 265–70. Graz: Akademische Druck- und Verlagsanstalt.

Krause, Walter. 1998. "Bekenntnisse eines alten Anatomie Lehrers." In: *Körper ohne Leben. Begegnung und Umgang mit Toten*, ed. Norbert Stefenelli, 40-43. Wien: Böhlau Verlag.

Lagerliste Sachsenhausen, 1939. "Zugänge vom 28.11.39." Accessed 15 August 2010 via International Tracing Service Arolsen, United States Holocaust Memorial Museum.

Langenfeld, Marian Stanislaw. 1993. "The Hundred and Twenty-Fifth Anniversary of the Birth of Henryk Ferdinand Hoyer Junior (1864–1989)." *Annals of Anatomy* 175: 101–3.

Lanz, Titus von. 1950. "Professor Dr. Harry Marcus zum 70. Geburtstag." *Münchner Medizinische Wochenschrift* 92 (11/12).

Laschke, Michael. 2003. *Das Oskar-Ziethen-Krankenhaus Berlin-Lichtenberg.* Altenburg: Leipziger Universitätsverlag.

Lippert, Herbert. 1967. "Titus Ritter von Lanz zum 70. Geburtstag." *Medizinische Klinik* 62(1): 28-29.

———. 1973. "Titus W.-H. Ritter von Lanz. 1897–1967." *Acta anatomica* 84: 465–74.

Lipphardt, Veronika. 2009. "'Investigation of Biological Changes': Franz Boas in Kooperation mit deutsch-jüdischen Anthropologen." In: *Kulturrelativismus und Antirassismus. Der Anthropologe Franz Boas (1858-1942)*, ed. Hans-Walter Schmuhl, 163–85. Bielefeld: transcript Verlag.

Lisowski, W. 1986. "Professor Kazimierz Kostanecki (1863–1940), Renowned Polish Anatomist." *Materia Medica Polana* 2(58): 115–18.

MacDonald, Callum, and Jan Kaplan. 1995. *Prague: In the Shadow of the Swastika.* London: Quartet Books Ltd.

Macleod, Malcolm. 1997. "Migrant, Intern, Doctor—Spy? Dr. Eric Wermuth in Second World War Newfoundland." *Newfoundland Studies* 13(1): 79–89.

Makarová, Jelena.1998. "Die Akademie des Überlebens." *Terezin Studies and Documents* 5:213–38. Accessed 31 January 2011. www.ceeol.com.

Maskar, Üveis. 1953. "Tibor Péterfi." *Acta Anatomica* 19: 1–7.

Matematica newsletter. 2011. "In Italia, 2011. Manifesto degli scienziati razzisti." *Matematica.* Accessed 26 January 2011. http://matematica-old.unibocconi.it/giornomemo/italiani.htm.

McCort, James J. 1957. "Doctors Afield: Franz Weidenreich (1873–1948)." *New England Journal of Medicine* 257(14): 670–71.

Medizinische Universität Wien. 2011. *Der Anschluss: Carla Zawisch-Ossenitz.* Accessed 6 April 2011. http://www.meduniwien.ac.at/geschichte/anschluss/an_zawisch.html.

Mehregan, David A. 2003. "Felix Pinkus: Dermatologist, Dermatopathologist, and Artist." *International Journal of Dermatology* 42: 44.

Mendes-Flohr, Paul, and Jehuda Reinharz. 2011. *The Jew in the Modern World. A Documentary History.* 3rd ed. New York: Oxford University Press.

Mottura, G. 1963. "Ettore Ravenna (1876–1963)." *Archivio Italiano di Anatomia e Istologia Patologica* 37: 148–51.

Mussgnug, Dorothee. 1988. *Die vertriebenen Heidelberger Dozenten: Zur Geschichte der Ruprecht-Karls-Universität nach 1933.* Heidelberg: Carl Winter.

Nathan, Otto, Heinz Norden, and Bertrand Russell, eds. 1960. *Einstein on Peace.* New York: Simon and Schuster.

NEJM. 1939. "Notes." *New England Journal of Medicine* 220(3): 124.

New York Times. 1956. "Obituary Dr. Oscar Benesi: 77, City Hospitals Aide." *New York Times* 20 January 1956.

Noack, Thorsten. 2008. "Begehrte Leichen. Der Berliner Anatom Hermann Stieve (1886–1952) und die medizinische Verwertung Hingerichteter im Natinoalsozialismus."

Medizin, Gesellschaft und Geschichte. Jahrbuch des Instituts für Geschichte der Medizin der Robert Bosch Stiftung 26: 9–35.

Notice. 1938. "Vacated Italian Chairs." *Irish Journal of Medical Sciences* 13(12): 770.

Novak, Emil, ed. 1949. *Autobiography of Dr. Robert Meyer (1864–1947)*. New York: Henry Schuman.

Ohm, Agnes. 2007. "Vergessene Vernichtung?" *Gedenkstättenforum-Rundbrief* 155: 26–35. Accessed 12 January 2011. http://www.gedenkstaettenforum.de/nc/gedenkstaetten-rundbrief/rundbrief/news/vergessene_vernichtung/.

Ortmann, Rolf. 1976. "Prof. Dr. Otto Veit. Ein Nachruf und ein Stück Geschichte der Kölner Anatomie." *Acta Anatomica* 94:161–68.

Ortmann. Rolf. 1986. *Die jüngere Geschichte des anatomischen Instituts der Universität Köln 1919–1984*. Köln: Böhlau Verlag.

Pauly, John E. 1987. "M. Hans Elias 1907–1985." *American Journal of Anatomy* 180: 123–125.

Peiffer, Jürgen. 1998. "Die Vertreibung deutscher Neuropathologen." *Nervenarzt* 69: 99–109.

Pross, Helge. 1955. *Die deutsche akademische Emigration nach den Vereinigten Staaten 1933–1941*. Berlin: Duncker & Humblot.

Purpura, Dominick P. 1998. "Berta Scharrer." In: *National Academy of Sciences US: Biographical Memoirs* 74:289–307. Washington DC: National Academic Press.

Redzik, Adam. 2011. *Polish Universities during the Second World War*. Accessed 2 September 2014. http://www.gomezurdanez.com/polonia/adamredzikpolishuniversitas.pdf .

Röder, Werner, and Herbert A. Strauss, eds. 1983: *Biographisches Handbuch der deutschsprachigen Emigration nach 1933. Volume II. Part 1: A-K*. München and New York: KG Saur.

Röhl, John Charles Gerald. 2002. *Kaiser, Hof und Staat. Wilhelm II. Und die deutsche Politik*. München: C.H. Beck.

Roelcke, Volker. 2007. "Trauma or Responsibility? Memories and Historiographies of Nazi Psychiatry in Postwar Germany." In: *Trauma and Memory: Reading, Healing and Making Law*, ed. Austin Sarat, Nadav Davidovitch, and Michael Alberstein, 225–42. Stanford: Stanford University Press.

Ro'i, Yaacov. 2009. "The Transformation of Historiography on the 'Punished Peoples.'" *History and Memory* 21(2): 150–76.

Roland, Charles. 1989. "An Underground Medical School in the Warsaw Ghetto, 1941–1942." *Medical History* 33: 399–419.

Rotblat, Joseph. 1979. "Einstein the Pacifist Warrior." *Bulletin of the Atomic Scientists* 35(3): 21–26.

Rudloff, Udo, and Hans Ludwig. 2005. "Jewish Gynecologists in Germany in the First Half of the Twentieth Century." *Archives of Gynecology and Obstetrics* 272: 245–60.

Rürup, Reinhard and Michael Schüring, eds. 2008. *Schicksale und Karrieren. Gedenkbuch für die von den Nationalsozialisten aus der Kaiser-Wilhelm-Gesellschaft vertriebenen Forscherinnen und Forscher*. Göttingen: Wallstein-Verlag.

Sachar, A. Leon. 1995. *A Host at Last*. Hanover, MA, and London: Brandeis University Press.

Saller, Karl. 1961. *Die Rassenlehre des Nationalsozialismus in Wissenschaft und Propaganda*. Darmstadt: Progress-Verlag.

Scheibe-Jaeger, Angela. 2008. "Das Leben des Prof. Dr. med. Harry Marcus. Ein geachteter Bürger und Soldat." In: *Ins Licht gerückt. Jüdische Lebenswege im Münchner Westen. Eine Spurensuche in Pasing, Obermenzing und Aubing*, ed. Bernhard Schossig, 107–15. München: Herbert Utz Verlag.

Schierhorn, Helmke. 1986. "Mitglieder der Anatomischen Gesellschaft im antifaschistischen Exil." *Verhandlungen der Anatomischen Gesellschaft* 80: 957–63.

Schwartz, Philipp. 1995. *Notgemeinschaft. Zur Emigration deutscher Wissenschaftler nach 1933 in die Türkei*. Marburg: Metropolis-Verlag.

Schwoch, Rebecca, ed. 2009. *Berliner Jüdische Kassenärzte und ihr Schicksal im Nationalsozialismus. Ein Gedenkbuch.* Berlin: Hentrich & Hentrich.

Seidler, Eduard, and Karl-Heinz Leven. 2007. *Die Medizinische Fakultät der Alberts-Ludwigs-Universität Freiburg im Breisgau. Grundlagen und Entwicklungen.* Freiburg: Verlag Karl Alber.

Singer, Ronald. 1993. *William Bloom, 1899–1972. A Biographical Memoir.* Washington, DC: National Academy of Sciences.

Spiel, Hilde. 1977. "Psychologie des Exils." In: *Österreicher im Exil 1934 bis 1945. Protokoll des internationalen Symposiums zur Erforschung des österreichischen Exils von 1934 bis 1945,* ed. Dokumentationszentrum des österreichischen Widerstandes und Dokumentationsstelle für neuere österreichische Literatur, XXII–XXXVI. Wien: Österreichischer Bundesverlag.

Sprague, Victor. 1975. "J. Richard Weissenberg: 1882–1974." *Anatomical Record* 183(1): 148–49.

Stieve, Hermann. 1926. *Unfruchtbarkeit als Folge unnatürlicher Lebensweise.* München: Verlag von J. F. Bergmann.

———. 1942a. "Der Einfluss von Angst und psychischer Erregung auf Bau und Funktion der weiblichen Geschlechtsorgane." *Zentralblatt für Gynäkologie* 66: 1698–1708.

———. 1942b. "Die Wirkung von Gefangenschaft und Angst auf den Bau und die Funktion der weiblichen Geschlechtsorgane." *Zentralblatt für Gynäkologie* 66: 1456.

———. 1952. *Der Einfluss des Nervensystems auf Bau und Tätigkeit der Geschlechtsorgane des Menschen.* Stuttgart: Thieme.

———. 1953. "Cyclus, Physiologie und Pathologie (Anatomie)." *Archiv für Gynäkologie* 183: 178–203.

Studnika, F. K. 1946. "Prof. Jan Florian." *Nature* 2 March 1946, 3983:257.

Tachau, Frank. 2002. "German Jewish Emigrés in Turkey." In: *Jews, Turks, Ottomans: A Shared History, Fifteenth through the Twentieth Century,* ed. Avigdor Levy, 233–29. Syracuse, NY: Syracuse University Press.

Ter Meulen, Volker. 2010. Rede des Präsidenten der Leopoldina bei der Einweihung der Gedenkstele am 1. Oktober 2009 in Halle (Saale). *Nova Acta Leopoldina* 22:7–10. Also online, accessed 9 September 2014. http://www.leopoldina.org/uploads/tx_leopublication/ProbeNALSu22.pdf

The Encyclopedia of Saskatchewan. *Rudolph Altschul.* Accessed 12 September 2014. http://esask.uregina.ca/entry/altschul_rudolph_1901–63.html.

Tochowicz, L., and S. J. G. Nowak. 1964. "History of Cracow School of Medicine." *Journal of the American Medical Association* 188(7):662–67.

Tonutti, Emil. 1977. "Harry Marcus +." *Anatomischer Anzeiger* 141: 97–105

Villiez, Anna von. 2009. *Mit aller Kraft verdrängt. Entrechtung und Verfolgung "nicht arischer" Ärzte in Hamburg 1933 bis 1945.* München: Dölling und Galitz Verlag.

Voogd, J. 2002. "Neuroanatomy." In: *History of Neurology in the Netherlands,* ed. J. A. M. Frederiks, G. W. Bruyn, and P. Eling, 123–64. Amsterdam: Boom.

Voswinkel, Peter. 2004. "Damnatio memoriae: Kanonisierung, Willkür und Fälschung in der ärztlichen Biographik." In: *Universitäten und Hochschulen im Nationalsozialismus und in der frühen Nachkriegszeit,* ed. Karen Bayer, Frank Sparing, and Wolfgang Woelk, 249–70. Stuttgart: Steiner Verlag.

Voswinckel, Peter, ed. 2002. *Fischer, Isidor: Biographisches Lexikon der hervorragenden Ärzte der letzten fünfzig Jahre. Band III.* Hildesheim: Georg Olms Verlag.

Wakefield, Edward. 1966. *Past Imperative: My Life in India 1937–1947.* London: Chatto and Windus.

Walczak, Marian. 1978. *Szkolnictwo Wyzsze i Nauka Polska w Latach Wojny i Okupacji 1939–1945.* Wroclaw: Zakład Narodowy im Ossolinskich.

Wallenberg-Chermak, Marianne. 1963. "Adolf Wallenberg (1862–1939)." In *Grosse Nervenärzte. Band 3*, ed. Kurt Kolle, 191–96. Stuttgart: Georg Thieme Verlag.

Watzka, Max. 1959. "Alfred Kohn." *Anatomischer Anzeiger* 106: 449–57.

Weindling, Paul J. 1992a. "German-Soviet Medical Co-operation and the Institute for Racial Research, 1927–c.1935." *German History* 10(2): 177–206.

———. 2009. "Medical Refugees and the Modernization of British Medicine: 1930–1950." *Social History of Medicine* 22(3): 489–511.

Wermuth, Eric G. 1939. "Anastomoses between the Rectal and Uterine Veins Forming a Connection between the Somatic and Portal Venous System in the Recto-Uterine Pouch." *Journal of Anatomy* 74(Pt1): 116–26.

Widmann, Horst. 1973. *Exil und Bildungshilfe. Die deutschsprachige akademische Emigration in die Türkei nach 1933*. Frankfurt: Peter Lang.

Wilde, J. and Volkmar Ellmauthaler. 2007. "Universitätsprofessor Dr. Walter Krause". Accessed 6 August 2010. http://www.medpsych.at/lehrer-krause.txt.

Winkelmann, Andreas. 2012. "The Anatomische Gesellschaft and National Socialism: A Preliminary Analysis Based on Society Proceedings." *Annals of Anatomy* 194: 243–50.

Winkelmann, Andreas, and Thorsten Noack. 2010. "The Clara Cell: A 'Third Reich Eponym'?" *European Respiratory Journal* 36:722–27.

Winkelmann, Andreas, and Udo Schagen. 2009. "Hermann Stieve's Clinical-Anatomical Research on Executed Women during the 'Third Reich.'" *Clinical Anatomy* 22(2): 163–71.

Wiskemann, Elizabeth. 1970. *Fascism in Italy: Its Development and Influence*. London: Macmillan.

Zeidman, Lawrence A., and Lauren Mohan. 2014. "Adolf Wallenberg: Giant in Neurology and Refugee from Europe." *Journal of the History of Neurosciences* 23: 31–44.

Chapter 6

ANATOMISTS WORKING
IN NS GERMANY

Es gibt keine unbelasteten Anatomen in Deutschland.
(There are no untainted anatomists in Germany.)
Hermann Stieve after the war, commenting
on the involvement of anatomists in the
use of bodies of executed persons[1]

The NS regime disrupted and indeed destroyed the careers and lives of many anatomists in Germany and the occupied countries. It also had a direct impact on those scientists who continued in their academic positions and anatomical work. There was a wide political spectrum among them, spanning from enthusiastic and sustained support of National Socialism to pragmatic acceptance of the political situation to, rarely, careful resistance. All of them came to an arrangement with the authorities that allowed them to maintain or even advance their careers, with the exception of Charlotte Pommer. Selected biographies will serve as illustrations of the personal backgrounds of anatomists and their professional environments that led to their political decisions. They will be presented here in groups of active National Socialists, pragmatists, winners of the system, and dissenters.

Active National Socialists: Pernkopf, Kremer, Hirt

Some anatomists forcefully supported the regime; they joined the NS movement early on or held important leadership roles in the political life of their universities. All of them were members of the NSDAP, some also of the SA or SS. Eduard Pernkopf and August Hirt had successful anatomical careers, while Johann Paul Kremer's life was one of academic mediocrity.

Following the war, discussions of their work and wrongdoings took a long time to emerge.

Eduard Pernkopf was born on 24 November 1888 in Austria, and received his medical degree from the Vienna Medical School in 1912, where he stayed for the rest of his professional life. In 1933 Pernkopf succeeded Ferdinand Hochstetter as director of the Second Anatomy Institute.[2] David J. Williams was the first to provide some insight into Pernkopf's political activities:[3] He had joined a nationalistic German student fraternity, became an NSDAP member in 1933 while the party was illegal under Austrofascist law, and signed up with the SA less than a year later.

Anatomy at the University of Vienna had been divided into two separate departments since 1870. The directors of the First Anatomy Institute, which was more systematically and clinically oriented, were liberal-democratic-minded Jewish scientists like Julius Tandler, who was dismissed in 1934 by the Austrofascist regime.[4] The chairmen of the Second Anatomy Institute tended toward nationalism and anti-Semitism. During the 1920s and 1930s violent battles including physical altercations over differences in politics were fought between students from the two institutes, which were finally merged under the chairmanship of Pernkopf in 1938.[5] Pernkopf was a meticulous and demanding supervisor, who developed new dissection and representation techniques with the illustrators for his anatomical atlas, establishing a personal routine of eighteen-hour workdays.[6] He was appointed dean of the medical faculty in April 1938, a few weeks after the annexation of Austria by NS Germany, and served as president of the University of Vienna from 1943 to 1945. He helped establish an office for "hereditary and racial biology" and made changes in the medical curriculum, which he had proposed in his first official speech as a dean.[7] In this lecture on 6 April 1938, Pernkopf proved his full alliance with the ideology of the new rulers.[8] He thanked Hitler for the integration of Austria into the German Reich, and formulated as the new goal of the medical school the education of German National Socialist doctors. He proposed a science of order and purpose. Specifically applied, this new concept would allow the anatomical sciences to explain human variation through concepts of constitution and "race." The new curriculum was to include racial physiology, psychology, pathology, and racial genetics. This knowledge could be used in such areas as sports, occupational counseling, and family counseling, serving not just the individual but the whole body of the people through supporting the healthy and worthy, their marriages and reproduction, as well as by eradicating the unworthy and bad, preventing "racial" mixing, and eliminating the genetically inferior by sterilization and other means.

At the end of the war Pernkopf was dismissed by the Allied Forces and detained in a prison camp from 1945 to 1948, stripped of his titles and appointments. For the time that his obituary called his "retirement," he was offered rooms by members of the medical faculty for work on his atlas. His manuscripts continued to be published in the *Wiener Klinische Wochenschrift,*[9] a journal whose editorial board he had joined in 1938.[10] Pernkopf achieved a revision of his legal status from "incriminated person" to "lesser incriminated person" and thus passed the official "denazification" procedure, even receiving his full pension from 1953 until his death.[11] Pernkopf was never questioned about the use of bodies of NS victims for the atlas and died from a stroke on 7 April 1955. The controversy around the atlas occurred many years after Pernkopf's death.[12]

Johann Paul Kremer's career was much less distinguished than Pernkopf's. He was born on 26 December 1883 to a family of farmers near Cologne, and received his PhD in zoology in 1914 and an MD in 1919. After several years of surgical work he joined the anatomical institute in Bonn in 1924, and accepted the position of senior assistant at the anatomical institute in Münster in 1927, where he remained for the rest of his career. In 1929 he attained his *Venia legendi,* and was promoted to *ausserplanmässiger* professor in 1939.[13] Apart from anatomy, Kremer taught sports medicine and hereditary biology.[14] He joined the NSDAP in July 1932 and the SS on 20 November 1934,[15] and was promoted to SS-*Obersturmführer* in 1943.[16] Despite his age and political position he was never seriously considered for a chair in anatomy.

Benninghoff thought Kremer's scientific work was "weak."[17] At that time, Kremer had studied the effect of hunger on tissues in amphibians,[18] and in 1941 he published a controversial manuscript claiming the inheritance of traumatic injuries in a litter of short-tailed cats.[19] This neo-Lamarckian theory brought him not only in conflict with his peers, but also with the NSDAP and the GeStapo, as news of a potential heritability of war injuries was highly inopportune for the government, and a colleague easily disproved Kremer's theory.[20] The controversy also precluded him from being considered for a planned chair in racial hygiene at the University of Münster,[21] and the lack of professional advancement greatly disappointed him. He considered himself a "victim of my sincere belief in scientific ideals and in the unlimited freedom of research."[22]

As an SS officer, Kremer had been detailed to the SS Main Sanitation Office in Berlin and the *Waffen-*SS frontline unit hospitals in Prague and Dachau from 20 August to 15 October 1941, before being ordered to Auschwitz on 30 August 1942.[23] In Dachau Kremer had been responsible for German personnel and had no contact with prisoners, but he may have been aware of medical experiments performed on them by other SS physicians, which had been authorized by Himmler in 1941.[24] In Auschwitz Kremer had direct access to prisoners through the selection procedures on the hospital wards, which were part of his duties in addition to the selection of prisoners at the train ramp.[25] According to his recollections during postwar interrogations by the Polish authorities, he realized shortly after his arrival in the camp that he had the opportunity to follow up on his hunger experiments on animals with studies on the prisoners. He presented this idea to his superior at Auschwitz, Dr. Eduard Wirths, who, in Kremer's words,

> told me that I could collect fresh living material (*lebensfrisches Material*) from prisoners who were put to death by phenol injection […]. I closely observed the prisoners in this group, and when one of them interested me because of a highly advanced state of starvation, I commanded the orderly to reserve that patient for me, and inform me of when that patient would be killed with the help of an injection. At the time indicated by the SS orderly, the patients I had chosen were led back to that last block and placed in a room there across the corridor from the examining room where they had been selected. There, the patient was placed on an autopsy table while still alive. I went to the table and asked the patient about various details that were important to my research. Thus: his weight before arrest, how much weight he had lost since then, any medicine he had taken and so on. After the collection of this information, an orderly approached the patient, and killed the patient with an injection in the vicinity of the heart. As far as I know, only phenol injections were used for killing. After such an injection death occurred immediately. I never administered the fatal injections myself. I stood far away from the autopsy table with jars prepared to collect samples of organs of the body that were necessary to examine for my work.[26]

Some prisoners were photographed before their murders in Kremer's presence. He took the collected tissue samples of spleen, liver, and pancreas together with the photographs back to Münster at the end of his Auschwitz assignment on 20 November 1942.[27] It is unlikely that Kremer performed any research on these tissues, as no publications of results exist, and he had refused to work at the institute after his return due to disagreements with his superior Hellmut Becher.[28]

After the war Kremer was arrested by the British military, who found his diary in his apartment.[29] This journal was the first document that proved the participation of physicians in medical experiments in concentration camps. Kremer was extradited to Poland and reported during his trial that he did not know where the photographs and tissue samples had ended up.[30] He was sentenced to death for sending at least 10,717 men, women, and children directly to their deaths. In 1948 the verdict was converted into a life sentence because of Kremer's "good conduct" and advanced age, and in 1958 he was extradited from Poland to Germany.[31] There he was again put on trial and received a verdict of ten years in prison for aiding and abetting murder in two cases. He walked away as a free man, as the court saw the punishment as already served through his incarceration in Poland.[32] Kremer died in January 1965 in Münster.[33]

Kremer's diary entries from 26 November 1940 to 11 August 1945 give insight into his work and personality.[34] Still in Münster in March 1942, he noted about the death of his canary Hänschen: "Hänschen ceased to suffer at 2 p.m. I was extremely sorry, as I had been so used to this poor little fellow, always so lively. Cremation." Upon arrival in Auschwitz he remarked on the details of his days, including his research:

August 31, 1942: Tropical climate with 28 centigrades in the shade, dust and innumerable flies! Excellent food in the Home. This evening, for instance, we had sour duck livers for 0.40 mark, with stuffed tomatoes, tomato salad etc. Water is infected. So we drink seltzer water which is free (Mattoni).[35] First inoculation for typhus. Had photo taken for the camp identity card.

September 1, 1942: Have ordered SS officer's cap, sword-belt and braces from Berlin by letter. In the afternoon was present at the gassing of a block with Cyclon B against lice.

September 2, 1942: Was present for the first time at special action at 3 a.m. In comparison with it Dante's Inferno seems to be almost a comedy. Auschwitz is justly called an extermination camp.

September 3, 1942: Was for the first time taken ill with the diarrhoea which attacks everybody in the camp here. Vomiting and colic-like paroxysmal pains.

[...]

September 5, 1942: This noon was present at a special action in the women's camp *Muselmänner*,[36] the most horrible of all horrors. *Hschf.* Thilo, military surgeon, is right when he said today to me we were located here in the '*anus mundi*'.[37] In the evening at about 8 p.m. another special action with a draft from Holland. Men compete to take part in such actions as they get additional rations then—1/5 litre vodka, 5 cigarettes, 100 grammes of sausage and bread. Today and tomorrow (Sunday) on duty.

[...]

October 3, 1942: Today I got quite living-fresh material of human liver, spleen and pancreas, also lice from a typhus case, fixed in absolute alcohol.

[...]

October 17, 1942: Was present at a punishment and 11 executions. Have taken living-fresh material of liver, spleen and pancreas after injection of pilocarpin.

[...]

October 18, 1942: In wet and cold weather was on this Sunday morning present at the 11th special action (from Holland). Terrible scenes when 3 women begged to have their bare lives spared.

[...]

November 13, 1942: Living-fresh material (liver, spleen and pancreas) from a Jewish prisoner of 18, extremely atrophic, who had been photographed before. As usual with liver and spleen, it was fixed in *Carnoy* and pancreas in *Zenker* (prisoner No. 68030).[38]

Asked at his Polish trial about the *"anus mundi"* entry he commented:

> I could deduce from the behaviour of these women that they realized what was awaiting them. They begged the SS-men to be allowed to live, they wept, but all of them were driven to the gas chambers and gassed. Being an anatomist I had seen many horrors, had to do with corpses, but what I then saw was not to be compared with anything seen ever before.[39]

In these writings Kremer comes across as a rather pedantic man full of self-pity, capable of clear and detailed observation as well as sympathy (for a dead pet canary). At the same time he was completely obtuse to the tragic irony and heartlessness of reporting on the quality of his daily meals at Auschwitz while investigating the effects of starvation on the bodies of freshly murdered emaciated prisoners. Auschwitz represented for him a terrible place (*"anus mundi"*) not because he felt sympathy with the suffering of the victim prisoners, but because he himself was uncomfortable. Any observations on the prisoners' desperate reactions came during his postwar reflections, not while he was present in Auschwitz.

August Hirt is the most notorious anatomist of the National Socialist period.[40] He was born in Mannheim, Germany, on 29 April 1898 to a Swiss salesman. In 1914, at age sixteen, the mediocre student volunteered for military service and later returned to high school in Mannheim in 1916 after receiving an Iron Cross Second Class and a gunshot wound to his face that left him badly scarred. In 1917 Hirt started his medical studies at the University of Heidelberg, applied for recognition of his German citizenship in addition to his Swiss one, and became a member of the nationalist fraternity "Normannia." After receiving his MD in 1922 and his *Venia legendi* in 1925, he was promoted to *ausserordentlicher* professor in 1930.

Hirt's work on the autonomic innervation of the human kidney brought him in contact with Philipp Ellinger, professor of pharmacology at Heidelberg. Between 1925 and 1932 the two collaborated on the development of an innovative technique using fluorescence with a new microscope, which allowed the observation of blood flow in living organs. The Ellinger-Hirt luminescence microscope is still regarded as an important historic step in the development of modern microscopy by its manufacturer Zeiss.[41] Even though Hirt was clearly the junior partner in this collaboration, he later claimed sole authorship of the invention and distanced himself from his Jewish colleague, who was forced to emigrate in 1933.

Hirt joined the *Kampfbund für deutsche Kultur* (a nationalistic anti-Semitic political society) in 1932, the SS on 1 April 1933 (No. 100414), the NSDAP on 1 May 1937 (No. 4012784), and the NSDDB on 27 July 1934. Despite the fact that Hirt's scientific reputation was inconsistent among his colleagues, the REM appointed him to the chair of anatomy at the University of Greifswald in 1936 without following the proper recruitment process. The faculty quickly reconciled itself with this decision by the authorities, and the dean expressed his grateful approval of Hirt in a letter to the REM after several meetings with the new appointee.[42]

In 1938 Hirt swapped chair positions with Wilhelm Pfuhl at the University of Frankfurt, whose health was suffering. However, Hirt did not spend much time there, as he performed war services as a physician from 1939 to 1941, but his research continued with the help of his assistants Karl Wimmer and Anton Kiesselbach. Hirt's idea of preventing both cancer and poison gas effects with the help of Vitamin A compounds was tested in animal models in Frankfurt and at the *Militärärztliche Akademie* (Academy for Military Physicians) in Berlin. However, his Berlin colleagues were not impressed with the quality of Hirt's research.[43]

Hirt was recruited as first chair of anatomy at the *Reichsuniversität* Strasbourg in 1941, and he moved there with Wimmer and Kiesselbach. During the celebrations for the inauguration of the university on 23 November 1941, Hirt met Sievers, who apparently mentioned the opportunity of gaining access to "prisoners and professional criminals" for Hirt's poison-gas experiments.[44] Sievers promoted Hirt's ideas, and the *Ahnenerbe* decided to support Hirt's experiments financially and organizationally (see chapter 3). He was the only anatomist whose name was mentioned during the Nuremberg Doctors' Trial in 1947, but he was not among those charged with war crimes there.[45] The fact that Hirt had committed suicide in June 1945 became known only much later.[46] In 1953 a court in Metz, still unaware of Hirt's death, sentenced him to death in absentia.[47]

The Pragmatists: Voss, Bargmann, von Hayek

Membership in the NSDAP did not necessarily mean approval of the NS regime,[48] and anatomists who disagreed with the new government often delayed their entry into the party as long as they thought their careers were not threatened. Benninghoff was such an example; he only joined the party when he feared obstacles for the completion of his textbook in 1940. Others who were firmly established never became members of the NSDAP, among them Stieve, who was recruited to Berlin in 1935,[49] and Stöhr, who attained the position of chairman of anatomy in Bonn in the same year. Stöhr was said to harbor a distinct aversion against the NS government and only joined minor NS political affiliations, like Stieve, and discontinued his participation at the meetings of the *Anatomische Gesellschaft* because of the active National Socialists among its members.[50] Anatomists who were hoping for promotion, among them Hermann Voss and von Hayek, decided to become party members when they felt that their careers were stalling, and there were young and ambitious scholars like Bargmann, who joined the party early. A comparison of the biographies of Voss, Bargmann, and von Hayek, who advanced through decisive steps of their careers during the NS period and had prosperous postwar careers, will point out the various pragmatic political decisions these anatomists made.

Hermann Voss's life is well documented in the personnel files from the various universities at which he worked during his long career, as well as through his diary, which he kept from 1932 to 1942. This diary, along with his chronicle of the department of anatomy at the University of Posen, was found in the anatomical institute by his Polish successor Stefan Rozycki in 1945.[51] Although excerpts from the diary had been published repeatedly in Polish media in the first decades after the war[52] and were mentioned in a German dissertation from 1964,[53] the text only became known to a wider audience in 1987, when author Götz Aly published it in a volume on NS medicine.[54] The authenticity of the text was disputed by friends of Voss; however, several independent evaluations all confirmed Voss as the author.[55] A comprehensive account of Voss' life can be found in Majewski.[56]

Voss was born on 13 October 1894 to a tenant farmer in Berlin, and studied medicine in Rostock from 1913 to 1919. He also served during WWI from 1917 to 1918. His first position in anatomy was as *Prosektor* in Rostock under Dietrich Barfurth, who retired in 1921. Voss stayed on under the new chairman Curt Elze, received his *Venia legendi* in 1923, and moved on to Leipzig in 1926. Voss became second *Prosektor* in 1929 and *ausserplanmässiger* professor in 1930.[57] His chief Hans Held had supported these promotions based on Voss's scientific work and ability as a teacher.[58] However, Held retired in 1934 and was succeeded by Clara, who was several years younger than Voss, and Voss's career seemed at an end when conflicts arose between them.[59] Voss decided to revive his chances for professional advancement by joining the party, according to his statement in 1948: "It is well known that any promotion during NS times was impossible without a party membership. So I joined the party in 1937. It was only in 1941 that Professor Clara succeeded in dumping me on Posen, the new *Reich* university."[60] Until then Voss had avoided a political commitment, even though he had initially viewed Hitler quite favorably, and harbored what he called a "national-social" conviction.[61] He also joined the NSDDB and the NS welfare organization.[62]

Clara recommended Voss for various anatomical positions, including a promotion to *ausserordentlicher* professor and first *Prosektor* in Leipzig. Clara believed that Voss was hardly a candidate for a chair in anatomy but that he was suitable for these other jobs, and thought that any bitterness Voss might have would be relieved with a promotion.[63] Voss was indeed resentful about the preferment of other colleagues, calling his competitor von Hayek—who was not of Jewish descent—a "little Viennese Jew."[64] Voss received his promotion on 15 February 1938[65] and his diary over the next few years shows a man accepting his fate and mostly content with his work and his family life with wife and daughter. His son, who had suffered from a chronic disease, had died in 1939.

It must have come as a surprise to Voss when he was finally offered a chair in anatomy in early February 1941, even if it was probably the least desirable anatomical position in Germany, at the new German medical school at the occupied University of Posen in Poland. His official appointment as full professor of anatomy in Posen was dated 7 July 1941.[66] He left his family behind in Leipzig because there was no suitable accommodation for them in Posen, and the ensuing loneliness and his utter disregard and outright hatred of the Polish people left their mark in his chronicle.

He succeeded in establishing a functional preclinical medical education as chair of anatomy and dean of the medical faculty before he had to flee Posen in January 1945. Bodies for teaching and research purposes were delivered from GeStapo execution sites as well as from nearby concentration camps. Voss' senior technician Gustav von Hirschheydt, an experienced dissector of Baltic-German origin, joined the institute on 1 April 1941. Hirschheydt was responsible for body acquisition and processing of the bodies. His production of skeletons, skulls, and casts of facemasks of the dead was so prolific that he sold them to other institutions, such as the anthropological department of the Museum of Natural History in Vienna and the anatomical institutes in Breslau, Leipzig, Königsberg, and Hamburg.[67] Hirschheydt died of typhus on 4 June 1942 after being bitten by a louse from the body of a Jewish concentration camp victim on 16 May 1942.[68]

Also responsible for body procurement was senior assistant Robert Herrlinger, who officially joined the institute in August 1941, but was frequently absent due to war service until the end of 1943. Voss mentored Herrlinger's research on the spleen, which was based on bodies of executed Polish citizens. While Voss had also worked with "material" from bodies of the executed before,[69] it seems that he had no time for anatomical investigations in Poland, as his list of scientific publications shows a gap between 1941 and 1948.[70]

In 1942 Voss reported on the status of the medical faculty in Posen, defining as its goal to educate doctors who help create "healthy Germans for the new Eastern territories as a prerequisite to the development of these regions in the German manner."[71] In reference to the anatomical dissection course, which was held for the first time for sixty students in the winter semester 1941–42, he pointed out that "due to the special situation in the *Wartheland* [NS administrative term for parts of occupied Poland], it was distinguished by a multitude and quality of demonstration material that can hardly be presented by any other German university."[72] In his diary Voss noted:

> March 10 1942: About 60 students took part in the dissection exercise. Nineteen corpses were dissected. Seven of them come from the Polish period, the rest had almost all been executed and injected a few hours after death with formol. [...] The dissections of the organs of the executed persons were the loveliest I have ever seen in a dissecting room.[73]

Voss' diary does not only give details of his academic work, but also about his attitude toward the people around him:[74]

> Posen, April 25, 1941: I quite like the city of Posen; one would only have to get rid of the Poles for it to be very pleasant here. Provisions are much better here than in Germany proper. You can get bread and rolls without any ration coupons!
>
> [...]
>
> Sunday evening, May 24, 1941: Here in the basement of the institute building, there is a crematorium for bodies. It now serves the Gestapo exclusively. The Poles they shoot are brought there at night and cremated. If one could only incinerate the whole Polish pack! The Polish people must be exterminated, or there will be no peace here in the East. It is terrible that we are still dependent on Polish labor here at the institute.
>
> [...]
>
> Whit Monday, June 2, 1941: I think one should look at this Polish question without emotion, purely biologically. We must exterminate them, otherwise they will exterminate us. And that is why I am glad for every Pole who is no longer alive.
>
> [...]
>
> Sunday, June 15, 1941: Yesterday I viewed the cellar for corpses and the cremation that is also located in the cellar. This oven was built to eliminate parts of bodies left over from dissection exercises. Now it serves to incinerate executed Poles. The grey cars with the grey men—that is, SS men from the Gestapo—come almost daily with material for the oven. [...] The Poles are quite impudent at the moment, and thus our oven has a great deal to do. How nice it would be if we could drive the whole pack through such ovens! Then there would be finally peace for the German people. [...] Today I wrote to Prof. Schoen in Göttingen and reminded him of my existence, in case they need an anatomist there. It won't do any good, but I want to take advantage of even the tiniest possibility of getting away from here.
>
> [...]
>
> Tuesday, September 30, 1941: Today I had a very interesting discussion with the chief prosecutor Dr. Heise, about obtaining corpses for the anatomical institute. Königsberg and Breslau also get corpses from here. So many people are executed here that there are enough for all three institutes.

[...]

Thursday, October 30, 1941: Tomorrow the anatomical institute will get its first bodies. Eleven Poles are being executed; I will take five of them, the others will be cremated.

[...]

Monday, April 27 1942: After lunch today, I sat upstairs for three-quarters of an hour right under the roof on our "bone whitener", soaking up the sun. To my right and left, Polish bones lay bleaching, occasionally giving off a slight snapping sound. This evening our air raid warden was here, to give me the pleasant news that night watches must now be set up in the building. They are expecting air raids on Posen. That's great. But in the end it doesn't matter where one croaks....

Tuesday, May 19, 1942: On Sunday, Herr von H[irschheydt] told me he had gotten lice on Saturday from a louse-ridden Jewish corpse. He has been making plaster casts of Jewish heads for the Vienna anthropological museum. That was wonderful news, since the Jewish corpses delivered here often died of typhoid fever. Because I was already in low spirits [English in the original] that day, as the English say, this had quite an effect on me. I thereupon decided to go home for Pentecost to see my family again. You never know when it will be the last time.

[...]

Monday, June 8, 1942: We just buried Herr von Hirschheydt. It all happened as I predicted three weeks ago. When I arrived at the institute on the morning of the 29th of May I was greeted with the news that Herr v.H. had become ill. I knew right away. It was good that Eva [Voss' wife] was here. I suffered through terrible hours. Our nerves are so on edge. These incidents don't seem to end.[...] Tomorrow Eva is returning home. Then I will be alone again with all my worries. What a miserable life.

Very much like Kremer's, Voss' diary reveals him as a self-pitying and somewhat pompous man. He cared for his family and students; however, when his close collaborator von Hirschheydt died, Voss felt sorry for himself rather than for his dead colleague.[75]

Voss' diary differs from Kremer's in that he unapologetically voices a visceral hatred for people of Polish and Jewish descent, including the Poles who worked at his institute. This is noteworthy, as testimonies from Voss' staff and colleagues given during his denazification procedures attested to his friendliness toward the Polish staff of the anatomical institute. Altogether seven character witnesses testified, more than in similar denazification procedures, possibly because of Voss' incriminating employment at a *Reichsuniversität*, and many claimed that he was not a convinced National Socialist.[76] Polish morgue technician Piotr Miklejewski testified in Voss' favor and reported of his superior's direct intervention with the German authorities in favor of Miklejewski's wife when she had been ordered to perform forced labor. Voss also increased Miklejewski's wages when he learned about his meager income. Finally, according to Miklejewski, Voss intervened to save him from the GeStapo in 1944 when they came to arrest him for withholding the body of a Polish assistant at the institute, Dr. Z. Meisnewowski, who had been murdered by the GeStapo in Fort VII, a concentration camp in Posen. Miklejewski had embalmed the body and stored it in the institute, hoping to give his colleague a dignified funeral after the war.[77] Three German friends of Voss' also affirmed his rejection of NS racism and anti-Semitism on scientific grounds.[78]

How can these declarations be reconciled with Voss' confessions in the diary? First, it should be mentioned that these types of denazification statements were popularly called "*Persilschein*" for good reason. They were meant to "clean" the guilty, scrubbing away their transgressions from the public record, as thoroughly as the well-known laundry detergent "*Persil*" rendered linen as white as snow. The simple fact of their multitude in this case seems to reveal Voss' need for his career to receive as meticulous a cleaning as possible. However, the testimony by the porter of the Posen anatomy institute, Jósef Jendykiewicz, documented by Halszka Szołdrska in 1948, tells a very different story. Jendykiewicz reported that the situation at the institute was difficult and inhumane for the Polish staff members, who were subject to Voss' and his assistant Schwarz's despotism. The anatomists abused the Poles verbally and sometimes even physically. In one incident Voss used his truncheon on the porter Kokciński, and Jendykiewicz had to mediate the situation.[79] This is at odds with Miklejewski's very specific narrative, suggesting that Voss indeed did help his Polish staff in certain circumstances. This apparent contradiction in Voss' character may be explained by his prevailing need to keep the institute functional. After all, his work was the most important thing to him next to his family, according to his friend Kalkbrenner, and he had already compromised his family life by agreeing to a temporary separation from his wife and daughter for the sake of his work.[80] So, if he had been dependent on Polish colleagues for the proper and seamless operation of the anatomical department, he probably would have done anything to keep the staff intact and willing to work. It is also possible that his general hatred of "Poles" as a group did not extend to certain individual Polish employees in his daily life.

On 20 January 1945, Voss fled from Posen to avoid the advancing Soviet army. He spent two months in Leipzig and then settled in the nearby city of Borna, where he volunteered as a physician at the local hospital. According to Voss, on 1 March 1946, the mayor of Borna found out about Voss' past and forced him to leave this job, in which Voss had hoped to "serve the new state." In the next two years Voss waited for his denazification and concentrated on writing textbooks for medical students, among them the final version of the short textbook known as Voss-Herrlinger.[81] During this time Voss started negotiating for his potential reemployment with various universities in the Soviet occupation zone. While the ministry of education in Berlin considered Voss as politically tainted, it had to admit by the fall of 1948 that he was the sole anatomist left in the Soviet zone who remained unemployed, and recommended his recruitment in a subordinate role.[82] Chairman of anatomy Günther Hertwig in Halle had previously reported to the ministry that all negotiations with colleagues from the Western zones had been futile,[83] and authorities in the Soviet zone apparently started to rehabilitate many academics with dubious political pasts at that time.[84] Voss was given a lectureship in anatomy at the University of Halle starting on 1 October 1948, which was upgraded later to a full position as professor.[85]

Voss continued negotiating for a chair position in anatomy in his hometown of Rostock,[86] but the chairman of anatomy at the University of Jena, Fritz Körner, died unexpectedly of a lung embolism at the age of only forty-six years on 20 December 1949, and this sudden loss necessitated an expedited recruitment process. Kurt Alverdes and von Hayek were the preferred candidates, but they ultimately declined; thus, the faculty was left with Voss,[87] who Benninghoff had recommended as the best academic in the Soviet zone.[88] Voss was appointed to the position in Jena on 1 January 1952, and stayed there until his retirement in September 1962.[89] He then moved to Greifswald and taught anatomy at the university there for many more years.

In 1959 the government of the GDR honored Voss with the title "Outstanding Scientist of the People."[90] Voss' work in histochemistry, which had not been appreciated by his colleagues previously, became relevant for new histological techniques in the 1950s and he was hailed as a pioneer in the field.[91] There was a proposal to bestow an honorary doctorate from the University of Halle upon Voss on his seventieth birthday, but he was not given the honor because of his activities in Posen.[92] However, this information was apparently not shared with other colleagues in the GDR at the time, as in 1978 negotiations began to rename the anatomical institute in Jena on Voss' eighty-fifth birthday to "Hermann-Voss Institute." The discussions were complicated by the fact that Voss was planning to leave the GDR and join family members in the Federal Republic of Germany, who were going to take care of him in his old age. The relatives were Fritz Körner's children, who had been orphaned and then raised by Voss and his wife, after their daughter had died.[93] Günther Geyer, a pupil of Voss and his successor as chair of anatomy in Jena, had written a letter of support for the tribute to Voss, but added as an afterthought that there had been a controversy surrounding Voss' wartime activities in Posen when Voss was nominated for an honorary membership in a professional society in 1974. Geyer recommended a reexamination of Voss' war record, as the institute in Jena had friendly relationships with Polish colleagues and did not want to endanger these.[94] After inquiries by the leadership of the University of Jena[95] and investigations by the ministry of education[96] the president of the university received a very short letter from the latter on 8 December 1978, which ordered the university to "immediately stop all proceedings concerning the affair of Hermann Voss."[97]

Voss moved to Hamburg in the FRG in 1979 and died there in 1987. He refused to answer any inquiries concerning the diary and his time in Posen.[98] The only indication of Voss' state of mind concerning Aly's revelations comes from the recollections of Professor Holstein, director of the anatomical institute in Hamburg, who had a conversation with Voss on the topic shortly before Voss' death. Holstein recalled Voss stating that "all of this is terrible,—that all of this was represented wrongly,—that he hadn't written this in this way,—that he could explain all of this,—that all of this had been quite different,— that he had done much for the Poles in his institute." Holstein counseled Voss to refuse any public comments.[99] A memorial plate in honor of Hermann Voss, which had been mounted on a wall at the anatomical institute of Jena, was removed during a renovation in the 1980s without official commentary.[100]

If Voss' career advancement was based to a large degree on pragmatism in political decisions, it came somewhat late in life. Wolfgang Bargmann[101] on the other hand is an example of a young anatomist whose political adaptability matched his ambitions in a way that enabled his professional success from the very beginning. Bargmann was born as Wolfgang Bardel on 27 January 1906 in Nuremberg and changed his name to Bargmann when he was adopted by his second stepfather in 1925. His early fascination with the structure and function of living organisms led him to study zoology and medicine. He received his medical degree in Frankfurt in 1932, where his teachers were Karl Zeiger, Franz Weidenreich, and Hans Bluntschli. Under Zeiger's guidance he studied the histology of the renal glomerulus in animals and humans, and his doctoral thesis was published in 1931. He described his human "material" as fourteen kidneys, which mostly came from executed men and had been "kindly provided by" colleagues.[102]

In 1933 Bluntschli offered Bargmann a position as assistant. During the following year Bargmann befriended Ernst and Berta Scharrer, who had recently moved from Munich, bringing their work on secretory nerve cells in the brain of invertebrates and vertebrates with them to the Ludwig Edinger Institute in Frankfurt. Ernst Scharrer had been hired as successor to paleoneurologist Tilly Edinger, also the daughter of the founder of the institute, after she had been forced to emigrate because of her so-called "non-Aryan" descent.[103] Bargmann and Ernst Scharrer probably first met during the anatomical dissection course in which Ernst was volunteering as a teacher.[104] The Scharrers shared their scientific findings with Bargmann, who was impressed by the quality of their work at a time when the concept of neurosecretion was still controversial.[105]

In 1934 Bargmann left Frankfurt for Freiburg, and followed von Möllendorff to Zurich in 1935. He received his *Venia legendi* from Freiburg in 1937.[106] Among his many publications from this period was another paper on the histology of human kidneys using "material" from the executed.[107] In 1938 Bargmann moved on to Leipzig for the next step in his career, a position as *Prosektor* at Clara's institute, and Clara strongly supported Bargmann's promotion to *ausserplanmässiger* professor, in 1941.[108] In the interim, Bargmann served as a physician in the military during the year of 1940.[109] Among the eight original publications produced by Bargmann during his time in Leipzig, two made use of "material" from bodies of the executed.[110] In addition, he mentored two students who studied such tissues.[111]

On 1 January 1942 Bargmann became section chief at the anatomical department of the University of Königsberg, which was directed by Robert Heiss.[112] Apart from gross anatomy and histology, he taught history of medicine and published a paper on the history of anatomy at the University of Königsberg.[113] His research interests still spanned a variety of subjects, and three of the four original publications from this time made use of "material" from the executed,[114] among them a study on "nuclear secretions" in the human pituitary gland,[115] and two of his doctoral students published studies on similar "material."[116] The anatomical department had collected tissues from many executed persons, such as "intestines from 18 executed persons aged 17–52, fixated immediately after death,"[117] and "10 hypophyses from executed persons ages 17–44, fixated while still warm from the body."[118]

During a period of illness of Robert Heiss, Bargmann served as acting director of the anatomical department from January to November 1943.[119] After the summer semester of 1944 regular teaching stopped at the University of Königsberg due to the impending arrival of the Soviet army.[120] Bargmann held exams for the last German medical students of Königsberg in January 1945 and then fled the city to join his family in Bavaria, taking some of the inventory of the institute with him.[121] Of the thirty original publications Bargmann wrote between 1929 and 1945, seven were based on studies of "material" from executed persons. In contrast to other anatomists, Bargmann did not publish any further studies from this "material" after the war.[122]

During the months in Bavaria Bargmann started work on his textbook of histology. After an intermezzo in the anatomical department at the University of Göttingen, he became director of the department of anatomy at the University of Kiel in February 1946, where he remained for the rest of his life despite several prestigious offers from other universities.[123] The city, including the buildings of the university, was almost completely destroyed by bombing raids, and Bargmann spent the first years after the war rebuilding the anatomical department. He also dealt with internal problems of the anatomical institute, among them inquiries in 1946 by relatives of executed political prisoners who were trying to trace the remains of their loved ones.[124] A critical situation of a different kind arose when Enno Freerksen asked to be reinstated as director of the institute. Despite his membership in the SS, Freerksen had been classified only as a "fellow traveler" in his denazification process (an official verdict for persons not directly involved in NS policies and crimes but cognizant of them) and thus had a legal right to reclaim his former position. The situation was resolved when Freerksen received a professional opportunity at the Tuberculosis Research Institute in Borstel.[125] Bargmann also endeavored to reconnect with the anatomical world community. He corresponded with emigrants like the Scharrers in the United States and Paul Glees in Oxford, England, and received a travel grant from the Rockefeller Foundation in 1950.

In his scientific work Bargmann experienced a breakthrough in 1948, when he was able to expand on the Scharrers' findings on the relationship between the hypothalamus and the neurohypophysis by applying a new staining technique, thus proving the previously unknown existence of a continuous neuronal system connecting the hypothalamus with the posterior hypophysis in mammals.[126] In the years to come he continued this line of inquiry but, deeply concerned about the comprehensive education of a new generation of physicians and scientists, took on various official roles in rebuilding German universities and their international contacts, the science of anatomy in Germany, and medical education in general.[127] As a man of principle, charm, and scientific insight he was much admired by his colleagues nationally and internationally,[128] but his refusal of political compromise later in life incited conflict with a younger generation.[129] Bargmann retired in 1974 and died suddenly of a cerebral hemorrhage on 20 June 1978 in Kiel.[130] Looking back on his work before 1945, Bargmann called his early scientific writings "literary sins I committed as a young man during the 1930's and early 1940's."[131] There is no indication of what exactly he considered as the "sin" in his early manuscripts: was it only the lack of scientific focus or possibly also the choice of "material" from executed persons, even if this was seen as accepted practice at the time? Bargmann firmly attributed his later success in the field of neuroendocrinology to his acquaintance with the Scharrers and their research.[132]

Bargmann's political record before 1945 speaks of conformity. He became a member of the NSDAP on 1 May 1933,[133] joined several other minor NS organizations,[134] and served as physician for the Hitler youth in Frankfurt and Freiburg.[135] There is no evidence that he was otherwise politically active during the Third Reich. Bargmann's joining of the NSDAP in 1933 could be interpreted as a sign of early consent with the party, but may have been influenced by his mentor Zeiger, who did the same and acted as regional physician for the Hitler youth.[136] Bargmann's NS memberships ensured his employability as a young academic, as he depended on recommendations by superiors and their assessment of his personal politics. Clara was known for his severe political judgment,[137] and it is unlikely that he would have supported Bargmann as decisively had he not at least superficially conformed to NS party standards.

The political memberships may have also helped with research funding (see chapter 3). Bargmann must have been keenly aware of the changes in his professional field brought about by NS discriminatory policies, as his teachers Bluntschli and Weidenreich had to leave Germany, and the Social Democrat von Möllendorff was forced to relinquish his position as president of the University of Freiburg to philosopher Martin Heidegger after only a few days of tenure in April 1933.[138] The Scharrers may have tried to adjust to the new regime in the beginning but left soon thereafter. According to Tilly Edinger, Ernst Scharrer had joined the SA on a suggestion by his superior—even though Scharrer himself stated in 1962 that he had been able to avoid any political involvement.[139] Bargmann did not leave any statements concerning his politics during these times, but his actions reveal his flexibility. He worked with the political dissident von Möllendorff in a foreign country, but had no scruples about returning to Germany and working for the active National Socialist Clara. Apparently he saw no reason to emigrate like his friends the Scharrers. His political convictions must have either not been particularly strong against the regime, thus enabling him to compromise with the political situation, or he might have been in mild acquiescence. In his denazification trial, which occurred after his recruitment to Kiel in 1947, Bargmann was categorized as "exonerated" despite his early party membership. The verdict was based on a statement by Professor von Mikulicz, chair of gynecology at the University of Königsberg, whom Bargmann cited as witness to his having been in trouble with the NS regime.[140] Unfortunately this document is not available. No other information on Bargmann's private or professional life mentioned any such political problems and the statement might have been of the "whitewashing" kind.[141] This seems likely, as the witness was Felix von Mikulicz-Radecki, the Königsberg chair of gynecology, an active supporter of NS policies and specialist in sterilization by irradiation.[142]

After the war Bargmann saw the long isolation of German science from international developments as fatal and necessitating tenacious efforts by him and his colleagues to reconnect with their international peers. He stated: "Germany, and especially her universities, had suffered materially, intellectually and morally."[143] It seems unlikely that he included his own political and professional activities during the Third Reich as having "suffered morally."

Political pragmatist Heinrich von Hayek's career was not quite as straightforward as Bargmann's. Von Hayek was born on 29 October 1900 in Vienna, Austria, to a physician and professor of botany. His older brother was Friedrich August von Hayek, Noble laureate in economics in 1973.[144] The family was of minor nobility, but after 1919 they were no longer allowed to carry the title "von" in Austria.[145] Von Hayek attained his medical degree in Vienna in 1924 and a PhD in zoology in 1929.[146] He published fourteen papers during his time as assistant to Hochstetter in Vienna, mostly on topics of comparative anatomy of the head and neck bones.[147] In 1929 von Hayek was appointed as a second *Prosektor* at the anatomical institute of the University of Rostock, Germany, which was directed by Curt Elze, another pupil of Hochstetter,[148] who justified the recruitment of an Austrian scholar with a lack of suitable German candidates.[149] In 1930 von Hayek received the *Venia legendi* for work on comparative embryology in vertebrates. Among his publications from Rostock was his first investigation based on the use of "material" from an executed person. In it he described how he injected "fresh" bodies of executed persons with formol-alcohol to cause the musculature to contract for a study on periarticular arteries.[150]

In 1935 von Hayek was recruited as professor of anatomy at the Tongji University in Shanghai, China. This university had originally been founded as a German medical school in 1907 and came under Chinese governance in 1919,[151] but the anatomy department was always directed by German professors.[152] In the summer of 1937 the institute was destroyed during the Second Sino-Japanese War, and von Hayek fled to inner China with other university members. He tried to teach there under primitive circumstances while his family stayed in Shanghai. After lengthy deliberations with the German and Chinese authorities, he returned to a position as senior assistant at the anatomical institute in Würzburg in fall 1938.[153] His publications from Shanghai included a racial anthropological study on the frontal lobes of Chinese brains.[154]

During a bombing raid on 16 March 1945, the city of Würzburg, including its university, was almost completely destroyed. Von Hayek, as deputy head of the anatomical department, was in charge of rebuilding the institute directly after the war,[155] but was dismissed on 10 August 1945 by the occupying military forces because of his membership in the NSDAP.[156] The seventeen papers he published between 1938 and 1945 mostly dealt with pulmonary histology, and nine explicitly referred to "material" from the bodies of executed persons. Such bodies were readily available, as 120 out of 910 bodies received by the anatomical institute in Würzburg between 1935 and 1945 had been delivered from execution sites.[157]

Von Hayek called his time of unemployment after the war "an externally difficult time, when practical scientific work was impossible in Germany,"[158] but he was able to write his textbook on the human lung during this period. He based this study in great part on his previous work with "material" from bodies of executed NS victims. Von Hayek attended the first German anatomy meetings after the war in 1946 and 1947.[159] He was finally reinstated at the University of Würzburg on 20 November 1947,[160] and in 1951 the Universities of Vienna, Rostock, and Jena each asked him to become chairman of an anatomical department. He accepted the Viennese offer in January 1952.[161] Von Hayek remained in this position until his death on 28 September 1969.[162] During his time there, one of his main tasks was exploring new sources for body procurement, as there was a dire shortage of bodies for anatomical dissection in Vienna after the war. According to Walter Krause, von Hayek focused on promoting a voluntary body-donation program. He successfully mobilized mass media to educate the general public of the need for body donation for the training of physicians, so that by the time of his death the body supply had again become sufficient.[163]

In the first years after the war von Hayek continued publishing studies based on "material" from the NS period. However, he avoided directly mentioning "bodies of the executed," Instead referring to "tissue as fresh as in life"[164] and "lungs of four well-built young men fixed in fresh condition."[165] He was nationally and internationally recognized as the leading authority on the anatomy of the lungs, and his book became a standard reference. He also received high honors from Austrian and foreign scientific societies and from the Danish government.[166] The fact that much of his research on lungs was based on work with "material" from executed persons was never publicly discussed.

Interestingly enough, von Hayek's political history seems to have been somewhat better known among his colleagues than Bargmann's.[167] In November 1933 he joined the SA and became a *Scharführer* (noncommissioned officer) in 1943. His membership in the NSDAP started in March 1938 (member number 5518677). Among other minor NS organizations he joined the NSDDB in 1939 and became its local manager.[168] He also was a member of the *Deutsche Akademie München*, an organization founded in the 1920s that promoted German culture in foreign countries and came under strong NS influence. It gained a reputation after the war as a propaganda institution of the NS regime.[169] From 1934 to 1935 he was a leader in the *Kampfring der Deutsch-Österreicher im Reich (Hilfsbund)*, an associa-tion of Austrians living in Germany that included a Swastika in its regalia. All of these memberships and minor positions seem to indicate that von Hayek was definitely a national conservative, supporting the NS move-ment at least pro forma and possibly in mild acquiescence with the regime. There are no statements by von Hayek concerning the politics of the time except for the defense statements in his two denazification trials in Decem-ber 1946 and July 1947. He declared to have joined the SA and NSDAP only to save his career and received a verdict of "fellow traveller."[170]

Voss, Bargmann, and von Hayek adapted to what they considered the political necessities of the Third Reich, joining the party and ultimately thriving professionally during this time and thereafter. However, their immediate postwar experiences differed quite a bit. While Voss and von Hayek both had to face years of unemployment due to their political past, Bargmann's career was only interrupted by his flight from Königsberg. He was recruited to Kiel even before his formal denazification trial had taken place, while Voss and von Hayek had to wait for their trials before they were considered for new positions years later. Even then, Voss was only re-admitted into his profession because there were simply no other anatomists left in the Soviet zone, and the education of medical students was in dan-ger. It is possible that these experiences were caused by variously stringent denazification practices in the British zone for Kiel, the American zone for Würzburg, and the Soviet zone for Jena. Evidence for this might lie in the fact that Bargmann, even though he joined the party in 1933, was ex-onerated on the basis of one positive witness statement, while von Hayek and Voss, who both had joined the party later and in Voss' case presented a multitude of supporting witnesses, were judged to be "fellow travellers." In Voss' case it is also possible that his work at a *Reichsuniversität* weighed strongly against him. However, the diary and its content were not public at that time in Germany and there was no official knowledge of his activities in Posen. Still, by the early 1950s all three anatomists were in chair posi-tions and stood at the beginning of significant postwar careers.

Anatomists Profiting from NS Policies: Stieve and Fischer

National Socialism created "special opportunities" for many anatomists through its ideology and practice, most decisively for Eugen Fischer and Hermann Stieve. Fischer's science was actively supported by the NS au-thorities and he and his colleagues became instrumental in establishing and enforcing discriminatory legislation that ultimately led to the murder of patients and prisoners. Similarly, high execution rates gave Hermann

Stieve opportunities to fulfill his research plans. Both Fischer and Stieve, who clearly profited from NS policies, were counted among the most prominent scientists of their time, and their work was quoted for a long time after the war.

Hermann Stieve is one of the most widely discussed German anatomists from the Third Reich era because of his work on the reproductive organs of women who were executed by the NS regime.[171] Stieve was born in Munich on 22 May 1886 as son of a professor of history. He had an early interest in zoology and used hunting expeditions at his family's summer home to procure material for scientific investigations. Hunting would remain a lifelong passion for him, and more often than not the slain animals served as objects of his research. He studied zoology and medicine at the University of Munich and joined the nationalist student organization *Corps Franconia*.

In 1912 Stieve received his MD and became assistant in anatomy to Johannes Rückert, who recognized Stieve's intense motivation and working capacity and suggested the area of reproduction as a promising field of research. Stieve's first independent study confirmed the influence of stress on a hen's ability to lay eggs and revealed the morphological equivalent of this phenomenon in the atresia of ovarian follicles.[172] This work was of immediate relevance, as a very public discussion had ensued on the topic of "war amenorrhea." Göttingen gynecologist Hans Albert Dietrich had published a first paper on several cases of apparently healthy women who had stopped menstruating, a symptom that he attributed to the stressful war situation.[173] While Stieve explained this "war amenorrhea" mainly as a consequence of malnutrition, he was also aware of the fact that a woman's menstruation might seize for a month or longer if she suffered emotional upsets.[174]

He pursued this question for the rest of his life in studies on various aspects of the influence of the nervous system on the reproductive system in animals and humans. Even during his war service as a physician from August 1914 to April 1918 he found time to hunt for jackdaws and describe their reproductive organs in detail, a work for which he received the *Venia legendi*. Investigations in the field of physical anthropology led to his PhD two years later under the mentorship of the influential Munich anthropologist Rudolf Martin. In 1918 he accepted a position as second *Prosektor* in Leipzig, and in 1921 he was hired as youngest chair of a German medical department at the anatomical institute at the University of Halle. His recruitment was controversial, as Stieve had made some enemies among his colleagues by unabashedly criticizing them.[175] During his time in Halle he developed a close friendship with gynecologist Hugo Sellheim, who encouraged Stieve's scientific interest in the female reproductive organs and their changes during pregnancy. Sellheim provided Stieve with many rare surgical specimens, so that the anatomist was able to study all stages of pregnancy, from embryonic development to enlarged uterus and mature placenta.

Stieve was very imaginative in his quest for new sources of research "material." Apart from the surgical specimens from the termination of pregnancies by hysterectomy in severely diseased patients,[176] he explored the traditionally available source of bodies of executed persons. He used this "material" for the first time in a 1919 study on the pyloric region of the stomach.[177] By the early 1920s he realized that the situation of prisoners on death row essentially mirrored the design of his anatomical experiments for the study of the influence of stress on reproductive organs. In the case of the prisoners, the chronic stress factor was the incarceration itself, and the acute stressor was the prisoner's notification of the execution date. By 1924 Stieve had collected fresh "material" from thirty-four bodies of men killed at the execution site in Halle, and found distinct changes from the normal tissue structure in their reproductive organs.[178] Similar studies of women were impossible then, as they were not subject to executions in the Weimar Republic. By the time of his recruitment in 1935 to the prestigious chair of the anatomical institute of the Friedrich-Wilhelms University Berlin, Stieve had published more than one hundred scientific papers.

Stieve held himself aloof from active involvement in National Socialism and never joined the NSDAP or any other major NS organization, even though he held some political convictions similar to those of the new rulers[179] and shared the racist and gender stereotypes as well as the cultural pessimism of his time.[180] He was a member of the DNVP (German national people's party) from 1918 to 1930, and joined the *Stahlhelm*, a paramilitary organization, in 1921. In 1934 he led a university section of the *Stahlhelm* in Halle. He was involved in skirmishes against communists and anarchists between 1919 and 1921 while a member of the paramilitary group *Freikorps Escherich*,[181] and defended his nationalism very eloquently in an official speech at the anniversary of the foundation of the German Empire at the University of Halle-Wittenberg on 18 January 1929.[182] In this address he criticized Germany's enemies as well as an overformal education and the physical and mental evils brought on by overexertion through modern life and urbanization. "All work and life of every German" should serve the goal to lead the fatherland to "its old power and new greatness."[183] This attitude may explain Stieve's 1933 welcome of Hitler "as having the potential to revive German pride after its defeat in World War I."[184]

He was promoted to president of the University of Halle in May 1933, but could neither agree with the NS government's curbing of the university's autonomy, nor tolerate the local students' National Socialist opinions, and was not reappointed in November 1933.[185] Just as he was known to voice his opinion loudly and clearly in scientific disagreements with colleagues, Stieve was not afraid to speak his mind when he deemed decisions necessary that did not entirely conform to the NS-regimes politics, especially in terms of personnel. He tried, however unsuccessfully, to hire von Lanz, who had been dismissed from his position in Munich. Also, in 1938, Stieve fired his assistant Horst Boenig, a NSDAP party member and leader of the NSDDB, because he thought him incompetent.[186] The conflict with Boenig, who had powerful friends in the political system, led to REM inquiries into Stieve's performance as an administrator and teacher.[187] Other disagreements with the university administration usually concerned the medical curriculum and interactions with students.[188]

After the war Stieve tried to portray himself as a political dissident because of his differences with university administrations,[189] but his Berlin recruitment alone spoke against such an interpretation. Stieve's behavior was most likely less motivated by a distinct ideological disagreement with the National Socialists than by the fact that he would not brook any compromise in the pursuit of excellence, be it in medical education or research. He also objected to any interference with his leadership of the anatomical institute. His character, which has been described as "great of willpower and determination, ruthless against himself" with "unflagging, tenacious diligence,"[190] would not allow for any decisions that promised less than the best results for the task at hand. The evidence of Stieve's decisions throughout his life makes it clear that he was politically passionate only about Imperial Germany, for which he fought with body and soul. After 1933 he managed to walk the fine line of political compromise with the NS regime, a feat he would repeat in the political atmosphere of the Soviet occupational zone and the young GDR. Even though he had fought communists in his youth, he aligned himself with the new rulers after the war. Throughout his life, his focus was first on his work and not on politics, and he was able to negotiate all four political systems he experienced in a manner that best enabled and supported his work. He was helped in this by a driven character, which was marked by superb self-esteem, if not grandiosity.

Given that much of his work focused on the female reproductive system, it is pertinent to explore Stieve's attitude toward women. The twenty-eight-year-old Stieve married his wife Maria when she was nineteen years old,[191] and they had three sons and two daughters. Maria Stieve followed the traditional pattern of taking care of husband, home, and children, which her husband considered the "true profession" of any woman.[192] He must have been conflicted when not only two sons but also his daughter Kriemhild decided to study medicine. He felt that the ongoing urbanization of society had led to unwanted infertility, brought on by high caffeine and alcohol consumption, exposure to other toxins, and nervous stress, to which he saw the healthy constitution of women succumbing.[193] It can be assumed that he was worried that his daughter would also suffer this fate. While Stieve conceded in 1942 that not every woman could expect to become a wife and mother, he still believed that intellectual pursuits, particularly university studies, led to an unhealthy overexertion of the female system and women's early "wilting," leaving them as "twitching bundles of nerves" in times of stress.[194]

Despite his frequent contention that women were ultimately less suited to the medical profession than men, Stieve collaborated with several female doctoral students and colleagues. He chose a female assistant for his histology course and had a respectful working relationship with her, even to the point that they "confessed" their disagreements with the NS regime to each other.[195] He was popular with male and female medical students, who enjoyed his theatrical style of lecturing and arrived as early as 7AM to get a seat in his 8AM lecture.[196] He would descend on the hall in a glistening black gown, present the Hitler salute with a certain lack of enthusiasm, and launch into a highly competent discourse on anatomy, illustrated by bimanual drawings on the blackboard and sometimes embellished with sexually suggestive jokes.[197] He was respected and feared by students for his insistence on excellence. In the dissection rooms Stieve strictly admonished his charges of the importance of respect for the dead and orderly conduct,[198] but sometimes condoned student "pranks" involving severed sexual organs, an attitude some considered cynical.[199] In all of this Stieve seems to have been the typical slightly misogynistic academic male of his time. He admired the idea of "woman as mother" and held a certain theoretical disdain for intellectually active women, even if he accepted them as respected partners in his daily work. "Woman as study object" became the center of his professional life, so much so that he apparently included his wife's anonymized gynecological history in his volume of case studies.[200]

The bodies of executed women were the main focus of Stieve's research. Berlin as the political center of Germany was the seat of the *Volksgerichtshof*, where most of the prominent political trials were held, and the exponentially increasing number of death sentences were performed at the Plötzensee execution site or in Brandenburg-Görden. Beginning in 1935 women were among the executed, and Stieve seized this "opportunity" immediately.[201] His collaboration with the prison authorities was a particularly close one. Apart from arranging execution times earlier in the day to allow a body transport compatible with the lab schedule, Stieve demanded in 1942 that the authorities cover the cost of coffins, as the anatomical institute accepted bodies in excess of its needs and could not otherwise guarantee their removal.[202] In 1942 alone more than 500 executions were performed at Plötzensee.[203] In early September 1943 the Plötzensee execution site was severely damaged by bombing raids, and as a result the prison became acutely overcrowded. The authorities decided to perform mass executions, which were carried out between 7 and 13 September, a period that became known as the "bloody nights" of Plötzensee when at least 258 persons were killed. Their bodies were stored in an open shed while further bombing prevented their collection by staff from the anatomical institute.[204] When the anatomy technicians Pachali and Schwalbe were finally able to perform their duty, the handling of the bodies had become so difficult due to the bodies' long exposure to the elements that the authorities decided to pay the technicians a bonus for their "heavy work".[205] In October 1943 the anatomical institute was no longer able to collect bodies from Brandenburg-Görden due to the fuel rationing.[206]

The bodies of executed persons and other victims of the NS system were used for the dissection course as well as for Stieve's research, and students and staff were sworn to secrecy.[207] Stieve and his colleagues had to cope with very high student numbers, especially during the war years, when they taught between 1,200 and 1,900 students per semester.[208] Despite these responsibilities Stieve published several studies based on "material" from hundreds of bodies of executed men and women.[209] He felt that it was his "duty" to store and use this "material of a kind that no other institute in the world can call its own".[210]

Stieve often referred to the bodies of executed persons as "bodies of criminals." During his conflict with the REM in 1941 he replied to the accusation of being "ruthless and always nervous" that he lived under the strain of being admonished of the need to keep the presence of 90 decapitated bodies of executed criminals in the dissection room a secret from the public and their families. Meanwhile, Stieve claimed, the families had "cunning ways" of discovering the fate of the "criminal's body" and had "crept" into the institute to retrieve their relatives' remains.[211] Even as late as 1952, Stieve asserted that his research was based on "material" from bodies of "common criminals".[212] This was an outright lie and he knew better. The documents from the office of the prosecutor in the case studies of Maria Diecker and Ewald Funcke contained information on the prisoner's verdict (see pages 69–70). And many of the women, whose bodies he had used for his research and whose names he had assembled in a list for the military authorities in 1946, were prominent members of German resistance groups. Stieve even changed some of the names of women he recognized on the list of his "research subjects," possibly to anonymize them for the general public.[213]

In several of his postwar interrogations Stieve contradicted himself when asked specifically about his dissection of political prisoners, often denying that he had ever used them in research.[214] Harald Poelchau, the former Protestant prison chaplain at Plötzensee, believed that Stieve had secretly handed bodies of the executed to their families, possibly because Stieve told him about returning ashes to the families.[215] There is no indication that Stieve ever gave undissected bodies to the families. It is only known that he identified and returned urns with the ashes of Mildred Fish-Harnack and Elisabeth von Thadden to their families.[216] The anatomical institute was damaged during bombing raids and many specimens were lost; any remaining ashes, bodies, or body parts of NS victims were interred shortly after the war.[217] Stieve had sent important histological specimens to Gösta Häggqvist, his Swedish friend in Stockholm, for safekeeping, but it is unclear what became of these specimens, as they cannot be located in Stockholm or in Berlin. It is also unknown what kind of specimens these were.[218]

After the war Stieve was questioned about his work by all military occupation forces as well as by the university administration. Given the fact that the body registers had "vanished" in 1945, Stieve was only able to put together a partial list of victims' names in 1946, which was based on his research notes.[219] The use of bodies of executed political prisoners did not sit too well with the minister of education of the GDR, Paul Wandel, and in 1949 he recommended avoiding a public discussion of the topic.[220] As Stieve had never joined the NSDAP he was eligible for continued employment as professor of anatomy in the Soviet occupation zone, and he was in demand given that he was one of the most prominent anatomists in Germany. Even without his prominent stature he would have been irreplaceable, as there were no other anatomists left in the Soviet zone and no one from outside wanted to move there. Thus the authorities decided to keep silent on the subject of Stieve's research,[221] and he stayed in this position until his sudden death following a stroke on 9 September 1952.

From the beginning of his professional career Stieve was recognized by those around him as an extraordinarily motivated and innovative scientist.[222] He wielded considerable power in the field of anatomy through his editorship of several important anatomical journals.[223] Many admired and supported his work, not only anatomists but also clinicians, especially gynecologists. Apart from Sellheim, he also developed a close working relationship with Walter Stoeckel, chair of gynecology at the University of Berlin, who named Stieve on the anatomist's sixtieth birthday "the gynecologists' anatomist."

Stieve's judgment of colleagues was entirely based on their quality of work and he criticized them freely and repeatedly when he was not convinced by their results. The most prominent controversy was his disagreement with Austrian gynecologist Hermann Knaus on the timing of ovulation and its interpretations.[224] Another example was the dispute with his younger colleague Spanner on the blood circulation in the human placenta. While he described Spanner as a competent anatomical technician, he also included a differentiated criticism of Spanner's scientific thinking in his postwar evaluation of candidates for the vacant chair of anatomy at the University of Kiel in 1945.[225] Despite this, Stieve was often a very loyal and helpful friend and could elicit similar feelings in others like Hans Elias.

Stieve also had a special friendship with the Swedish anatomist Gösta Häggqvist.[226] Their relationship was based on the mutual personal respect of their work and high regard for each other's country. Stieve had probably first become acquainted with Sweden and its people through his brother Friedrich, who was married to a Swedish woman and served as press attaché for the German consulate in Stockholm during WWI.[227] In his obituary of Stieve, Häggqvist praised his colleague as the most important researcher in reproductive medicine in Germany.[228] On at least three occasions, Häggqvist and Stieve nominated each other for prestigious awards or membership in academic societies. Stieve became a member of the Swedish academies of science of Stockholm and Uppsala, as well as of the German academic societies of Halle, Berlin, and Munich, and was awarded an honorary membership in the German society of gynecology.[229]

Stieve's larger-than-life personality attracted friend and foe alike among colleagues, pupils, and students, and he became subject of many admiring tributes as well as several accusations about his activities brought before the authorities during WWII and thereafter.[230] Even his friend Benno Romeis mentioned in Stieve's obituary that his combativeness as well as his fixed opinions could burden a friendship. However, Stieve's scientific work was lasting and often quoted without hesitation by anatomists and clinicians, as late as 1995.[231] Even though the sources of Stieve's "material" had been discussed widely in Berlin after the war and colleagues were aware of Stieve's activities, his studies were not closely scrutinized until a first critical mention by Aly in his essay on Voss.[232] Since then medical historians have investigated the topic,[233] but the first evaluation by anatomists did not come before 2007.[234] So in many ways, Stieve has managed to stay "aloof" until very recently.

Eugen Fischer also profited from NS policies. While he was already a leading scientist in Germany before 1933, his work in racial biology made him one of the most important representatives of NS German science nationally and internationally. He promoted the application of his eugenic ideas by the NS state, and this application ultimately led to the destruction of all perceived "enemies" of the German people. The following note appeared on 25 October 1946 in the weekly journal *Science*:

Eugen Fischer, professor of physical anthropology, University of Berlin, and director of the Kaiser Wilhelm-Institut für Physische Anthropologie ... was one of the leading Nazi anthropologists who are morally responsible for the persecution and extinction of the peoples and races the Nazis considered "inferior". He was the first Nazi rector of the University of Berlin and took over the post when decent scholars withdrew or refused to do business with the Nazis. His address delivered at the inauguration ceremonies (29 July 1933) was entitled: "The Conception of the 'Völkisch' (Nazi) State in the View of Biology." This address foreshadowed the official execution of the principles of "racial hygiene" as taught by the Nazis. If anyone, he is the man who should be put on the list of war criminals.[235]

The letter was signed by Franz Weidenreich, who in one short paragraph succinctly drew an outline of the facts that made anatomists like Eugen Fischer and other racial hygienists main contributors to the catastrophe of the Third Reich. This truth, so clear in Weidenreich's mind as early as 1946, and extensively elaborated on by Max Weinreich in his 1946 publication, *Hitler's Professors*,[236] was long ignored in postwar medicine. Benno Müller-Hill touched on it in 1984,[237] but first critical analyses of Fischer's life were published only many years later.[238]

Fischer was born on 5 June 1874 and grew up in Freiburg, remaining there for much of his life. Early on he adopted the national-conservative attitude common in imperial Germany. He studied medicine and befriended Theodor Mollison and Erwin Baur.[239] Mollison shared Fischer's interest in zoology and morphology and became chair of anthropology in Munich, while Baur became Fischer's coauthor of the handbook on racial hygiene. Fischer received his MD in 1898 and joined his mentor Robert Wiedersheim at the anatomical department, receiving his *Venia Legendi* in 1900.[240] In 1901 he was offered a position in Strasbourg, but stayed with Wiedersheim, who promised him a better professional advancement. However, his next major promotion only happened many years later, when he succeeded his mentor as chair of the department in 1918.

Wiedersheim had set Fischer in charge of lectures on anthropology and of the skull collection assembled by anthropologist Alexander Ecker, Wiedersheim's predecessor. Fischer visited anthropologist Rudolf Martin in Zurich to learn the techniques of anthropometry and became involved in local archeology and history. One of his first anthropological studies was on the heads of two decapitated men from Papua New Guinea, provided through the services of the imperial governor of this German colony in 1905.[241] Racial anthropology became his main subject of research and, together with the medical student and future collaborator Fritz Lenz, he founded the local branch of the "German Society for Racial Hygiene" in 1910.[242] He developed his idea of the heritability of "racial" characteristics based on August Weismann's germ plasma theory, as well as the rediscovery of Mendel's laws for botany and rare human diseases, and Gertrude and Charles Davenport's claim of the Mendelian inheritance of normal human traits, such as hair form, hair color, and eye color.[243]

Political debates in 1907 about the financing of German colonialism prompted him to study the literature on racial anatomy in the German colonies, and he came across an account of a so-called "population of bastards" in the Rehoboth region of Namibia, which he believed to be superbly suited for a field study on Mendelian inheritance of "racial" traits. Fischer arrived in Namibia in 1908 and collected data on 310 descendants of an inhomogeneous group of white men and their African female partners. He determined hair color and shape, eye color and shape, skin color, and skull shape, and observed the social interactions of this population. The results of his study were not published before 1913, and even though his raw data and statistics were ambiguous, Fischer claimed that they proved the inheritance of "race" by Mendelian laws.[244] At the same time, he commented extensively on the inferiority of Africans in terms of character and intelligence, conclusions that were not based on any scientific data but on his own Eurocentric and stereotypic thinking. In terms of the relationship between the "white race" and the Africans he declared:

> One should give them [the Africans] shelter to the extent they need it as a race inferior to ours, so that they can survive as long as they are of use to us—or else free competition, which from my point of view means extinction! This may sound nearly brutally egotistical—but anybody who considers the term race consequently [...] cannot be of a different opinion. [...] I do not need to dwell here on the ethical or legal aspects of the question how all this should be regulated—and I say so in full awareness of my words—the survival of our race is at stake, this has to be in every instance the highest consideration, and ethical and legal norms have to accommodate this.[245]

Fischer presented a racist view of mankind in which certain "races" were inferior to others, with Africans occupying the lowest rung. With his report on the "Rehoboth Bastards" Fischer became known as the founder of a new science called racial biology. A closer analysis of his data shows that the statistics do not hold up to Fischer's claim.[246] Indeed, they could not, as Fischer did not differentiate between phenotype and genotype, and the parent populations were not homogenous, a fundamental requirement for Mendelian analysis. Mendelian inheritance in man can be phenotypically observed in some cases of monogenic disease, but the characteristics Fischer had chosen were multigenic. He could have known that the latter was a problem had he read the paper published by geneticist Wilhelm Weinberg in 1908, in which Weinberg had pointed out that Mendelian inheritance in humans could only be proved in rare, exceptional traits.[247]

Such rare human characteristics had first been described and linked to Mendelian inheritance by Archibald Garrod in 1902, in his work on the monogenic disease alkaptonuria and later on other inborn errors of metabolism.[248] A critical reading of the Davenports' papers reveals the same weaknesses in their data, i.e., shaky statistics and arbitrary definitions of traits. Despite the flawed data, the concept of Mendelian inheritance of "racial" characteristics, both physical and psychological, was seen as proven by the scientific community thereafter, and Fischer became known as a preeminent racial hygienist and expert on racial genetics. Through Fischer's work, inheritable "races" became the central paradigm of anthropology in Germany, opposed only by a few colleagues like Weidenreich and Saller.[249] In 1927 Fischer was recruited as the founding director of the KWI for Anthropology, Human Heredity and Eugenics, which played a central role in NS eugenic policies after 1933.[250]

As president of the *Landesverein "Badische Heimat e.V."* (regional association homeland of Baden) Fischer had ample opportunity to share his eugenic visions with the public, encouraging the reproduction of the healthy population while speaking against the procreation of "degenerated" families and those with inheritable diseases.[251] Together with his colleagues at the KWI Anthropology, Otmar von Verschuer (department of human inheritance) and Heinrich Muckermann (department of eugenics), he developed the theoretical basis for eugenic propaganda and professional education in the Weimar Republic, with first training courses established in 1929 and 1932.[252] In 1932 Fischer declared that the NSDAP was the only party whose program in eugenics he could fully support, even if he did not agree with them on the topic of "foreign races,"[253] and he welcomed Hitler's appointment as chancellor in January 1933.[254]

On 1 February 1933 Fischer gave a public speech on "crossbreeding and intellectual performance," in which he presented his concept of modern "races" as the result of crossbreeding. The National Socialists in the audience were appalled, as they understood Fischer's remarks as meaning that "bastards" (offspring from crossbreeding), even Jewish ones, were more capable and intelligent than the "pure races," especially the "Nordic race." Fischer tried to explain his Rehoboth findings, but the public outrage was great and additionally fuelled by anthropological colleagues from Munich, who were vying for political favoritism and preeminence in the Society for Racial Hygiene.[255] This event, which advertised his differences with the NSDAP, may actually have positively influenced his election as *Rektor* (president) of the University of Berlin. At that time the university senate was deeply divided between supporters and opponents of the new regime, and Fischer was seen as a candidate of compromise. The election was often considered the last "free" one, as from then on the *Rektor* was appointed directly by the REM.[256] However, shortly thereafter Fischer had to declare his alliances when he was called to the Ministry of the Interior for several discussions in June 1933. Given his perceived "soft" attitude toward the elimination of hereditarily diseased persons and his controversial opinion about Jews, the ministry demanded that he remove all collaborators from his institute who did not conform to NS standards, foremost the Catholic Muckermann. Otherwise, the ministry threatened, all interaction between the KWI Anthropology and the government would cease, especially Fischer's role as consulting specialist on racial biology.[257] Fischer chose to fulfill the ministry's conditions, even though his employment at the KWI was in no way dependent on the government. His next major speech on the founding day anniversary of the university, 29 July 1933, was fully in line with the party:

> It is the essence of the nationalist idea of the state to emphasize the unity and common bloodline of the entire people, and to create laws and administrative rules based on these to exclude foreign elements. [...] Regardless if they are good or bad, if bloodlines are different or foreign, they have to be rejected. [...] The fact that National Socialist politics is especially aimed against the Jews is easily explained by the circumstance that they are the only racially different element of a large enough number to matter for our country and people. This statement is not meant as a value judgment. [...] And as we know today, it is impossible to render an individual bloodline healthy or diseased, but we have to increase the hereditarily healthy and decrease the hereditarily diseased. [...] The new National Socialist Germany is the first state to have taken on this task on a large scale. The recently announced law for the preservation of the hereditarily healthy family is the first significant step.[258]

At the end of this speech Fischer declared that his interpretation of the *völkisch* people's state was not based on a political program but on the pure science of biology.[259] Schmuhl argues that Fischer was here not so much bowing to the NS state but trying to plead for a state that fully promoted his concept of eugenics.[260] What Fischer failed to foresee was that his ideas could justify not only sterilization and positive eugenic measures like marital counseling, but also the murderous policies of the NS regime.

In November 1933 he agreed to be reappointed as *Rektor* of the university directly through the REM,[261] and he signed the dismissal of all Jewish colleagues in 1933. In April 1935 he retired from this post to focus on his other duties, and he continued to represent NS Germany internationally. In November 1939 Fischer applied for membership in the NSDAP, together with Lenz, who had replaced Muckermann at the KWI in 1934. Their applications were personally evaluated and approved by Himmler because of their great scientific contributions to NS ideology.[262] In the eyes of Werner Weisbach, an acquaintance of Fischer's who emigrated for "racial" reasons in 1933, Fischer showed a facile adaptability that made him especially useful to the regime.[263]

By 1935 Fischer's KWI Anthropology was focused on services for the regime, among them continued professional education in hereditary biology, for example for the SS, and expert statements funded by the government.[264] Fischer and his colleagues wrote the expert opinions on the so-called "*Rheinland-Bastarde,*" descendants of German mothers and French colonial soldiers born after the occupation of the Rhineland in the 1920s. At least 385 of these young persons were subjected to illegal forced sterilizations between 1935 and 1937—illegal, because the 14 July 1933 Law for the Protection of Hereditary Health covered only hereditarily diseased persons, but not persons of "mixed race."[265] The law was enforced by newly created hereditary health courts, which were composed of lawyers and medical experts—Fischer served as a senior judge on the hereditary health court in Berlin. All documentation from individual cases was collected as material for future research on the health status of the German people.[266]

Fischer retired in 1942 and was succeeded by von Verschuer. Fischer's statements on Jews became overtly anti-Semitic during the war. In 1941 in Paris he gave a speech in French, in which he declared Jews to be of inferior value and even of a different species, and in 1944 he agreed to attend a planned anti-Jewish congress in Cracow that never materialized because of the war events. In 1943 he contributed to an openly anti-Semitic book produced by Gerhard Kittel called "*Das antike Judentum*" (*The Ancient Jewry*). Fischer received a doctorate of honor from the University of Freiburg and the Goethe-Medal for the Arts and Sciences awarded by Hitler in 1939. In 1943 he was given the "*Adlerschild des Dritten Reiches*" (eagle shield of the Third Reich), Hitler's substitute for the Nobel Prize, which he had forbidden German scientists to accept. The KWI Anthropology was renamed to *Eugen-Fischer-Institut.*[267]

In 1944 Fischer moved to Sontra in the state of Hessen, and lived with his daughter for the next six years. After the war he had lost his bearings, as so many other Germans, and noted in 1946:

> Broken were the German front lines, broken the heretofore blind trust in the leadership, broken the walls erected in front of inhumanities and acts of insanity committed by the previous leaders, inhumanities that cry to heaven. Broken was the pride to be a German, broken the pure feeling of having served the ideal of my people's distant future, having served to the best of my knowledge and without suspicion of any wrongdoing. [...] I honestly admit my great guilt of blindness, unquestioning confidence, unworldliness, of complete ignorance of all this evil—but only these—and am ready to atone for these.[268]

Fischer's claim of ignorance about the realities of the NS system's murderous eugenic measures does not ring true. Where was his "unworldliness" when he signed the dismissal of his Jewish colleagues in 1933? Where was his "blindness" when he ruled on the "genetically inferior"? Where was any atonement at all? Even his limited confession in his unpublished manuscript was never shared publicly, when an open admission of guilt by Fischer, even if only a partial one, might have promoted a postwar discussion of scientists' involvement in NS atrocities.

In 1947 Fischer received his denazification verdict as "fellow traveller." He was surprised and delighted by the mild ruling, but taken aback by Weidenreich's letter in *Science,* calling it "an act of meanness. But it does not surprise me at all, as we know the author" in a letter to von Verschuer.[269] It was not before 1948 that Fischer started publishing again.[270] Like most of his colleagues he now called his science "human genetics," and he facilitated through connections with old friends like Hellmut Becher the recruitment of von Verschuer and Lenz as chairs of that discipline at the Universities of Münster and Göttingen, respectively. Of the nine institutes of human genetics or genetics-oriented anthropology existing in the FRG after the war, four were led by students of Fischer. In 1954 Fischer achieved the legal status of professor emeritus at the University of Freiburg, even though many at the faculty there questioned his behavior during the Third Reich. Fischer stayed in close communication with his students. They in turn ensured that he became known to the postwar public as the man who discovered Mendelian genetics in humans, and not as the founder of racial biology.[271] He was named an honorary member of the German Society of Constitution Research in 1951, the German Anthropological Society in 1952 and the *Anatomische Gesellschaft* in 1954.[272] His book on the "Rehoboth Bastards" was reprinted with a nearly unchanged text in 1961. On 9 July 1967 Fischer died at the age of ninety-three.[273]

So, was Weidenreich right in his verdict of Fischer? Was Fischer "the man who should be put on the list of war criminals"? He gave the National Socialists the scientific and legal tools that enforced NS ideology and policies. Even if he claimed his "pure science" had been "abused" by the regime, he had collaborated willingly in providing the scientific basis for discriminatory laws that ultimately led to the Holocaust. It is highly unlikely that Fischer had no direct knowledge of the activities in the concentration camps before 1945, given that he was well integrated in high political circles. These included the *Mittwochsgesellschaft* (Wednesday Society), a politically diverse group of well-informed academics in Berlin which he attended from 1927 to 1943,[274] of whom four were executed as traitors after the assassination of Hitler on 20 July 1944.[275]

Fischer had personally profited from his collaboration with the regime that ensured his influential academic position throughout the NS period.

Max Weinreich put it succinctly in his comment on the collaboration be-tween zealously aggressive politicians and scientists: "There were in the memory of mankind Jenghis Khans and Eugen Fischers but never before had a Jenghis Khan joined hands with an Eugen Fischer. For this reason the blow was deadly efficient."[276] Weidenreich explained his understanding of the term "war criminal" in a memorandum after the war: "But war crim-inals are certainly those people who recommended, on the basis of those theories, the physical extermination of individuals, races, and nations only because they considered them as physically and mentally unfit and danger-ous for the persistency of the Nationalistic State."[277] In this sense, Fischer had indeed become a most dangerous man, and was not the "harmless opportunist" described by one of his emigrated colleagues from the *Mitt-wochsgesellschaft*.[278] What is truly astounding and reprehensible is the fact that Fischer and his students took over much of postwar German human genetics and influenced its future course. What's more, Fischer's idea of the genetic aspects of "race" has become pervasive in global modern genet-ics and medicine and needs to be questioned.[279] The roots of these thought concepts must be critically reconsidered.

Discreet Political Dissenters

Open political or practical opposition against the NS regime was not com-patible with continued professional life at a university in NS Germany.[280] Several anatomists who disagreed politically with the regime chose to emi-grate. However, discreet political dissent was possible, if not always without danger, as the following biographies of the anatomists Ferdinand Wagen-seil, Dietrich Starck, and Charlotte Pommer illustrate.

Ferdinand Wagenseil was born in Augsburg on 7 September 1887[281] and was a sickly child who spent much of his time reading. In school he met the future neuroanatomist Hugo Spatz, with whom he shared a lifelong friend-ship. Wagenseil studied medicine in Munich and received his MD in 1914. In May 1915 Wagenseil joined the Red Cross, which was home to many of those deemed physically unfit for military service, but still willing to serve their country.[282] Wagenseil was sent to Istanbul, were he worked in a military hospital and collected anthropological "material" that became the basis for his research. His superior Theodor Zlocisti, a former leader of the Zionist youth movement in Berlin, shared Wagenseil's love of literature and sparked his interest in the physical anthropology of the Jewish "race." Wagenseil was not interested in racial ideology, but in the exact science of anthropometry, and some of his observations ran counter to the common contemporary "race" stereotypes.[283]

During his time in Turkey Wagenseil gained experience in the building and organization of medical institutions, tasks he would be confronted with several times throughout his life. In February 1919 Wagenseil became Stieve's direct successor at Rückert's anatomical institute in Munich, and in April 1921 he moved to Freiburg as assistant to Fischer, who supported his anthropological work for the *Venia legendi*. In January 1923 Wagenseil was recruited as director of the anatomical institute at Tongji Medical School in Shanghai, which he rebuilt after WWI. Main research topics during his ten years in China included anthropometric studies of Chinese eunuchs and investigations of the musculature in Chinese persons. Two of his publications were based on "material" from bodies of the executed.[284] Despite several attempts to return to Germany,[285] Wagenseil wasn't offered a position until 1931, as second *Prosektor* in Bonn. For the rest of his life Wagenseil felt a deep connection with the Chinese people and culture. During his first years in Bonn Wagenseil was able to finish several of his Chinese anthropological studies.

The year 1935 brought a profound crisis for Wagenseil, as the NS regime threatened his siblings and mother, who were at the center of the bachelor's life. His much younger brother Kurt, for whom he felt a fatherly love, had been interred for political reasons at the concentration camp Dachau on 23 January 1935 and was held there for nearly a year. Much of Wagenseil's energy was taken up by trying to obtain his brother's release and by his mother's subsequent two-year hospitalization due to the mental strain. At the same time Wagenseil realized that the NS regime was making use of the science of anthropology in a perverted manner and he did not want to be part of it. Thus he decided to establish himself in a new field of research and began studying cell biology, first with tissue cultures of embryonic chicken cells and then, starting on 31 July 1935, with cells from an executed man. This work was supported by the DFG.[286] In 1946, during his denazification proceedings, Wagenseil explained that he could have had an easy career in the NS system with his work in "racial" anthropology, but he did not want to risk having to speak out against his convictions. Because he chose to change course to a new and complicated scientific field, his professional development suffered.[287] Despite his mediocre publication record Wagenseil was promoted to department head in Bonn in 1935, most likely because of his patriotic service in China.

Wagenseil never joined the NSDAP. Instead, he became a supporting member of the SS in 1933, a candidate for the NS physicians association in 1936, and a member of the NSDDB in 1941. Apparently he saw fit to enlist in all these minor NS groups in order to appear politically acceptable while he was trying to help his brother, who had again been incarcerated in 1939, as well as to promote his own career. The position of "supporting member of the SS" involved the payment of a monthly financial contribution, but not an oath to Hitler or any other commitment. His colleague Stöhr, who shared Wagenseil's skepticism of the NS regime, also chose to join this group to satisfy the need for a formal commitment to the government. Once Wagenseil was recruited as director of the anatomical institute at the University of Giessen in 1940, he cancelled his membership as supporter of the SS. He accepted the position reluctantly as the situation in Giessen was less than ideal for financial and organizational reasons, but he may have realized that this was the only chair position he might ever be offered given his short publication list and his family's political problems.

Wagenseil quickly became known as a liberal and freethinking academic. In Bonn he had assisted a Jewish student with a letter of recommendation and had supported emigrating scientists. From 1940 on he helped medical student Renate Röse repeatedly, who was put on trial for her role in a local resistance group in 1943.[288] Due to Wagenseil's advocacy Röse was able to continue her medical studies in Bonn after the end of her prison sentence. On Röse's recommendation, the young physician Werner Schmidt, who was of so-called "mixed race," sought Wagenseil's assistance. Wagenseil helped Schmidt find a position for his dissertation work in Hamburg. He was also not afraid to be seen with Schmidt in public, which meant a great deal to the ostracized young man. At the same time Wagenseil must have been quite aware of the political situation: the GeStapo delivered bodies of executed forced laborers directly to the anatomical institute. Other victims of the NS regime also ended up in the anatomical department. Röse reported that Wagenseil and his colleague Ernst von Herrath, who had joined the institute in 1941, were appalled when they realized that many bodies delivered to the institute were those of persons who had been killed for "racial" and political reasons. In one incident Wagenseil showed Schmidt the bodies of several emaciated persons who had been found along the train tracks, presumably thrown from train carriages that transported concentration camp inmates.[289] Another time, Wagenseil complained to Schmidt that life was unbearable after he had received a letter from Stieve, in which he related details on his research of executed women.[290] However, while Wagenseil was aware of the plight of the persons whose bodies arrived at the institute, and while he grieved for them, he still used these bodies for teaching purposes.

Wagenseil continued in his position after the war, as he was the only member of the Giessen faculty without a NSDAP membership, and taught anatomy in Marburg until his own institute was reopened in 1948. In his position as dean of the medical faculty he was responsible for rebuilding. He also caused his friend Hugo Spatz, whose neuroanatomical institute and collection had been evacuated from Berlin to Dillenburg, to move permanently to the University of Giessen.[291] Wagenseil retired in 1955 and the university elected him as a "senator of honor" in 1957. He moved back to Munich and returned to his original field of anthropology by going on an expedition of the Bonin Islands south of Japan in 1956. In Munich he was supported by his colleagues von Lanz and Saller, who shared Wagenseil's approach to anthropology. In 1963 Wagenseil was awarded the Great Cross of Merit of the Federal Republic of Germany for his reconstruction work after the war. He died on 28 February 1967.

Dietrich Starck was another anatomist who discreetly dissented during the NS period.[292] He was born on 29 September 1908,[293] and while his childhood was marked by war, hunger, and deprivation, he enjoyed a supportive teacher and a liberal family environment. They fostered his early interest in the observation of animals, which led to his study of medicine in Jena, Vienna, and Frankfurt—he enriched these studies with visits to lectures by prominent zoologists and embryologists. In 1931 he received his MD, mentored by Bluntschli and Hans Schreiber in Frankfurt, and became assistant to Veit in Cologne in 1932. There he pursued craniological studies that led to his *Venia legendi* in 1936. During a 1998 interview Starck emphasized repeatedly that the difficulties of daily life occupied him more than any political or philosophical theories.[294] He had to cope with the fact that Veit was dismissed from his position in 1937[295] and he had to act as substitute director of the anatomical institute. He was without any chance of promotion, as he was considered "not dependable in terms of National Socialism"; Hans Böker and later Franz Stadtmüller were instead recruited as chairman in Cologne.[296]

Starck joined the NSDAP in 1939 and also became a member of the NSDDB and the NS welfare organization. During his denazification trial, where he was exonerated, he explained that the curator of the university had threatened him with dismissal if he did not enroll in the party. While in Cologne, he was responsible for the morgue and had noticed that since 1934, bodies with terrible injuries were delivered by the GeStapo and the SS. Starck recorded the status and names of these bodies and gave them to Ullrich Hecht, the middleman of a resistance group, who helped transmit this information to contacts outside Germany. Also, Professor Veit testified that Starck had supported him and other dismissed faculty in many ways during the time of their persecution.[297] After the war Starck reported that his exam privileges had been taken away in 1940 because of his rejection of the NS system. Furthermore, in 1942, von Verschuer had offered Starck the opportunity to build a new department of embryology at the KWI Anthropology, Heredity and Eugenics in Berlin. However, Starck did not accept this offer as he did not want to be associated with an institute that was deeply involved with the NS "racial" policies.[298]

In late 1944 Starck assumed the position of *Prosektor* in Frankfurt. When the director of the institute, Schreiber, voluntarily resigned from his position at the end of the war, he once again took on the position of substitute director. In 1949 he was formally recruited as chair of the anatomical institute in Frankfurt. He rebuilt the department and became one of the leading figures in comparative anatomy. He established contacts with his international colleagues, specifically with the paleoneuroanatomist Tilly Edinger, and negotiated the independence of the Edinger Institute of Neuroanatomy after the war. He received many honors and died on 14 October 2001.

In contrast to her dissenting colleagues who still had distinguished careers in anatomy, Charlotte Pommer was the only anatomist who left the discipline during the NS period because the regime used its victims for anatomical purposes. Little is known about her life, even though her biographer Barbara Orth pursued all possible sources in addition to an undated manuscript by Pommer herself, named *Air Mail to Lexi in Elysium*. "Lexi" was Alexandra von Alvensleben, cousin of Libertas Schulze-Boysen. Pommer first met Lexi on 15 August 1944 and subsequently befriended her.[299]

Martha Brigitte Charlotte Pommer was born in Berlin on 9 November 1914, the daughter of a bookseller. In 1936 she started her medical studies at the Friedrich-Wilhelms University in Berlin and was licensed as a physician in 1941. She finished her doctoral thesis in the same year, which had been mentored by the pharmacologist Wolfgang Heubner. Pommer admired Heubner for his uncompromising stance against National Socialism.[300]

In the fall of 1941 she began work as an assistant at Stieve's institute. It was one of her tasks to assist Stieve in his work of retrieving tissues from the bodies delivered to the department. On the night of 22 December 1942 she entered the dissection room and was confronted with the bodies of three decapitated women and several men who had been hanged. She recognized Libertas Schulze-Boysen and Libertas' husband Harro Schulze-Boysen, as well as Arvid von Harnack and Rudolf von Scheliha. All of them had been members of the *Rote Kapelle* resistance group and had been executed for treason. Pommer was so shocked by this encounter that she was hardly able to perform her duties for Stieve as he examined the bodies "as always with great care and unusual diligence."[301] In the background, the anatomy technician complained how much work he had to do when people were executed in three-minute intervals. Pommer was profoundly shaken by the experience and wrote to Lexi many years later: "This night has darkened the Christmas day of 1942 and many more that came after. [...] Since I saw your cousin I have given up any claim on working on purely scientific questions, even though I am specially suited to the theoretical field. After the impressions of that night I resigned from my position."[302]

This decision meant for Pommer the end of her free control over further professional training, as she now had to perform mandatory service for the government. So, after leaving her anatomical position voluntarily on 31 March 1943, she was assigned to the surgical department at the state hospital of the police in central Berlin on 1 April 1943. Pommer was content with this decision, as her anatomical knowledge helped in her surgical duties, and some of the senior physicians at the hospital were not National Socialists. The execution numbers were quickly increasing, and she was deeply touched by the fate of the members of the *Weisse Rose* resistance group in Munich, whose young members were also executed. On her first day at the state hospital she met Thure von Uexküll, the prominent founder of psychosomatic medicine, who had also been detailed for war duty to this hospital. When he learned that she had worked at the anatomical institute he remarked that she had seen his childhood playmate there. He was most likely referring to his cousin Libertas.[303]

Pommer was responsible for the care of SS men and resistance fighters alike. Among her patients were prisoners from various detention facilities, including concentration camps, and she was given the oversight of the surgical outpatient department in March 1944. Lexi's husband, Wilhelm Roloff, was involved in the 20 July 1944 conspiracy against Hitler. He tried to escape after the failed assassination attempt, but was imprisoned and admitted to Pommer's hospital after he tried to end his life with cuts to his wrists. Pommer's superior refused to treat this prisoner properly, but she was able to ensure Roloff's correct care and passed messages between him and his wife. After the recovery from his wounds Roloff was transferred to the internal medicine department due to cardiac problems. The depressed man suffered additionally when he learned about the prisoners around him who were to be executed. Pommer felt much the same. She perceived herself as powerless against the system, having to witness fates like the one of a fifteen-year-old boy who was executed for arson and whose body she had encountered during her time in the anatomy department.[304] She tried to alleviate the suffering of prisoners and shield them from the authorities, often against the determined efforts of some of her superiors. In September 1944 Pommer noticed that she herself had come under GeStapo surveillance for the first time, when she was followed by two men on her way to work. Roloff was finally retransferred to prison in November 1944. Meanwhile Lexi's father had also been incarcerated for political reasons. Pommer stayed in contact with the family and other members of the resistance movement and helped them in any way she could, often by enabling the transfer of messages. She even collected information on ways to commit suicide to help her friends avoid execution by hanging.[305]

In early 1945 Pommer took care of Lexi, who was suffering from influenza. On 11 March 1945 both women were taken into custody by the SS on suspicion of collaboration with resistance fighters. They were repeatedly interrogated and transferred between prisons. In a bunker during a bombing raid Pommer encountered a six-months-pregnant Polish prisoner who was forced to stand throughout the entire length of the alarm. The woman had been sentenced to death for the murder of her rapist. Pommer remarked that she had never seen the body of an executed pregnant woman during her time in the anatomy institute.[306] However, Stieve had dissected at least two pregnant women, but not at the time when Pommer was working with him.[307] Near the end of the war Pommer had become prison secretary and was able to write discharge letters for herself, Lexi, and other friends, which they used for their escape on 21 April 1945 during the confusion of the disintegrating NS system.

Pommer spent the last days of the war and the ensuing chaotic months with Lexi and her family. She then decided on a new start in Hamburg. The rest of her life is marked by discontinuities and what seems to be a permanent quest for meaning. She held short-term positions in surgery, pathology, anatomy, and histology in various German cities and as far away as Ethiopia, India, and the United States. Her last employment was as a pathologist at the Red Cross hospital in Munich. She converted to Catholicism in 1953, never married, had no children, and entered a retirement home near Munich in 1995 where she died on 23 April 2004. Charlotte Pommer had never joined the NSDAP or any other NS organization. Years later she was still ruminating on what else she could have done to succeed against the violent NS regime. She tried to resign herself to never finding an answer, while hoping for a better future.[308]

The biographies of Wagenseil, Starck, and Pommer illustrate the narrow path that academics in the Third Reich had to navigate if they did not agree with the regime and tried to stand up for their own values. Wagenseil and Starck were careful in their dissent and thus able to continue their careers. Also, they did not refuse to work with bodies of NS victims for anatomical purposes, even if they objected to the murder of these persons. Pommer went a step further. She felt unable to pursue an anatomical career based on the work with bodies of NS victims and finally ended up actively taking part in resistance activities, which resulted in her imprisonment. Her life was acutely in danger, and only the end of the war set her free from prison. Pommer never found her way back to the academic pursuit to which she had felt so particularly well suited. It is very likely that this was caused by the trauma she suffered during the Third Reich. Charlotte Pommer survived, but the price she had to pay for her upright moral stance was exceedingly high. Wagenseil and Starck, on the other hand, thrived in their postwar careers.

Notes

1. Quoted after Viebig 2002.
2. Hayek 1955.
3. Williams 1982; Williams 1988; Williams 2004.
4. Goetzl and Reynolds 1944; Sablik 1983; Gisel 1988.
5. Angetter 1999.
6. Williams 2004.
7. Pernkopf 1938; Neugebauer 1998b.
8. Pernkopf 1938.
9. E.g., Pernkopf 1955a; Pernkopf 1955b.
10. Weissmann 1985.
11. Malina and Spann 1999; Angetter 2000.
12. See chapter 9 and Holubar 2000.
13. UAM Bestand 10 Nr.12531 Bd.1.
14. Thamer et al. 2012.
15. NSDAP member number 1265405; SS member number 262703.
16. Equivalent to First Lieutenant, Landgericht Münster 1960.
17. Letter from Benninghoff to Geheimrat Schwoerer, 31 January 1935; estate Benninghoff.
18. Kremer, 1930, 1932, 1933, 1937, 1938a, 1938b, 1939, 1941/42, 1942/43, all results of animal experiments, effects of hunger.
19. Kremer 1942.
20. Weinert 1942.
21. Landgericht Münster 1960.
22. Höss et al. 1984, 235.
23. Strzelecka 2011; Höss et al. 1984.
24. Mitscherlich and Mielke 1960.
25. Landgericht Münster 1960.
26. Strzelecka 2011, 161.
27. Ibid., 161, 162.
28. Höss et al. 1984, 246.
29. Rawicz 1972, 3.
30. Strzelecka 2011, 162.
31. Rawicz 1972, 8.
32. Landgericht Münster 1960.
33. Klee 2003.
34. Höss et al.1984, 213–30.
35. Brand name of a Czech mineral water.
36. Literally "Muslims," NS term for prisoners in the final stages of starvation.
37. Latin for "anus of the world."
38. The prisoner was later identified as Hans de Yong, or de Gong, born 18 February 1924 in Frankfurt, Höss et al. 1984, 230.
39. Höss et al. 1984, 215.
40. Lachman 1977; Kasten 1991; Bauer 1996; Wechsler 2005; Lang 2007; Uhlmann and Winkelmann 2015.
41. Zeiss 2004.

42. UAG Med Fak I 88, letter from the dean of Greifswald to REM, 6 May 1936.

43. Klee 2004, 356–58.

44. Letter from Sievers to Hirt, 3 January 1942; quoted after Lang 2013, 375.

45. Mitscherlich and Mielke 1960.

46. Kasten 1991, 195.

47. Lang 2007, 212.

48. Giles 1993, 88.

49. Noack 2008; Winkelmann and Schagen 2009.

50. Forsbach 2006, 79; Fleischhauer 1981.

51. Aly 1994.

52. Pospielszalski 1992.

53. Goguel 1964.

54. Aly 1987.

55. Scharf 1991; Fanghänel et al. 1988; Zimmermann 1988; Zimmermann 1990; Pospielszalski 1992.

56. Majewski 2013, 130–53.

57. UAL PA 1648, 1.

58. Ibid., 16.

59. Letter from Voss to Professor Brugsch, 14 June 1948, UAJ Bestand D, Nr.598, 29.

60. See note 55.

61. Aly 1994, 108–9.

62. UAL PA 1648, 89.

63. Letter from Clara to the ministry for national education, Dresden, 4 August 1937, UAL PA 1648, 64–65.

64. Aly 1994, 112.

65. UAL PA 1648, 69.

66. Ibid., 77.

67. Goguel 1964, 125.

68. Aly 2003, 145–54.

69. Voss 1932, 1937a, 1937b, 1940, anatomical publications.

70. UAJ, Bestand D, Nr 597, 23.

71. Voss 1942, 357.

72. Ibid., 356.

73. Aly 1994, 135.

74. Ibid., 121–45.

75. Aly 1987, 55. Aly's translator used here the English term "typhoid fever" for the original German term "*Flecktyphus*," correctly translated as typhus. A further distinction must be made. Those with typhus are infected with Rickettsia prowazekii, which is transmitted, for example, by louse bites. Typhoid fever, confusingly called *Typhus* in German, is transmitted orally. Sufferers are infected by food contaminated with Salmonella typhi. Charles G. Roland (Roland 1992, 120) has pointed to this common confusion of the two terms in English and German; the author thanks Bill Seidelman for this information.

76. Copy of testimony, Elisabeth Eckardt, Leipzig, 18 March 1947; UAL PA 1648, 91.

77. Copy of "*Mein Bekenntnis*," Piotr Miklejewski, 19 March 1947; UAL PA 1648, 95.

78. UAL PA 1648, 96–99.

79. Majewski 2013, 140; Szołdrska 1948, 47.

80. Copy of "*Eidesstattliche Erklärung*," 1 January 1947; UAL PA 1648, 98.

81. Curriculum vitae, Voss, 14 December 1948; UAL PA 1648, 16r, 16v.

82. Note ministry of education September 24, 1948; UAJ Bestand D, Nr. 597, page 33r.

83. Letter from Hertwig, 20 September 1948; UAJ Bestand D, Nr. 597, 32-32v.

84. Schumacher 2013, 76.

85. Curriculum vitae, Voss, 5 May 1952; UAJ Bestand D, Nr. 597, 19r-20.

86. Letter from Voss, 11 November 1950; UAJ Bestand D, Nr. 597, 55.

87. Letters from the dean of the medical faculty in Jena, UAJ Bestand D, Nr. 597, 57r-58.

88. Letter from Benninghoff to Professor Guleke, Jena, 20 February 1951; UAM-EB.

89. UAJ Bestand D, Nr. 597, 61–65, 127.

90. Ibid., 128–29.

91. Scharf 1991.

92. UAJ Bestand D, Nr. 597, 136, 167.

93. Ibid., 148–59.

94. Geyer to university leadership Jena, 2 November 1978; UAJ Bestand D, Nr. 597, 162.

95. UAJ Bestand D, Nr. 597, 167, 169-70.

96. Ibid., 171.

97. Ibid., 147.

98. Wróblewska 2000, 120.

99. Letter from Professor A.F. Holstein to Professor Linss, 7 July 1987; file anatomy institute Jena, personal collection Redies.

100. Redies, personal communication via electronic mail, 10 March 2014.

101. Biographical data on Bargmann are taken from Bargmann 1967; Bargmann 1975; UAL, PA 1255, f. 21; Fleischhauer 1979; Hildebrandt 2013c.

102. Bargmann 1931, 85, 97–101.

103. Peiffer 1998.

104. Doerr and Korf 1995.

105. Fleischhauer 1979; Bargmann 1975.

106. UAL, PA 1255, f12.

107. Bargmann 1938.

108. UAL, PA 1255, ff15 and 26.

109. Ibid., f21.

110. Bargmann 1941a; Bargmann1941b.

111. Schneider 1939.

112. UAL, PA 1255, ff 27 and 28.

113. Bargmann 1943.

114. Bargmann and Scheffler 1943a; Bargmann and Scheffler 1943b.

115. Bargmann 1942.

116. Ziesche 1945; Steege 1945.

117. Bargmann and Scheffler 1943b, 7.

118. Bargmann 1942, 395.

119. BA, PK, Bargmann, Wolfgang, Prof., 27.1.06, UBS1/1000041204.

120. Bruns 2009, 113.

121. Bargmann 1975; Sano 1979.

122. Hildebrandt 2012b.

123. Universitätsarchiv Göttingen, Personalvorgang des Rektors, Brief des Rektors vom 16. Oktober 1945 an den Kurator der Universität; Bargmann 1975; Fleischhauer 1979.

124. Dr. Ratschko, personal information; Buddecke 2011.

125. Buddecke 2011.

126. Bargmann 1949.

127. Bargmann 1951, 1953, 1955, 1957, articles on medical education.

128. Scharrer and Bern 1979.

129. Fleischhauer 1979.

130. Scharrer and Bern 1979; Fleischhauer 1979; Sano 1979.

131. Bargmann 1967, 5.

132. Bargmann, 1965, 1975.

133. BA, NSDAP-Gaukartei, Bargmann, Dr. Wolfgang, 27.1.06.

134. Hildebrandt 2013c.
135. UAF B24/104.
136. Bussche 1989; Aumüller et al. 2001; Klee 2003.
137. Winkelmann and Noack 2010.
138. Schulze-Baldes 2002.
139. Kohring and Kreft 2003; Doerr and Korf 1995.
140. LA Schl-H, Abt 460, Nr. 4392, Fragebogen from 1 January 1947 and German denazification panel—categorization in accordance with Z.E.I. No 54, 3 December 1947.
141. Remy 2002.
142. Zimmermann 1994.
143. Bargmann 1967, 6.
144. Platzer 1971; Ebenstein 2001.
145. RIS 2012.
146. UAW, Signatur PH RA 10184 von Hayek, Heinrich, 1929, curriculum vitae.
147. Platzer 1971.
148. Buddrus and Fritzlar 2007.
149. UAR, Personalakte Heinrich von Hayek, 1929–38, f1 letter from Elze to University of Rostock.
150. Hayek 1935.
151. Unger 1998.
152. Tongji 2012.
153. BayHst, Signatur MK 43720, Lebenslauf 26.9.38; Der Reichs- und Preussische Minister 23 Dezember 1937; letters from von Hayek to Professor Petersen, 28 January 1938 and 4 April 1938; and Dienst-Vertrag.
154. Hayek 1937.
155. UWü PA272, Bestätigung Rektorat der Universität Würzburg, 6. Juni 1945.
156. BayHst, Signatur MK 43720, letter from Hipp to Hayek, 24 August 1945; UWü PA272, Abschrift von der Abschrift, Nr. V 18445, der Bayer. Staatsminister für Unterricht und Kultus, 24 August 1945.
157. Blessing 2012.
158. Hayek 1953, iv.
159. Hayek 1946; Hayek 1948b.
160. BayHst, Signatur MK 43720, letter from Bayerisches Ministerium für Unterricht und Kultus an das Rektorat der Universität Würzburg 20 November 1947.
161. BayHst, Signatur MK 43720, letters from von Hayek An seine Magnifizenz, den Rektor der Universität Würzburg 2 Mai 1951, Hayek An den Herrn Dekan der Medizinischen Fakultät Würzburg 21 Januar 52.
162. Platzer, 1971; UAW MED PA 189, ff25, 32, letters from Dekan An das Bundesministerium für Unterricht 10 März 1952, and 28 Januar 1957.
163. Walter Krause in UAWMED PA 189, ff 40, 41, Nachruf Univ. Prof. DDr. Heinrich Hayek; also, Platzer, 1971.
164. Hayek 1948a, 123.
165. Hayek 1950a, 88, with reference to Hayek 1941; similarly Hayek 1950b.
166. Krause, see above; Platzer 1971.
167. LA Schl-H Abt. Acc 59/11, Nr.265; UAWü PA272, Fragebogen 1945; BayHst, Signatur MK 43720, Abdruck Überleitung in die neue Besoldungsverordnung 1 September 1938.
168. Hildebrandt 2013c.
169. Historisches Lexikon Bayerns 2012.
170. Hildebrandt 2013c.
171. Examples of press reports and obituaries: Bazelon 2013; Hoffmann 1951; Grosser 1951; Romeis 1953; Kirsche 1953.

172. Stieve 1918.
173. Stukenbrock 2008.
174. Stieve 1918, 568, 570.
175. Romeis 1953, 405.
176. Schagen 2005, 42.
177. Stieve 1919.
178. Stieve 1924.
179. Noack 2008.
180. Stieve 1926.
181. Schagen 2005.
182. Stieve 1929.
183. Stieve 1929, 26.
184. Winkelmann and Schagen 2009, 164.
185. Heiber 1994, 461; Noack 2008.
186. BA R4901/1329, fol. 1, 2, 8, 114, 6, 14.
187. Schagen 2005; Zimmermann 2007; Noack 2008.
188. Schagen 2005, 48.
189. Ibid., 57.
190. Heiss 1952, 1729.
191. Schagen 2013, 336.
192. Stieve 1929, 19.
193. Stieve 1926.
194. Ibid., 47; Stieve 1929, 20; Stieve 1942c, 260.
195. Personal communication from Dr. Ingeborg Lötterle, former student of Hermann Stieve.
196. Ibid.
197. Bräutigam 1998, 8–9; Ditfurth 1993, 166; Richter 1986, 39; Podszus 2000, 21.
198. Lötterle, personal communication.
199. Richter 1986, 39.
200. Stieve 1952, 107–8.
201. Hildebrandt 2013a.
202. Waltenbacher 2008, 221.
203. Winkelmann and Schagen 2009, 165.
204. Gostomski and Loch 1993, 34, 36
205. Oleschinski 1997, 63.
206. Waltenbacher 2008, 221.
207. Noack 2008.
208. Romeis 1953, 406.
209. Stieve 1939, Stieve 1942a, Stieve 1942b, Stieve 1942c, Stieve 1943, Stieve 1946.
210. Stieve in 1938, quoted in Winkelmann and Schagen 2009, 165.
211. Zimmerman 2007, 33.
212. Stieve 1952, III.
213. Hildebrandt 2013a.
214. Schagen 2005; Noack 2008.
215. Gostomski and Loch 1993, 36.
216. Wulfert 2010; Noack 2008; BA N2506/281, letter from Stieve to Walter Hammer 4 June 1952.
217. Noack 2008.
218. Hansson and Hildebrandt 2014.
219. Hildebrandt 2013a.
220. Zimmermann 2007, 38.
221. Zimmermann see note 206.

222. Heiss 1951.
223. Kirsche 1953.
224. E.g., Stieve 1942c, 261; Marx 2003.
225. Landesarchiv Schl. Holst. Abt. Acc 59/11, Nr. 265, letter from Hermann Stieve to Professor Netter Kiel, 19 November 1945.
226. Hansson and Hildebrandt 2014.
227. Kraus 2005; Haar 2000, 206.
228. Häggqvist 1953.
229. Romeis 1953, 406.
230. Zimmermann 2007.
231. Bettendorf 1995, 568–72; Wischmann 2008; Wischmann 2011.
232. Aly 1987, 61.
233. Oleschinski 1992; Seidelman 1996; Schagen 2005; Noack 2008.
234. Winkelmann 2007.
235. Weidenreich 1946.
236. Weinreich 1946.
237. Müller-Hill 1984.
238. Lösch 1997; Gessler 2000; Ferdinand and Maier 2002.
239. Gessler 2000, 154.
240. Ibid., 13.
241. Fischer 1905.
242. Gessler 2000, 144.
243. Davenport and Davenport 1907; Davenport and Davenport 1908; Davenport and Davenport 1909.
244. Fischer 1913.
245. Ibid., 303.
246. Mai and Bussche 1989; Lösch 1997; Gessler 2000, 68.
247. Lösch 1997, 70.
248. Garrod 1902; Garrod 1908.
249. Lösch 1997, 155; Lipphardt 2008a; Lipphardt 2008b.
250. Schmuhl 2008.
251. Gessler 2000, 156, 166.
252. Ibid., 218–19.
253. Weingart et al. 1992, 385.
254. Gessler 2000, 170.
255. Lösch 1997, 231.
256. Ibid., 256.
257. Schmuhl 2008, 124.
258. Fischer 1933, 14–15.
259. Ibid., 17.
260. Schmuhl 2005, 176.
261. Lösch 1997, 264.
262. Ibid., 276.
263. Scholder 1982.
264. Lösch 1997, 316.
265. Ibid., 344; Lusane 2002.
266. Lösch 1997, 349.
267. Gessler 2000, 171–72; Lösch 1997, 292, 417.
268. Quoted after Lösch 1997, 440, 445.
269. Lösch 1997, 458.
270. Ferdinand and Maier 2002.
271. Lösch 1997, 461, 483.

272. Ferdinand and Maier 2002.
273. Gessler 2000, 173.
274. E.g., Hassell 1988, 339–40.
275. Scholder 1982.
276. Weinreich, 1946, 240–41.
277. Quoted from Lipphard 2008a, 277.
278. Weisenberg 1956, 339.
279. E.g. Whitmarsh and Jones 2010, *What's the use of race?*.
280. For a first systematic discussion on physicians' resistance against the regime, see Kudlien 1985, 210.
281. Unger 1998; Tonutti 1971.
282. Ibid., 39, 49.
283. Ibid., 42.
284. Wagenseil 1934, 1937.
285. Tonutti 1971, 362.
286. Unger 1998, 96.
287. Ibid., 97.
288. Oehler-Klein 2007.
289. Ibid., 367.
290. Schmidt 1993, 114.
291. Tonutti 1971.
292. Korf and Winkler 2002; Fischer and Hossfeld 2002; Kreft 2008; Hossfeld and Junker 1998.
293. Lebenslauf Dietrich Starck, UAE Personalakte Bauer, Karl Friedrich, Blatt 97.
294. Hossfeld and Junker 1998.
295. Golczewski 1988.
296. Hossfeld and Junker 1998, 135, 136; Kaiser 2013.
297. Spruchkammer Frankfurt/Main 8. Dezember 1947, UAG Akte Greifswald 1948 Med Fak II-52 Blatt 113–15.
298. Lebenslauf Dietrich Starck, UAE Personalakte Bauer, Karl Friedrich, Blatt 97.
299. Orth 2013.
300. Ibid., 78.
301. Ibid., 25.
302. Ibid., 26.
303. Ibid., 27; Otte 2001, 43.
304. Orth 2013, 47.
305. Ibid., 78, 79.
306. Ibid., 96.
307. Hildebrandt 2013a.
308. Orth 2013, 117.

Archival Sources

Bundesarchiv
– BA, PK, Bargmann, Wolfgang
– BA, NSDAP-Gaukartei
– BA R4901/1329
– BA N2506/281
Bayerisches Hauptstaatsarchiv (BayHst)
– Signatur MK 43720

Landesarchiv Schleswig-Holstein
 – LA Schl-H, Abt 460, Nr. 4392
 – LA Schl-H Abt. Acc 59/11, Nr. 265
Universitätsarchiv Erlangen
 – UAE Personalakte Bauer, Karl Friedrich
Universitätsarchiv Freiburg
 – UAF B24/104
Universitätsarchiv Göttingen, Personalvorgang des Rektors
Universitätsarchiv Greifswald
 – UAG Med Fak I 88
 – UAG 1948 Med Fak II-52
Universitätsarchiv Jena
 – UAJ Bestand D, Nr. 598
Universitätsarchiv Leipzig
 – UAL PA 1648
 – UAL, PA 1255
Universitätsarchiv Marburg:
 – Estate (*Nachlass*) Benninghoff (UAM-EB)
Universitätsarchiv Münster (UAM)
 – UAM Bestand 10 Nr. 12531 Bd. 1
Universitätsarchiv Rostock (UAR)
 – Personalakte Heinrich von Hayek
Universitätsarchiv Wien (UAW)
 – Signatur PH RA 10184
 – Signatur MED PA 189
Universitätsarchiv Würzburg (UAWü)
 – PA272

Bibliography

Aly, Götz. 1987. "Das Posener Tagebuch des Anatomen Hermann Voss." In: *Biedermann und Schreibtischtäter: Materialien zur deutschen Täter-Biographie*, ed. Götz Aly, Peter Chroust, and Christian Pross, 15–66. Berlin: Rotbuch Verlag.
———. 1994. "The Posen Diaries of the Anatomist Hermann Voss." In: *Cleansing the Fatherland. Nazi Medicine and Racial Hygiene*, ed. Götz Aly, Peter Chroust, and Christian Pross, 99–155. Baltimore: Johns Hopkins University Press.
———. 2003. *Rasse und Klasse. Nachforschungen zum deutschen Wesen*. Frankfurt am Main: S. Fischer Verlag GmbH.
Angetter, Daniela C. 1999. "Die Wiener anatomische Schule." *Wiener Klinische Wochenschrift* 111(18): 764–74.
———. 2000. "Anatomical Science at University of Vienna 1938–45." *Lancet* 355: 1445–57.
Aumüller, Gerhard, Kornelia Grundmann, Esther Krähwinkel, Hans H. Lauer, and Helmuth Remschmidt. 2001. *Die Marburger Medizinische Fakultät im "Dritten Reich."* München: K.G. Saur.
Bargmann, Wolfgang. 1931. "Über Struktur und Speicherungsvermögen des Nierenglomerulus." *Zeitschrift für Zellforschung* 14:73–137.
———. 1938. "Über die Gitterfasern des Nierenglomerulus." *Zeitschrift für Zellforschung* 28: 99–102.
———. 1941a. "Zur Kenntnis der Hülsenkapillaren der Milz." *Zeitschrift für Zellforschung* 31(4): 630–47.

———. 1941b. "Neuere morphologische Untersuchungen zum Thymusproblem. Eine kritische Betrachtung." *Zentralblatt für Innere Medizin* 62: 713–20.

———. 1942. "Über Kernsekretion in der Neurohypophyse des Menschen." *Zeitschrift für Zellforschung* 32(3): 394–400.

———. 1943. "Zur Geschichte der Anatomie in Königsberg (Pr.)." *Anatomischer Anzeiger* 94, 161–208.

———. 1949. "Über die neurosekretorische Verknüpfung von Hypothalamus und Neurohypophyse." *Zeitschrift für Zellforschung* 34: 610–34.

———. 1951. *Das Bild der modernen Anatomie.* Kieler Universitätsreden. Kiel: Kommissionsverlag Lipsius & Tischer.

———. 1953. "Die Aufgaben des World University Service in Deutschland." *Universitas* 8: 1101–2.

———. 1955. "Wiederaufbau und Gestaltung der deutschen Universitäten 1945–1955." *Universitas* 10: 649–59.

———. 1957. "Medical Education in Germany." *Journal of Medical Education* 32(6): 422–26.

———. 1967. "Wolfgang Bargmann." In: *Reflections on Biological Research,* ed. Gulio Gabbiani, 4–7. St. Louis: Warren H. Green Inc.

———. 1975. "A Marvelous Region." In: *Pioneers in Neuroendocrinology,* ed. Joseph Meites, Bernard T. Donovan, and Samual M. McCann, 36–43. New York and London: Plenum Press.

Bargmann, Wolfgang, and A. Scheffler. 1943a. "Zur Frage der parthogenetischen Furchung menschlicher Ovarialeizellen." *Anatomischer Anzeiger* 94: 97–100.

———. 1943b. "Über den Saum des menschlichen Darmepithels." *Zeitschrift für Zellforschung* 33(1–2): 5–13.

Bauer, Axel. W. 1996. "Die Universität Heidelberg und ihre medizinische Fakultät 1933–1945. Umbrüche und Kontinuitäten." *1999: Zeitschrift für Sozialgeschichte des 20. und 21. Jahrhunderts* 11(4): 46–72.

Bazelon, Emily. 2013. "The Nazi Anatomists." *Slate.* Accessed 28 August 2014. http://www.slate.com/articles/life/history/2013/11/nazi_anatomy_history_the_origins_of_conservatives_anti_abortion_claims_that.html.

Bettendorf, Gerhard. 1995. *Zur Geschichte der Endokrinologie und Reproduktionsmedizin.* Berlin: Springer-Verlag.

Blessing, Tim, Anna Wegener, Hermann Koepsell, Michael Stolberg. 2012. "The Würzburg Anatomical Institute and Its Supply of Corpses (1933–1945)." *Annals of Anatomy* 194: 281–285.

Bräutigam, Hans H. 1998. *Beruf: Frauenarzt. Erfahrungen und Erkenntnisse eines Gynäkologen.* Hamburg: Hoffmann und Campe.

Bruns, Florian. 2009. *Medizinethik im Nationalsozialismus.* Stuttgart: Franz Steiner Verlag.

Buddecke, Julia. 2011. *Endstation Anatomie: Die Opfer nationalsozialistischer Vernichtungsjustiz in Schleswig-Holstein.* Hildesheim: Georg-Olms Verlag.

Buddrus, Michael, and Sigrid Fritzlar. 2007. *Die Professoren der Universität Rostock im Dritten Reich. Ein biographisches Lexikon.* München: K. G. Saur.

Bussche, Hendrik van den, ed. 1989. *Medizinische Wissenschaft im "Dritten Reich." Kontinuität, Anpassung und Opposition an der Hamburger Medizinischen Fakultät.* Berlin/Hamburg: Dietrich Reimer Verlag.

Davenport, Gertrude C., and Charles B. Davenport. 1907. "Heredity of Eye-Color in Man." *Science, New Series* 26(670): 589–92.

———. 1908. "Heredity of Hair-Form in Man." *American Naturalist* 42(497): 341–49.

———. 1909. "Heredity of Hair-Color in Man." *American Naturalist* 43(508): 193–211.

Ditfurth, Hoimar von. 1993. *Innenansichten eines Artgenossen. Meine Bilanz.* München: Deutscher Taschenbuch Verlag.

Doerr, Hans W., and Horst-Werner Korf, eds, 1995. *Berühmte Ärzte und Forscher in Frankfurt am Main*. Lampertheim: Alpha.

Ebenstein, Alan. 2001. *Friedrich Hayek. A Biography*. New York: Palgrave.

Fanghänel, Jochen, H. Spaar, and Achim Thom. 1988. "Die Schatten der Vergangenheit. Zur Veröffentlichung des Tagebuches des Anatomen Hermann Voss aus den Jahren 1932–1944." *Zeitschrift der gesamten Hygiene* 34(12): 715–17.

Ferdinand, Horst, and Kurt Erich Maier. 2002. "Eugen Fischer." In: *Baden-Württembergische Biographien*, ed. Kommission für geschichtliche Landeskunde in Baden-Württemberg, 78–85. Stuttgart: Kohlhammer.

Fischer, Eugen. 1905. "Anatomische Untersuchungen zu den Kopfweichteilen zweier Papua." *Correspondenzblatt der deutschen Gesellschaft für Anthropologie, Ethnologie und Urgschichte* 36: 118–23.

———. 1913. *Die Rehobother Bastards und das Bastardisierungsproblem beim Menschen*. Jena: Verlag Gustav Fischer.

———. 1933. *Der Begriff des völkischen Staates, biologisch betrachtet. Rede bei der Feier in Erinnerung and den Stifter der Universität Berlin, König Friedrich Wilhelm III. In der Alten Aula am 29. Juli 1933*. Berlin: Preussische Druckerei und Aktiengesellschaft Berlin.

Fischer, Martina, and Uwe Hossfeld. 2002. "Professor Starck in Memoriam (29. September 1908–14. Oktober 2001)." *Verhandlungen der Geschichte und Theorie der Biologie* 9: 377–95.

Fleischhauer, Kurt. 1979. "In Memoriam Wolfgang Bargmann." *Anatomischer Anzeiger* 146: 209–34.

———. 1981. "In Memoriam Philipp Stöhr Jr." *Anatomischer Anzeiger* 150: 239–47.

Forsbach, Ralf. 2006. *Die Medizinische Fakultät der Universität Bonn im "Dritten Reich."* München: R. Oldenbourg Verlag.

Garrod, Archibald. 1902. "The Incidence of Alkaptonuria: A Study in Chemical Individuality." *Lancet* 160(4137): 1616–20.

———. 1908. "The Croonian Lectures on Inborn Errors of Metabolism." *Lancet* 172(4427): 1–7.

Gessler, Bernhard. 2000. *Eugen Fischer (1874–1967): Leben und Wirken des Freiburger Anatomen, Anthropologen und Rassenhygienikers bis 1927*. Frankfurt: Peter Lang.

Giles, Geoffrey J. 1993. "Die Tätigkeit Hamburger Hochschullehrer in der NS-Bewegung." In: *Akademische Karrieren im Dritten Reich. Beiträge zur Personal- und Berufungspolitik an Medizinischen Fakultäten*, ed. Günter Grau and Peter Schneck, 83–88. Berlin: Pegasus Druck und Verlag.

Goetzl, Alfred, and Ralph Reynolds. 1944. *Julius Tandler: A Biography*. San Francisco: Privately printed.

Goguel, Rudi. 1964. "Über die Mitwirkung deutscher Wissenschaftler am Okkupationsregime in Polen im zweiten Weltkrieg, untersucht an drei Institutionen der deutschen Ostforschung." Dissertation. Medical Faculty Berlin: Humboldt-Universität.

Golczewski, Frank. 1988. *Kölner Universitätslehrer und der Nationalsozialismus. Personengeschichtliche Ansätze*. Köln: Böhlau Verlag.

Gostomski, Victor von, and Walter Loch. 1993. *Der Tod von Plötzensee. Erinnerungen, Ereignisse, Dokumente*. Frankfurt: Bloch Verlag.

Grosser, Otto. 1951. "Hermann Stieve zum 65. Geburtstag." *Anatomischer Anzeiger* 98: i–iv.

Haar, Ingo. 2000. *Historiker im Nationalsozialismus*. Göttingen: Vandenhoeck & Ruprecht.

Häggqvist, Gösta. 1953. *Svenska akademiens årsbok för år 1953*. Stockholm: Almqvist & Wiksell.

Hansson, Nils, and Sabine Hildebrandt. 2014. "Swedish-German Contacts in Anatomy 1930–1950: The Example of Gösta Häggqvist and Hermann Stieve." *Annals of Anatomy* 196: 259–67.

Hassell, Ulrich von. 1988. *Die Hassell-Tagebücher 1938–1944. Herausgegeben von Friedrich Frhr. Hiller von Gaertringen.* Berlin: Jobst Siedler Verlag GmbH.

Hayek, Heinrich von. 1935. "Das Verhalten der Arterien bei Beugung der Gelenke." *Zeitschrift für Anatomie und Entwicklungsgeschichte* 105(1): 25–36.

———. 1937. "Die Stirnlappen des Chinesengehirnes und seine Beziehungen zum Schädel." *Zeitschrift für Morphologie und Anthropologie* 37: 114–18.

———. 1946. "Die Umstellung des Lungenkreislaufs nach der Geburt durch epitheloidzellige Sperrarterien." *Ärztliche Wochenschrift* 1: 251.

———. 1948a. "Über die Beziehung der Alveolarepithelien zu den Capillaren." *Klinische Wochenschrift* 26: 123–24.

———. 1948b. "Über die Alveolarepithelzellen und Alveolarkapillaren." *Ärztliche Wochenschrift* 3: 381.

———. 1950a. "Die Muskulatur im Lungenparenchym des Menschen." *Zeitschrift für Anatomie und Entwicklungsgeschichte* 115(1): 88–94.

———. 1950b. "Zur Frage der Lungenmuskulatur." *Klinische Wochenschrift* 28: 268–69.

———. 1953. *Die menschliche Lunge.* Berlin: Springer-Verlag.

———. 1955. "Prof. Dr. Eduard Pernkopf +." *Wiener Klinische Wochenschrift* 67(19): 350.

Heiber, Helmut. 1994. *Universität unterm Hakenkreuz. Teil II: Die Kapitulation der Hohen Schulen. Das Jahr 1933 und seine Themen. Band 1.* München: KG Saur.

Heiss, Robert. 1952. "Lebensbild. Hermann Stieve zum Gedächtnis." *Medizinische Klinik* 52: 1729–30.

Hildebrandt, Sabine. 2013a. "The Women on Stieve's List: Victims of National Socislism Whose Bodies Were Used for Anatomical Research." *Clinical Anatomy* 26: 3–21.

———. 2013b. "Research on Bodies of the Executed in German Anatomy: An Accepted Method that Changed during the Third Reich. Study of Anatomical Journals from 1924 to 1951." *Clinical Anatomy* 26: 304–26.

———. 2013c. "Wolfgang Bargmann (1906–1978) and Heinrich von Hayek (1900–1969): Careers in Anatomy Continuing through German National Socialism to Postwar Leadership." *Annals of Anatomy* 195: 283–95.

Historisches Lexikon Bayerns. 2012. *Deutsche Akademie 1925–1945.* Accessed 30 October 2012. http://www.historisches-lexikon-bayerns.de/artikel/artikel_44466.

Höss, Rudolf, Pery Broad, and Johann Paul Kremer. 1984. *KL Auschwitz Seen by the SS: Selection, Elaboration and Notes by Bezwinska, Jadwiga and Czech, Danuta.* New York: Howard Fertig.

Hoffmann, Auguste. 1951. "Hermann Stieve zu seinem 65. Geburtstag am 22. Mai 1951." *Zeitschrift für Mikroskopisch-Anatomische Forschung* 57: 117–28.

Holubar, Karl. 2000. "The Pernkopf Story: The Austrian Perspective of 1998, 60 Years after It All Began." *Perspectives in Biology and Medicine* 43(3): 382–88.

Hossfeld, Uwe, and Thomas Junker. 1998. "Dietrich Starck zum 90. Geburtstag." *NTM Zeitschrift für Geschichte der Wissenschaften, Technik und Medizin* 6:129–47.

Kaiser, Stephanie. 2013. "Tradition or Change? Sources of Body Procurement for the Aatomical Institute of Cologne in the Third Reich." *Journal of Anatomy* 223: 410–18.

Kasten, Frederick H. 1991. "Unethical Nazi Medicine in Annexed Alsace-Lorraine: The Strange Case of Nazi Anatomist Professor Dr. August Hirt." In: "Historians and Archivists." In: *Essays in Modern German History and Archival Policy,* ed. George O. Kent, 173–208. Fairfax Virginia: George Mason University Press.

Kirsche, Walter. "Hermann Stieve +." *Gegenbaurs Morphologisches Jahrbuch* 93: 1–13.

Klee, Ernst. 2003. *Das Personenlexikon zum Dritten Reich: Wer war was vor und nach 1945?* Frankfurt am Main: S. Fischer.

———. 2004. *Auschwitz, die NS-Medizin und ihre Opfer.* 3rd ed. Frankfurt am Main: Fischer Verlag.

Kohring, Rolf, and Gerald Kreft, eds. 2003. *Tilly Edinger. Leben und Werk einer jüdischen Wissenschaftlerin.* Frankfurt: E. Schweizerbart'sche Verlagsbuchhandlung.

Korf, Horst-Werner, and Jürgen Winckler. 2002. "Zum Gedenken an Dietrich Starck." *Hessisches Ärzteblatt* 1: 701.

Kraus, Karl. *Briefe an Sidonie Nádherný von Borutin.* Vol. 1. Göttingen: WallsteinVerlag.

Kreft, Gerald. 2008. ""...nunmehr judenfrei...." Das Neurologische Institut 1933 bis 1945." In: *Frankfurter Wissenschaftler zwischen 1933 und 1945,* ed. Jörn Kobes and Jan-Otmar Hesse, 125–56. Göttingen: Wallstein Verlag.

Kremer, Johann. 1930. "Die histologischen Veränderungen der quergestreiften Muskulatur der Amphibien im Hungerzustande." *Zeitschrift für Mikroskopisch-Anatomische Forschung* 21: 184–349.

———. 1932. "Die fortlaufendenVeränderungen der Amphibienleber im Hungerzustande." *Zeitschrift für Mikroskopisch-Anatomische Forschung* 28: 81–156.

———. 1933. "Die morphologische Gestaltung der Gallensekretion." *Zeitschrift für Mikroskopisch-Anatomische Forschung* 33: 486–524.

———. 1937. "Zur Frage der Lokalisation des Hungerpigmentes in der Kaltblüterleber." *Anatomischer Anzeiger* 83: 316–30.

———. 1938a. "Das Problem der Pigmentablagerung in der Leber und Milz der Kaltblüter und seine Beziehungen zur Frage des Blutabbaues und Eisenstoffwechsels." *Zeitschrift für Mikroskopisch-Anatomische Forschung* 44: 234–323.

———. 1938b. "Über das Wesen und die Bedeutung der pegmentierten Zellen in der Leber hungernder Kaltblüter." *Anatomischer Anzeiger* 85: 310–12.

———. 1939. "Die zentrale Bedeutung der Leber im Pigment- und Eisenstoffwechsel der Reptilien." *Anatomischer Anzeiger* 88: 119–29.

———. 1941/42. "Neue Fundamente der Zellen- und Gewebeforschung." *Mikrokosmos* 51–52.

———. 1942/43. "Das Wesen und die Herkunft der mit der Zerstörung roter Blutkörperchen in Verbindung gebrachten eisenpigmenthaltigen Zellen der Milz." *Mikrokosmos-Jahrbuch* 36(6/7): 77–80.

———. 1942. "Ein bemerkenswerter Beitrag zur Frage der Vererbung traumatischer Verstümmelungen." *Zeitschrift für menschliche Vererbung und Konstitutionslehre* 25: 535–70.

Kudlien, Fridolf. 1985. *Ärzte im Nationalsozialismus.* Köln: Kiepenheuer und Witsch.

Lachman, Ernest. 1977. "Anatomist of Infamy: August Hirt." *Bulletin of the History of Medicine* 51: 594–602.

Landgericht Münster. 1960. "Das Urteil gegen Dr. Johann Paul Kremer." In: *Justiz und NS-Verbrechen.* Band XVII. Accessed 16 December 2013, http://web.archive.org/web/20081207153240/http://www1.jur.uva.nl/junsv/Excerpts/Kremer.htm.

Lang, Hans-Joachim. 2007. *Die Namen der Nummern: Wie es gelang, die 86 Opfer eines NS-Verbrechens zu identifizieren.* Überarbeitete Ausgabe. Frankfurt am Main: S. Fischer Verlag.

———. 2013. "August Hirt and 'Extraordinary Opportunities for Cadaver Delivery' to Anatomical Institutes in National Socialism: A Murderous Change in Paradigm." *Annals of Anatomy* 195: 373–80.

Lipphardt, Veronika. 2008a. *Biologie der Juden: Jüdische Wissenschaftler über "Rasse" und Vererbung 1900–1935.* Göttingen: Vandenhoeck & Rupprecht.

———. 2008b. "Das "schwarze Schaf" der Biowissenschaften: Marginalisierungen und Rehabilitierungen der Rassenbiologie im 20. Jahrhundert." In: *Pseudowissenschaft,* ed. Dirk Rupnow, Veronika Lipphardt, Jens Thiel, and Christina Wessely, 223–50. Frankfurt am Main: Suhrkamp.

Lösch, Niels C. 1997. *Rasse als Konstrukt: Leben und Werk Eugen Fischers.* Frankfurt: Peter Lang GmbH.

Lusane, Clarence. 2002. *Hitler's Black Victims: The Historical Experiences of Afro-Germans, European Blacks, Africans, and African Americans in the Nazi Era.* New York and London: Routledge.

Mai, Christoph, and Hendrik van den Bussche. 1989. "Die Forschung." In: *Medizinische Wissenschaft im "Dritten Reich." Kontinuität, Anpassung und Opposition an der Hamburger Medizinischen Fakultät*, ed. Hendrik van den Bussche, 165–266. Berlin/Hamburg: Reimer Verlag.

Majewski, Olaf Edward. 2013. "Medizin an der Reichsuniversität Posen (1941–1945) und der polnischen Untergrunduniversität der westlichen Gebiete U. Z. Z. (1942–1945)." Medizinische Dissertation, Universität Heidelberg.

Malina, Peter, and Gustav Spann. 1999. "Das Senatsprojekt der Universität Wien 'Untersuchungen zur Anatomischen Wissenschaft in Wien 1938–1945.'" *Wiener Klinische Wochenschrift* 111(18): 743–53.

Marx, Jörg. 2003. "'Der Wille zum Kind' und der Streit um die physiologische Unfruchtbarkeit der Frau. Die Geburt der modernen Reproduktionsmedizin im Kriegsjahr 1942." In: *Biopolitik und Rassismus*, ed. Martin Stingelin, 112–59. Frankfurt am Main: Suhrkamp Verlag.

Mitscherlich, Alexander, and Fred Mielke. 1960. *Medizin ohne Menschlichkeit. Dokumente des Nürnberger Ärzteprozesses.* Frankfurt: Fischer Taschenbuch Verlag. 1997 printing.

Müller-Hill, Benno. 1984. *Tödliche Wissenschaft.* Reinbek bei Hamburg: Rowohlt Taschenbuch Verlag.

Neugebauer, Wolfgang. 1998. "Zum Umgang mit sterblichen Resten von NS-Opfern nach 1945." In: *Senatsprojekt der Universität Wien: Untersuchungen zur anatomischenWissenschaft in Wien 1938–1945*, ed. Akademischer Senat der Universität Wien, 459–65. Unpublished manuscript.

Noack, Thorsten. 2008. "Begehrte Leichen. Der Berliner Anatom Hermann Stieve (1886–1952) und die medizinische Verwertung Hingerichteter im Natinoalsozialismus." *Medizin, Gesellschaft und Geschichte. Jahrbuch des Instituts für Geschichte der Medizin der Robert Bosch Stiftung* 26: 9–35.

Oehler-Klein, Sigrid. 2007. *Die Medizinische Fakultät der Universität Giessen im Nationalsozialismus und in der Nachkriegszeit: Personen und Institutionen, Umbrüche und Kontinuitäten.* Stuttgart: Franz Steiner Verlag.

Oleschinski, Brigitte. 1992. "Der 'Anatom der Gynäkologen': Hermann Stieve und seine Erkenntnisse über die Todesangst und weiblichen Zyklus." *Modelle für ein deutsches Europa- Ökonomie und Herrschaft im Grosswirtschaftsraum*, ed. Horst Kahrs, Ahlrich Meyer, and MG Esch, 211–18. Berlin: Rotbuch Verlag.

———. 1997. *Gedenkstätte Plötzensee.* Berlin: Möller Druck und Verlag.

Orth, Barbara. 2013. *Gestapo im OP: Bericht der Krankenhausärztin Charlotte Pommer.* Berlin: Lukas-Verlag.

Otte, Rainer. 2001. *Thure von Uexküll. Von der Psychosomatik zur integrierten Medizin.* Göttingen: Vandenhoeck & Ruprecht.

Peiffer, Jürgen. 1998. "Die Vertreibung deutscher Neuropathologen." *Nervenarzt* 69: 99–109.

Pernkopf, Eduard. 1938. "Nationalsozialismus und Wissenschaft." *Wiener Klinische Wochenschrift* 51(20): 545–48.

———. 1955a. "Die Wegbereitung und der Charakter der modernen Anatomie." *Wiener Klinische Wochenschrift* 67(18): 312–15.

———. 1955b. "Gedanken, die einer allgemeinen, funktionellen Anatomie des Nervensystems vorangestellt werden können." *Wiener Klinische Wochenschrift* 67(19): 341–44.

Platzer, Werner. 1971. "Heinrich v. Hayek 1900–1969." *Anatomischer Anzeiger* 128: 409–17.

Podszus, Werner. 2000. *Grosse Charitéärzte in Krieg und Frieden.* Norderstedt: Libri-Books on demand.

Pospieszalski, Karol Marian. 1992. "Zur Frage der Echtheit des Tagebuchs von Hermann Voss." In: *Modelle für ein deutsches Europa- Ökonomie und Herrschaft im Grosswirtschaftsraum*, ed. Horst Kahrs, Ahlrich Meyer, and Michael G. Esch, 207–11. Berlin: Rotbuch Verlag.

Rawicz, J. 1972. "Foreword." In: *KL Auschwitz seen by the SS*, ed. K. Smolén, J. Bezwińska, Danuta Czech, T. Iwaszko, I. Polska, and I. Oświęcimiu, 5–32. Oświęcimiu: Państwowe Muzeum w Oświęcimiu.

Remy, Stephen P. 2002. *The Heidelberg Myth: The Nazification and Denazification of a German University*. Boston: Harvard University Press.

Richter, Horst-Eberhard. 1986. *Die Chance des Gewissens. Erinnerungen und Assoziationen.* Hamburg: Hoffmann und Campe.

RIS. 2012. *Bundeskanzleramt: Rechtsinformationssystem. Bundesrecht konsolidiert: Gesamte Rechtsvorschrift für Adelsaufhebungsgesetz, Fassung vom 25.10.2012.* Accessed 25 October 2012. http://www.ris.bka.gv.at/GeltendeFassung.wxe?Abfrage=Bundesnormen&Gesetzesnummer=10000036.

Roland, Charles. 1992. *Courage under Siege: Starvation, Disease and Death in the Warsaw Ghetto*. New York and Oxford: Oxford University Press.

Romeis, Benno. 1953. "Hermann Stieve +." *Anatomischer Anzeiger* 99: 401–40.

Sablik, Karl. 1983. *Julius Tandler. Mediziner und Sozialreformer. Eine Biographie.* Wien: Verlag A. Schendl.

Sano, Yutaka. 1979. "In Memoriam Wolfgang Bargmann (1906–1978)." *Archivum Histologicum Japonicum* 42(3): 196–200.

Schagen, Udo. 2005. "Die Forschung an menschlichen Organen nach 'plötzlichem Tod' und der Anatom Hermann Stieve (1886–1952)." In: *Die Berliner Universität in der NS-Zeit. Band II: Fachbereiche und Fakultäten*, ed. Rüdiger von Bruch and Rebecca Schaarschmidt, 35–54. Stuttgart: Franz Steiner Verlag Wiesbaden GmbH.

———. 2013. "Stieve." In: *Neue Deutsche Biographie*, ed. Bayerische Akademie der Wissenschaften, 335–37. Berlin: Duncker & Humblodt.

Scharf, Joachim-Hermann. 1991. "Prof. Dr. med. Dr. med. h.c. Hermann Voss." *Acta Histochemica* 90: 1–3.

Scharrer, Berta V., and H. A. Bern. 1979. "Wolfgang Bargmann 1906–1978." *General and Comparative Endocrinology* 38(3): 389–91.

Schmidt, Werner. 1993. *Leben an Grenzen. Autobiographischer Bericht eines Mediziners aus dunkler Zeit.* Baden-Baden: Suhrkamp/Nomos Verlagsgesellschaft.

Schmuhl, Hans-Walter. 2005. *Grenzüberschreitungen: Das Kaiser-Wilhelm-Institut für Anthropologie, menschliche Erblehre und Eugenik 1927–1945*. Göttingen: Wallstein Verlag.

———. 2008. *The Kaiser Wilhelm Institute for Anthropology, Human Heredity and Eugenics, 1927–1945: Crossing Boundaries*. Dordrecht: Springer Science and Business Media B.V.

Schneider, H. J. 1939. "Über die Speicherung von Vitalfarbstoffen im Thymusretikulum. (Mit Bemerkungen über die Architektur des Thymus)." *Zeitschrift für Anatomie und Entwicklungsgeschichte* 113(1): 187–203.

Scholder, Klaus. 1982. *Die Mittwochsgesellschaft. Protokolle aus dem geistigen Deuschland 1932–1944.* 2nd ed. Berlin: Severin und Siedler.

Schulze-Baldes, Annette. 2002. "Das Jahr 1933. Die medizinische Fakultät und die 'Gleichschaltung' an der Universität." In: *Medizin und Nationalsozialismus. Die Freiburger Medizinische Fakultät und das Klinikum in der Weimarer Republik und im "Dritten Reich"*, ed. Bernd Grün, Hans-Georg Hofer, and Karl-Heinz Leven, 139–60. Frankfurt am Main: Peter Lang Verlag.

Schumacher, Gert-Horst. 2013. *Unzeitgemäss in den Zeiten*. Rostock: Redieck& Schade GmbH.

Seidelman, William E. 1996. "Nuremberg Doctors' Trial: Nuremberg Lamentation: For the Forgotten Victims of Medical Science." *British Medical Journal* 313: 1463–67.

Steege, Helmut. 1945. "Über den Histotopochemischen Nachweis von Vitamin C in der menschlichen und tierischen Schilddrüse." *Zeitschrift für Zellforschung* 33(3): 412–23.

Stieve, Hermann. 1918. "Die Entwicklung des Eierstocks der Dohle." *Archiv für mikroskopische Anatomie* 92: 137–288.

———. 1919. "Der Sphincter pylori des menschlichen Magens." *Anatomischer Anzeiger* 51: 513–34.

———. 1924. "Untersuchungen über die Wechselbeziehungen zwischen Gesamtkörper und Keimdrüsen. III. Beobachtungen am menschlichen Hoden." *Zeitschrift für Mikroskopisch-Anatomische Forschung* 1: 491–512.

———. 1926. *Unfruchtbarkeit als Folge unnatürlicher Lebenweise.* München: Verlag von J.F. Bergmann.

———. 1929. *Gedanken über die geistige und körperliche Ausbildung des akademischen Nachwuchses auf der Schule. Hallische Universitätsreden 40.* Halle (Saale): Max Niemeyer Verlag.

———. 1942a. "Der Einfluss von Angst und psychischer Erregung auf Bau und Funktion der weiblichen Geschlechtsorgane." *Zentralblatt für Gynäkologie* 66: 1698–1708.

———. 1942b. "Die Wirkung von Gefangenschaft und Angst auf den Bau und die Funktion der weiblichen Geschlechtsorgane." *Zentralblatt für Gynäkologie* 66: 1456.

———. 1942c. "Der Einfluss des Nervensystems auf Bau und Leistungen der weiblichen Geschlechtsorgane de Menschen." *Zeitschrift für Mikroskopisch-Anatomische Forschung* 52: 189–266.

———. 1943. "Über Follikelsprung, Gelbkörperbildung und den Zeitpunkt der Befruchtung beim Menschen." *Zeitschrift für Mikroskopisch-Anatomische Forschung* 53: 467–582.

———. 1946. "Über Wechselbeziehungen zwischen Keimdrüsen und Nebennierenrinde." *Das deutsche Gesundheitswesen* 18(1): 537–45.

———. 1952a. *Der Einfluss des Nervensystems auf Bau und Tätigkeit der Geschlechtsorgane des Menschen.* Stuttgart: Thieme.

Strzelecka, Irena. 2011. *Medical Crimes: Medical Experiments in Auschwitz.* Oświęcim, Poland: International Center for Education about Auschwitz and the Holocaust.

Stukenbrock, Karin. 2008. "Der Krieg in der Heimat: 'Kriegsamenorrhoe' im Ersten Weltkrieg." *Medizinhistorisches Journal* 43: 264–93.

Szołdrska, Halska. 1948. *Walka z kultura Polską.* Poznan, Poland: Odbito w drukarni Uniwersytetu Poznanskiego.

Teichmann, W. 1940. "Über die Gitterfasern des Thymus." *Zeitschrift für Zellforschung* 30(5): 689–701.

Thamer, Hans-Ulrich, Daniel Droste, and Sabine Happ. 2012. *Die Universität Münster im Nationalsozialismus. Kontinuitäten und Brüche zwischen 1920 und 1960.* Vols. 1 and 2. Münster: Aschendorff Verlag.

Tongji. 2012. *Chronology of Tongji University.* Accessed 25 October 2012. http://www.tongji.edu.cn/english/themes/10/template/abouttongji/Chronology.shtml.

Tonutti, Emil. 1971. "In Memoriam Ferdinand Wagenseil." *Anatomischer Anzeiger* 129: 361–65.

Uhlmann, Angelika, and Andreas Winkelmann. 2015. "The Science Prior to the Crime: August Hirt's Career before 1941." *Annals of Anatomy,* DOI: http://dx.doi.org/10.1016/j.aanat.2014.10.001.

Unger, Michael. 1998. *Ferdinand Wagenseil (1887–1967). Integrer Forscher und Bewahrer der Medizinischen Fakultät Giessen.* Giessen: Wilhelm Schmitz Verlag.

Viebig, Michael. 2002. "Zu Problemen der Leichenversorgung des Anatomischen Institutes der Universität Halle vom 19. bis Mitte des 20. Jahrhunderts." In: *Beiträge zur Geschichte der Martin-Luther-Universität 1502–2002,* ed. Hermann-Josef Rupieper, 117–46. Halle: MDV, Mitteldeutscher Verlag Halle.

Voss, Hermann. 1932. "Die Beobachtungen eines drüsenartigen Lumens mit Sekret in der Nebennierenrinde des Menschen." *Zeitschrift für Mikroskopisch-Anatomische Forschung* 28: 158–82.

———. 1937a. "Vergleichende Untersuchungen über den Aufteilungsgrad der kontraktilen Masse in den Skelettmuskeln." *Zeitschrift für Mikroskopisch-Anatomische Forschung* 42: 418–32.

———. 1937b. "Untersuchunge über Zahl, Anordnung und Länge der Muskelspindeln in den Lumbricalmuskeln des Menschen und einiger Tiere." *Zeitschrift für Mikroskopisch-Anatomische Forschung* 42: 509–24.

———. 1940. "Vergleichende Histotopochemische Untersuchungen über das Verhalten der Nebenniere zur Plasmalreaktion." *Zeitschrift für Zellforschung* 31(1): 43–53.

———. 1942. "Die Medizinische Fakultät der Reichsuniversität Posen." *Deutsches Ärzteblatt* 72: 356–57.

Waltenbacher, Thomas. 2008. *Zentrale Hinrichtungsstätten. Der Vollzug der Todesstrafe in Deutschland von 1937–1945. Scharfrichter im Dritten Reich.* Berlin: Zwilling.

Wechsler, Patrick. 2005. "La Faculté des Medicine de la 'Reichsuniversität Strassburg' (1941–1945) a l'heure nationale-socialiste." Dissertation. Faculté de Medicine de Strasbourg.

Weidenreich, Franz. 1946. "Letter to the Editor: On Eugen Fischer." *Science* 104: 399.

Weinert, Hans. 1942. "Eine Erklärung auf die Frage der 'Vererbung traumatischer Verstümmelungen.'" *Zeitschrift für menschliche Vererbung und Konstitutionslehre* 25: 780–85.

Weingart, Peter, Jürgen Kroll, and Kurt Bayertz. 1992. *Rasse, Blut und Gene: Geschichte der Eugenik und Rassenhygiene in Deutschland.* Frankfurt am Main: Suhrkamp.

Weinreich, Max. 1946. *Hitler's Professors.* New Haven and London: Yale University Press. Repr. 1999.

Weisbach, Werner. 1956. *Geist und Gewalt.* Wien-München: Verlag Anton Schroll & Co.

Weissmann, Gerald. 1985. "Springtime for Pernkopf." Repr. 1987 in: *They All Laughed at Christopher Columbus,* ed. Gerald Weissmann, 48–69. New York: Times Books.

Whitmarsh, Ian, and David S. Jones. 2010. *What's the Use of Race? Governance and the Biology of Difference.* Boston: The MIT Press.

Williams, David J. 1982. "The Work of Eduard Pernkopf, et al." Presentation at 37th Annual Meeting of the Association of Medical Illustrators, Anaheim, CA, 13 October 1982. Manuscript.

———. 1988. "The History of Eduard Pernkopf's Topographische Anatomie des Menschen." *Journal of Biocommunication* 15(2): 2–12.

———. 2004. "Is a Picture Worth a Thousand Lives? The Truth about the Most Beautiful Book You Will Never See." The Twenty-Third Annual Charles Henry Hackley Distinguished Lecture in the Humanities. Muskegon, MI, 20 May 2004. Manuscript.

Winkelmann, Andreas. 2007. "Die menschliche Leiche in der heutigen Anatomie." In: *Grenzen des Lebens. Beiträge aus dem Institut Mensch, Ethik und Wissenschaft.* Band 5, ed. Sigrid Graumann and Katrin Grüber, 62–74. Berlin: LIT Verlag Dr. W. Hopf.

Winkelmann, Andreas, and Thorsten Noack. 2010. "The Clara Cell: A 'Third Reich Eponym'?" *European Respiratory Journal* 36: 722–27.

Winkelmann, Andreas, and Udo Schagen. 2009. "Hermann Stieve's Clinical-Anatomical Research on Executed Women During the 'Third Reich.'" *Clinical Anatomy* 22(2): 163–71.

Wischmann, Tewes. 2008. "'Paracyclische Ovulationen' und 'Schreckblutungen.' Zur Rezeption der Arbeiten Hermann Stieves in der psychosomatischen Gynäkologie." *Gynäkologische Praxis* 32: 709–13

———. 2011. "'Paracyclische Ovulationen' und 'Schreckblutungen'— eine Fortsetzung." In: *Nichts ist unmöglich!? Frauenheilkunde in Grenzbereichen. Beiträge der 39. Jahrestagung* der Deutschen Gesellschaft fur Psychosomatische Frauenheilkunde und Geburtshilfe (DG-PFG e. V.), ed. Susanne Ditz, Brigitte Schlehofer, Friederike Siedentopf, Christof Sohn, Wolfgang Herzog, and Martina Rauchfuss, 147–52. Frankfurt am Main: Mabuse-Verlag.

Wróblewska, Teresa. 2000. *Die Reichsuniversitäten Posen, Prag und Strassburg als Modell Nationalsozialistischer Hochschulen in den von Deutschland besetzten Gebieten.* Torun: Marszalek.

Wulfert, Tatjana. 2010. "Margarete von Zahn (geb. 1924)." *Der Tagesspiegel* 18 November 2010. Accessed 26 February 2014. http://www.tagesspiegel.de/berlin/nachrufe/marga rete-von-zahn-geb-1924/2892332.html.

Zeiss, Carl. 2004. "Das Phänomen Fluoreszenz." *Innovation* 14: 4–10.

Ziesche, Karl T. 1943. "Zur Histologie des Tuber cinereum des Menschen." *Zeitschrift für Zellforschung* 33(1–2): 143–50.

Zimmermann, Susanne. 1988. "Was die Tagebücher des Hermann Voss offenbaren." *Universitätszeitung* Nr. 5:7.

———. 1990. "Jenaer Medizin im Spannungsfeld von faschistischer Ideologie und ärztlicher Ethik." In: *Judenhass und Judenmord: unerklärlich, unbegreiflich. Jenaer Reden und Schriften,* ed. Friedrich-Schiller-Universität Jena, 76–80. Jena: Universitätsverlag.

———. 2007. ""… er lebt weiter in seinen Arbeiten, die als unverrückbare Steine in das Gebäude der Wissenschaft eingefügt sind"- Zum Umgang mit den Arbeiten des Anatomen Hermann Stieve (1886–1952) in der Nachkriegszeit." In: *Täterschaft — Strafverfolgung — Schuldentlastung. Ärztebiographien zwischen nationaler Gewaltherrschaft und deutscher Nachkriegsgeschichte,* ed. Boris Böhm, and Norbert Haase, 29–40. Leipzig: Leipziger Universitätsverlag.

Zimmermann, Volker. 1994. "Mikulicz-Radecki, Felix." *Neue Deutsche Biographie* 17: 499. Accessed 18 April 2014. http://www.deutsche-biographie.de/pnd105684007.html.

Chapter 7

NS VICTIMS AND THE USE OF THEIR BODIES FOR ANATOMICAL PURPOSES

> They all had names, faces, hopes and longings […] and for that
> reason the suffering of the least of them is no less than of the
> first, whose names will be preserved. I only wish you felt close to
> all of them, as though you knew them, as though they were of
> your own family, or even you yourself.
>
> Julius Fuchik (1903–43)[1]

Instead of focusing on the victims and survivors of medical atrocities in the Third Reich, research on the history of medicine during National Socialism has focused on the perpetrators and their crimes. The sufferers were rarely mentioned, and at times were even victimized again as some attributed the victims' fates to their own being and actions.[2] This was in part due to the fact that Germany as a society depended on the work of physicians and other professionals who had been involved in the NS system but who continued in their old positions and provided essential services. Other factors contributing to the neglect of research into the victims' histories include the relatively small number of perpetrators compared to the great number of victims, the scarcity of documentation on victims' lives, and the silence of perpetrators as well as victims after the war. Recent approaches concentrate on factual research into the history of the ethically unthinkable in medicine, and detailed presentations of the victims as well as the perpetrators of iniquities,[3] and the focus has shifted to an appropriate memorialization of these victims. Weindling published a first approach toward an overview of all victims and survivors of Nazi experiments in 2015.[4] Among others, the biographies of psychiatric patients have been investigated,[5] and the exploration of the lives of NS victims whose bodies were used for anatomical purposes has begun.

Giving Them Back Their Names

The publication of NS victims' names has been discussed controversially within the disciplines of psychiatry and anatomy.[6] An argument against identification is the potential suffering of surviving members of victims' families; however, it seems particularly important to give a story to all the victims who have remained anonymous since their bodies were used for anatomical teaching and research during the Third Reich. Only with a name and a biography is it possible to make these persons visible as individuals with full lives and hopes for a future they were denied. Even in the case of the "euthanasia" victims the problems of "naming victims' names" are not insurmountable any longer[7]. With respect to the NS victims in anatomy, the archival laws on statutes of limitations for publication of personal data usually do not apply any longer, as these persons died more than sixty years ago.[8] Authors concerned with the victims of Stieve's activities in Berlin have handled the problem of naming the victims in various ways. Oleschinski[9] referred to several of the women by name, while Zimmermann[10] refrained from naming victims with the exception of two, thereby following wishes of family members. The contentious issue of the publication of names is compounded in the cases of persons who had been convicted of activities considered criminal and deserving the death penalty under NS legislation. While it is now seen as highly honorable to have died as a member of a dissident group such as the *Rote Kapelle*, relatives of other victims might not want to know or have it published that their loved ones were convicted for murder or arson. Therefore the reasons for the death penalty will here be connected only with names in individual biographies. So far, publications with the names of victims in anatomy have not elicited any negative response, such as historian Ralf Forsbach's list of execution victims in Bonn,[11] Lang's naming of August Hirt's victims,[12] or the names on Stieve's list.[13] On the contrary, publishing the names and stories of NS victims dissected by Stieve resulted in a wide and positive echo from the media.[14] The identification of these victims and the restoration of their full biographies is necessary to remember their individuality, to aid the realization of their humanity, to honor their memory, and to acknowledge the iniquities committed against them not only by a criminal regime but also by members of the scientific community.[15]

Provenance of Bodies for Anatomy: Tradition and Change

All German anatomical departments used the bodies of victims of the National Socialist regime for education and research during the Third Reich (see appendix, table 3). Bodies came from traditional legal sources: hospitals, communal retirement homes and nursing facilities, psychiatric institutions, suicides, prisons, and execution sites. Very rarely were bodies donated, such as those reported for Göttingen and Vienna.[16] The traditional sources changed under the influence of the NS regime in that they included increasing numbers of NS victims. Benigna Schönhagen, author of the Tübingen study, estimated that two-thirds of all bodies used for anatomical purposes during this time stemmed from what she called "a context of injustice."[17] Of the 623 bodies delivered to the Tübingen anatomical institute during the war alone, 429 were victims of the NS system: ninety-nine had been decapitated, forty-four hanged, and six executed by shooting.[18] Schönhagen also counted among the victims those whose lives were destroyed by the living conditions in camps, prisons, psychiatric hospitals, or other communal facilities. In Tübingen, as well as in many other anatomical departments where body registers were still existent, it was relatively easy to identify the executed by their cause of death, as they were listed as "decapitated," "shot," or "hanged." However, in the case of those destroyed by their living conditions, the cause of death was generally given as a "natural cause," for example as an infectious disease like "pneumonia," or as "cardiac failure," or as "unknown."[19]

The exact number of bodies delivered to anatomical departments during the Third Reich is not known. A first estimate can be calculated from the information listed in table 3 in the appendix, which summarizes the currently available information on the body supplies of anatomical departments at German Universities from 1933 to 1945. However, it should be kept in mind that even in those institutes where the body registers still exist, these books do not always reflect the whole truth—for example, in Würzburg, the bodies of the "euthanasia" victims were not listed in the register,[20] and a study of forced laborers whose bodies were delivered to the anatomy department in Marburg showed that not all of the names were officially registered.[21]

In all, 14,956 bodies were delivered to the investigated departments. Given that these thirteen institutes represent roughly a third of all German anatomical departments at the time, a conservative estimate of the total number of bodies used for anatomical purposes in the Third Reich arrives at 35,000 to 40,000 bodies. The percentage of NS victims within this number is not yet known. Systematic identification efforts so far have concentrated on the executed victims, as they often left a trail of documentation throughout the legal system. It is much more difficult to find information on victims from psychiatric institutions and the various camps, especially in the cases of anatomical departments where body registers no longer exist.

It is easy to discern the changes in the traditional sources of anatomical body procurement between 1933 and 1945 if one considers the sources themselves. Psychiatric hospitals have delivered bodies of their unclaimed dead to anatomical institutes for centuries. In times of crisis, such as WWI, psychiatric patients were among the vulnerable groups who suffered most and perished in great numbers.[22] However, under the NS regime they became victims of an unprecedented organized killing campaign, the so-called "euthanasia" program, in which malnutrition, neglect and over- or undermedication contributed to their deaths.[23]

The results of this program affected the delivery of bodies differently from one anatomical department to the other. Those in Tübingen and Marburg (1939–41) and Jena (1940–42) experienced an increase in the delivery of bodies from psychiatric institutions.[24] These numbers, which included children, dropped off later in Jena, after the dissolution of the Blankenhain psychiatric hospital in September 1940. More than 200 patients were transferred from there to Sonnenstein/Pirna, where they were murdered.[25] A similar development may have arisen at other German anatomical departments, e.g., Göttingen and Berlin, which saw an overall decline in the delivery of bodies from psychiatric hospitals during WWII.[26] There were no records of "euthanasia" victims delivered to the anatomy department in Halle, despite the fact that the central extermination facility of Bernburg was close by.[27]

An unusual situation ensued in Würzburg, when in 1941 or 1942 Werner Heyde, a leading organizer of the "euthanasia" program, asked Curt Elze, chairman of the anatomical department, whether the body supply for anatomy was sufficient. When Elze replied that there was a lack of bodies due to decreased delivery from psychiatric hospitals, Heyde promised help. Soon after, the anatomical institute secretly received the bodies of eighty "euthanasia" victims from a psychiatric institution. These victims showed obvious signs of carbon monoxide poisoning, and their names were not entered in the body register.[28]

The delivery of bodies from prisons, GeStapo and police custody and other detention camps increased as the reasons for imprisonment escalated during the NS period from merely criminal offenses to political ones, and prisoners were treated violently. The detailed study from Würzburg revealed that the number of bodies from prisons rose in an unprecedented fashion between 1941 and 1944, and these bodies showed clear indications of violence. They also included persons killed by the GeStapo.[29] In Jena the situation was similar, if not quite as pronounced, from 1938 to 1942.[30] The direct delivery of bodies by the GeStapo or the police was reported for many other anatomical departments (see appendix, table 3). A completely new source for anatomical body procurement was the extensive network of prison camps developed by the NS regime. These facilities included concentration camps, prisoner-of-war camps, and forced-labor camps, and they existed in great numbers and a variety of sizes. Because there were so many of them[31] one or more camps likely existed within the geographical procurement area of each anatomical institute. The camps shared harsh living conditions, nutrition levels equaling starvation, and excessive violence toward the inmates that often resulted in premature death, including murder by camp officials, and executions without legal proceedings.[32] All detailed studies of anatomical institutes show that they received bodies from at least one of the various types of camps, with the possible exception of Vienna, for which no documentation of body procurement from camps exists. Rothmaler found evidence for the delivery of 666 bodies of inmates from the concentration camp of Neuengamme to the anatomical institute of Hamburg.[33]

The bodies of those who had committed suicide represented another traditional source for anatomical dissection.[34] Only the studies of Jena, Innsbruck, and Marburg specifically mention suicides as a source,[35] but it has to be assumed that the other anatomical institutes did not change their traditional acquisition patterns and also received these bodies. The population of persons committing suicide changed during the Third Reich. While overall numbers of suicides did not increase during the war years, as documented for Berlin (with the exception of 1945, when overall suicide rates soared), the percentage of Jewish citizens among all those ending their own lives in Berlin rose: in 1941 one out of ten persons committing suicide was Jewish, and in 1942 six out of ten suicides were committed by Jews.[36] These suicides came in the wake of the increasing number of deportation orders for Jewish citizens to concentration camps. Similar developments have been reported for Vienna.[37] But even before the war Jewish citizens committed suicide in reaction to discriminatory and often violent treatment by officials and colleagues.[38] Prisoners who faced inhumane conditions in camps or prisons also committed suicide at a high rate, as exemplified by Tadeusz Wesolewski, a twenty-year-old forced laborer who killed himself directly after his second commitment to the Breitenau work camp. His body arrived in Marburg on 23 September 1941 to be used for a dissection of muscles.[39]

Following the increasingly harsh and unjust legislation of the NS regime, execution numbers soared, especially during the war. A third of all civilian executions took place in the facilities of Berlin Plötzensee and Brandenburg-Görden, more than 1,400 in Dresden, and 1,381 in Munich.[40] A total number of bodies of the executed then delivered to anatomical departments is unknown. The studies listed in table 3 of the appendix document the anatomical use of 3,887 bodies of the executed. The true number will prove to be much higher, as for Berlin alone potentially 3,000 bodies may have been transported from Plötzensee to the anatomical department. The institute also had access to bodies from Brandenburg-Görden.[41]

The Victims

The change of body supply for anatomical departments during the Third Reich was twofold. It was quantitative, as the amount of bodies increased considerably for most institutes during the war, and it was qualitative, as an increasingly higher number of victims of the NS regime were among the bodies delivered. While Jews as defined by NS laws were the primary targets of the regime, many other groups of people were persecuted too.[42] Previous historical analyses of victims of the Holocaust proposed categorizations of victims into "those who were targeted because of what they did" and those "because of what they were".[43] This approach attempts to separate persons pursued for their so-called "race" and supposed biological make-up—e.g., Jews, Sinti and Roma, Eastern Europeans, psychiatric patients, and homosexuals—from those who actively opposed the regime for political and other reasons. The reality, however, is more complex, as it includes, for example, Polish and German Jewish political dissidents and others who belonged to more than one of the categories. Documentation exists for the transport of bodies of members from all groups of victims to anatomical departments, except for Sinti, Roma, and homosexuals.[44] However, persons from the latter groups could have been among the unclaimed bodies.

First inquiries into the potential continued use of anatomical specimens stemming from NS victims focused initially on the remains of Jewish citizens. Studies since have shown a much wider spectrum of NS victims in anatomy.[45] As Jews suffered from increasing persecution, their percentage among the general population decreased, at first due to emigration, and later following transportation to extermination camps, where they were murdered and their bodies destroyed. Thus Jewish citizens became less often part of the traditional sources for anatomical body procurement, with notable exceptions. One of them was the 1943 transport of eighty-six Jews from Auschwitz to the concentration camp Natzweiler/Struthof, where they were killed and delivered to the anatomical department of Strasbourg on the orders of anatomist and SS-officer Hirt.[46] Bodies of Jewish citizens were also delivered from concentration camps in or near Posen to the anatomical institute, where the severed heads of these persons, together with those from executed Polish political prisoners, were taken, and plaster casts of the heads as well as the skulls themselves were sold to the anthropological collection of the Vienna Museum of Natural History.[47] In Berlin, female members of the "Baum" group in Berlin, an alliance of mostly German Jewish political dissidents named after its leader Herbert Baum and associated with the *Rote Kapelle*, were executed at Berlin Plötzensee and their bodies used for research by Stieve.[48] In Bonn, one of the 191 executed persons was identified as Jewish,[49] and in Vienna seven Jewish citizens were among the executed persons delivered to the anatomical institute.[50]

The political dissidents who died in prison or were executed were a heterogeneous group in terms of their nationality and motivation. Among them were Germans, Poles, Czechs, Belgians, Austrians, and French citizens. The national makeup of the prison population differed depending on the geographical location of the facility and the status of the ruling court. A sizable number of NS victims were imprisoned or executed because of dissent based on religious reasons. Among them were Christians of Catholic and Protestant faith,[51] and also Jehovah's Witnesses.[52] The latter saw "themselves as citizens of another state and members of another army" and thus refused to swear allegiance to Hitler.[53] The NS regime sentenced 253 Jehovah's witnesses to death, and 203 of these sentences were carried out, eighty of whom were male conscientious objectors tried by military courts.[54] The bodies of four female believers were dissected by Stieve.[55] Prisoners of war made up another large group of victims. In 2013, Russian prisoners whose bodies were delivered to the anatomical department in Strasbourg could finally be identified.[56]

The groups of conscientious objectors and deserters were also heterogeneous. The political spectrum included pacifists and social democrats, and the religious backgrounds spanned from Jehovah's Witnesses to Quakers, Catholics, baptists, and Seventh Day Adventists.[57] Information on objectors and deserters is often only sporadic, as they were defamed as "traitors" for a long time after the war. This stigma was only removed in 2009, when Germany and Austria published official statements and legally rehabilitated these men.[58]

One of these deserters was Austrian Franz Dollnig.[59] Born on 21 July 1923 in the Styrian city of Rosenthal/Voitsberg, Dollnig had to join the Hitler youth like all other Austrian boys following his country's annexation in 1938. He was then drafted into the military and trained as a machine gunner. When he received the order to move with his company to the front in late 1943, he decided, against his family's advice, to desert. He explained to his aunt: "How can these people be enemies, we have never met them before. [...] Nobody has attacked us. Why should I kill 3000 human beings per day?"[60] The aunt believed his decision was based on his Christian faith, as Dollnig had no political agenda. He was able to hide in the forest with his family's aid, but his brother-in-law finally denounced him. Dollnig was imprisoned on 21 November 1943, sentenced to death on 23 December 1943, and executed on 12 May 1944 in Graz, and his body was transported to the anatomy department at the university there. The Dollnig family spent 230 *Reichsmark* for an urn with his ashes, but after the war they discovered that these were not his remains. During an investigation into the activities of Anton Hafferl, who was chairman of anatomy in Graz, authorities uncovered that the bodies of forty-four resistance fighters, among them Dollnig's, had been buried in an anonymous mass grave in 1946 without anyone notifying their families.

Another changing source of bodies for anatomical purposes during the Third Reich was the prison population. "Criminals" now constituted not only thieves, embezzlers, and rapists, etc., but also an increasing number of persons who had violated the new and expanding NS legislation. New "major crimes" reached from membership in leftist parties over listening to enemy radio, robbing cars on highways, black-marketeering, looting, mail theft, so-called "racial" defilement, and speaking deprecatingly about the NS regime, to aiding and abetting the enemy and high treason.[61] All of these crimes could lead not only to imprisonment but also to a death verdict and execution. In the Weimar Republic the only crimes demanding capital punishment had been murder and high treason.[62]

Forced laborers comprised another new source from which anatomical bodies were procured. Between 1939 and 1945, 13.5 million persons were forced to perform labor for the NS regime. Of those, 8.4 million civilians were employed in the private industry, including farming; 4.7 million were prisoners of war; and 1.7 million were inmates of concentration camps. Among the slave laborers were 4.7 million citizens of the USSR, 2.3 million from France, 1.9 million from Poland, 1.4 million from Italy, and a number of citizens of the Baltic states, Belgium, Bulgaria, Denmark, Greece, the UK, Croatia, the Netherlands, Switzerland, Serbia, Slovakia, the Czech Republic, Hungary, and many other countries.[63] Here the term "forced laborer" is used for the group of *Fremdarbeiter* (foreign laborers) who worked in the civilian industry, to keep them distinct from prisoners of war and inmates of concentration camps.[64]

Early on some of these laborers had been recruited on a voluntary basis, whereas later most were rounded up in their home countries and deported to Germany and its occupied territories.[65] Forced laborers had a high mortality rate because of the bad hygienic and nutritional conditions in camps, the lack of safety in their workplaces, and the violence they had to endure from their guards. This was especially true for children born to female laborers, who often died in their infancy due to starvation and lack of medical care.[66] There are also reports that hospitalized children of forced laborers were intentionally given spoiled food to hasten their deaths.[67] A study from Freiburg documented that the bodies of ten Russians, nine Poles and one Romanian had been delivered to the Freiburg anatomical department, among them fifteen children. One of these children was Anatoli Morudowa, whose mother, a forced laborer working in Emmendingen, had gone into childbirth after a blow to her abdomen and died three days after his birth.[68] In Göttingen, three children of forced laborers were delivered to the anatomical department in 1944, in addition to over seventy adult laborers who had been executed at the Wolfenbüttel execution site between 1937 and 1945.[69] In Tübingen, the bodies of sixty-four foreigners were sent from forced laborer camps to the anatomical department.[70] Eight women and two men on Stieve's list were forced laborers executed at Plötzensee.[71] And for the anatomical institute of Marburg there is documentation on the delivery of bodies of twenty-four forced laborers, mostly from the work camp in Breitenau.[72]

History of the Identification of Victims

The identification of the bodies of NS victims received at anatomical departments can be grouped in three distinct phases since WWII. Phase one lasted from 1945 to 1948 and followed inquiries by government authorities and families, and phase two in 1989 came after persistent probing from students and historians outside the field of anatomy. Since 2000 anatomists have increasingly spearheaded a third phase of investigations into their own history, often in collaboration with historians.

During phase one, between 1945 and 1948, many people were seeking to find out the fate of their loved ones, and this quest frequently led them to anatomical departments. In addition, the military governments in the four occupation zones, as well as German organizations, were trying to locate the bodies of foreign nationals and political dissidents. Anatomists at many universities were asked about the identity of the bodies remaining in their institutes' storage spaces.[73] The discovery of such bodies by the Military Forces liberating these cities led to detailed investigations by the new authorities in Posen, Danzig, Greifswald, and Strasbourg.[74] In many cases it was impossible to identify the bodies due to loss of records or advanced dissection status.

The Tübingen study recorded in great detail several lists drawn up by the anatomical institute for various authorities during the first years after the war. Among them was an inquiry from 21 September 1945 by the attorney general in Stuttgart demanding information on the fate of persons executed for political reasons and ordering their respectful interment. Most of the bodies of the executed had been buried already, but some remained at the institute.[75] In 1946–47 a "list of foreigners" was created for the French occupational authority with the names of 273 foreign nationals, mostly Russians and Poles, whose bodies were used for anatomical purposes. Another list of Polish prisoners executed at camp Welzheim was assembled following a demand by the US military forces. In 1947 the anatomical institute was asked to inform the French authorities about all the bodies that were still held at the department. It appears that even after previous identification many of these bodies were still used for anatomical purposes until 1947 and 1948. As late as 1952, the Netherlands Tracing Mission located the remains in Tübingen of a Dutch citizen named Mr. Moonen and had them buried.[76] Interestingly, the Tübingen institute shared the stored bodies with its French medical colleagues, as eleven bodies were handed over to the French military hospital #421 for surgical training in 1947.[77] Certain specimens derived from the bodies of NS victims stayed in the anatomical collection for several decades longer.[78]

In Vienna, the anatomical institute was partially destroyed by a bomb on 7 February 1945, but the storage facilities for bodies were left intact. Between 1945 and 1946, assistant Franz Bauer, who was left in charge of the institute, organized for hygienic reasons the interment of at least 216 coffins filled with anonymous unembalmed body parts at the Vienna Central Cemetery. However, the storage facilities were not completely emptied, and bodies of NS victims were kept at the Vienna anatomical department for several years thereafter and used for teaching purposes.[79] In July 1945 Hanika, a resistance fighter and foster mother to the orphaned children of executed political dissident Dr. Jacob Kastelic, heard that 250 bodies of the executed were still held at the anatomical institute.[80] She searched for members of the resistance groups to which Kastelic, Karl Lederer, and Roman Karl Scholz belonged,[81] but it took until October 1945 before Bauer allowed access to several storage containers with bodies and heads of victims.

Other groups had also asked for the surrender of bodies of political prisoners, and several bodies of NS victims stayed at the institute until their relatives discovered their location and claimed them.[82] In 1952 von Hayek, the new director of the institute, noticed the continued presence of twenty-eight bodies of executed NS victims and had them buried over the next five years.[83] All names of the persons interred were known, but some of the bodies were buried in the manner customary for the anatomical department, that is, in anonymous graves at the Vienna Central Cemetery, whereas others were interred in marked locations. When von Hayek was criticized for the anonymous burials in 1962, he countered that nobody else had claimed these bodies before he finally took care of them.[84] A similar situation existed in the two other Austrian anatomical departments.

In Innsbruck the French occupation authority sent official inquiries concerning the bodies of victims to the anatomical institute in 1948.[85] It was discovered that when Gustav Sauser took over as new chair of anatomy in 1946, thirty bodies from the NS period had still been in storage and were used regularly for teaching purposes until February 1948, when thirteen bodies were found in the vats. Former chair Felix Sieglbauer declared that he had always seen these bodies as "scientific material" and never considered them to present "human remains."[86] The French authorities could not find grounds for criminal prosecution, but Sieglbauer was reprimanded. Some of the bodies were interred.[87] In Graz it took the efforts of the Styrian regional association of political dissidents to uncover the fate of some of their friends in August 1946.[88] They discovered that chairman of anatomy Anton Hafferl had ordered the secret interment of forty-four bodies from the NS period earlier that year. The following exhumation of the remains allowed the identification of only twenty-one of the bodies. Hafferl was imprisoned but released after claiming that he had informed the authorities about these bodies as early as May 1945. However, he had not provided this information to families and friends who had been searching for loved ones. Many of the victims from Graz and Innsbruck have been identified by name through the work of Herwig Czech.

For Bonn, Forsbach's detailed list of names of executed persons whose bodies were used for anatomical purposes showed burial dates in a majority of cases, many of them from 1946 and 1947.[89] In 1947 the anatomical institute of Bonn produced a list for the German civil registry office with the names of dead foreigners.

The medical historian Thorsten Noack reported that the anatomical institutes in Bavaria were inspected by a commission of the United Nations Relief and Rehabilitation Administration (UNRRA) after the war in order to locate and identify missing foreigners. However, this search did not include German NS victims.[90] In 1946 Philipp Auerbach, the Bavarian federal commissioner for the "racially, religiously and politically persecuted," received an anonymous tip about the presence of bodies of German resistance fighters at the institute in Würzburg. A search revealed not only the remains of the Munich communists Engelbert Kimberger, Otto Binder, and Wilhelm Olschewski,[91] but also those of three executed persons from Austria. The men from Munich were interred and honored with a public memorial service on 14 September 1947.[92] In Erlangen, Karl Bauer, the new director of the anatomical institute, informed the authorities on 11 September 1947 about the continued presence and anatomical use of bodies of the executed from the war. He also mentioned that due to the dissection process these bodies could no longer be identified. Neither the American nor the German authorities objected to this argument.[93] In Munich, bodies of persons executed in NS times were used for anatomical dissection at least until the 1947–48 winter semester, when they made up most of the dissection "material" for students. This only ended after Auerbach informed the Bavarian government and the UNRRA commission. UNRRA then confiscated twenty-two bodies in October 1947.[94]

In Heidelberg a list of 130 names was put together directly after the war registering persons whose bodies were delivered to the anatomical institute from places other than hospitals and other care facilities.[95] In 1945 the remains of twenty-seven executed persons were removed from the institute and buried in a collective grave at the cemetery *Bergfriedhof* in Heidelberg at the behest of the director of the anatomical institute. A memorial plaque was installed bearing a general inscription, but without names of individuals.[96]

The anatomical institute at the University of Danzig had been abandoned by Spanner, who had escaped from the advancing Soviet military with the help of his technician Eduard von Bargen on 30 January 1945.[97] The institute was first inspected by the new Polish authorities in March 1945.[98] The writer Zofia Nałkowska was part of the Main Commission for the Investigation of German Crimes in Poland. She gave a vivid description of her impressions of the victims' bodies during a second visit with the Polish commission at the anatomical institute in May 1945, days after an inspection by Soviet experts, among them the chief forensic physician of the Second Belorussian Front:[99]

> We entered the gloomy spaciousness of the basement first. [...] Two vats contained only decapitated, shaved heads piled one on top of the other—human faces like potatoes poured onto the ground, some on their side as though they were resting on a pillow, others facing down or up. They were yellowish, smooth, perfectly preserved, evenly severed at the neck, as if they, too, had been cut from stone.[100]

Among the bodies delivered to the anatomical department had been those from the psychiatric institution in Conradstein (Kocborowo), from the prisons in Elbing (Elbląg), Königsberg, and Danzig, some bodies of homeless persons, executed Poles, and ethnic Germans, as well as two or four bodies of inmates from the concentration camp in Stutthof.[101] From 1940 on the German director of the psychiatric hospital in Conradstein, Waldemar Schimansky (later, Siemens) from Danzig, sold the bodies of psychiatric patients for nine *Reichsmark* each to Spanner.[102] Apparently this was a rather low remuneration, as an official from the local NS administration pointed out that anatomical institutes in the *Reich* paid up to twenty *Reichsmark* per body.[103] Spanner actively lobbied for the bodies from psychiatric institutions, as letters from 1941 show.[104] In October 1941 he also signed the death certificates of thirty Polish citizens from Bydgoszcz, who had been executed by firing squad because of dissident political activities on 19 May 1941. It is highly likely that these bodies were delivered to the anatomical department.[105]

During the postwar inspections at Danzig, the remains of 147 persons were found, among them eighteen women, four children and 126 men. Eighty-two bodies were without heads. Autopsies were performed to ascertain cause of death, but there is currently very little information on the identity of these bodies. The remains were buried at the Holy Trinity Cemetery close to the anatomical institute in June 1945.[106] However, this was not the end of the immediate postwar history of the anatomical institute in Danzig. During the May inspection, Nałkowska noticed that "pieces of rough, white soap and a pair of metal molds stained with dried soap lay on a high table".[107] The inspecting committee's conclusion that this was soap made from human fat led to the rumor that the NS regime manufactured soap from Jewish bodies in concentration camps. This rumor entered the "collective memory of the Polish people"[108] through Nałkowska's short story "Profesor Spanner," written in 1945 and first published in 1946.[109] During several interrogations between 1946 and 1948, Spanner conceded the production of small amounts of soap from human fat for anatomical purposes, but was not prosecuted. All other evidence supports the fact that the alleged industrial production of soap from human fat, specifically the fat of Jews, never happened.[110]

After the war press reports revealed the continued existence of bodies and body parts from the executed of the NS period at the anatomical department of the Charité in Berlin, which led to the interment of these remains. Until then the bodies had still been used for educational and research purposes.[111] Stieve had kept urns with the ashes of dissected persons in the institute. At least some of them must have been identifiable, as he returned them to the families and handed a group of forty-two urns over to the French authorities. Apparently eighty urns remained unclaimed and were buried during the dedication of the memorial for the execution site Plötzensee at the Altglienicke cemetery.[112] In 1946 Stieve also identified the names of 182 persons whose bodies he had used for his research.

In Kiel, a first inquiry into the fate of bodies of the executed was launched by the "committee of former political prisoners" and relatives in July 1945.[113] On police orders, Enno Freerksen had the remains of bodies from the destroyed anatomy building buried in an anonymous mass grave at the Eichhof Cemetery in June 1945, shortly before his imprisonment by British occupying forces. All relevant documentation on the identity and disposal of the bodies was lost during the war. Bargmann responded to the relatives' questions by pointing to this lack of information and did not actively support outside efforts to investigate this issue.[114] It took an exhumation of the graves in 1947 to identify some of these victims. The remains were subsequently reinterred in a grave of honor in Hamburg. Inquiries by relatives and French authorities continued into the 1950s.[115]

The anatomical department in Marburg also received bodies of NS victims, but there is no information on any identification efforts directly after the war. At this point it is only known that any "material" left over from war times was collectively and anonymously cremated in 1948 and 1953.[116] Similarly in Hamburg, new chair of anatomy Helmut Ferner oversaw the cremation of bodies and body parts remaining from NS times at the end of 1946. There is no information on who requested these cremations.[117]

In November 1944 in Strasbourg, the advancing Allied Forces discovered the remains of at least 150 bodies in the storage facilities of the anatomical institute, sixty of which belonged to Soviet prisoners of war from the detention camp in Mutzig. The French authorities investigated in December 1944, and performed autopsies on the remaining bodies, but further identification seemed impossible.[118] Compounding matters, the body registers from the anatomical institute were missing. It was only known that in August 1943, SS sources delivered eighty-six bodies under great secrecy. The unidentified bodies were interred at the Strasbourg North Cemetery on 23 October 1945 and later transferred to the Jewish cemetery in Cronenbourg, a suburb of Strasbourg.[119] Tissues from prisoners of the Natzweiler-Struthof concentration camp were not all buried in these graves, but they were also used for microscopic studies, and histological slides were discovered after the war.[120] More tissues from Hirt's victims were found at the Forensic Institute of the University of Strasbourg in 2015.[121]

The liberators of Posen found bodies of forty-eight NS victims in the storage spaces of the anatomical department. Autopsies and forensic studies revealed clear signs of violent abuse, torture, execution by hanging in thirty-seven victims, and removal of gold teeth.[122] There is no information on the fate of these bodies after the autopsies; it is also unclear if identification was attempted or if the bodies were buried in a mass grave. The porter of the Posen institute, Jósef Jendykiewicz, reported after the war that there were women among the victims and many showed signs of intense physical abuse, including torture by electrocution.[123]

In Greifswald the Soviet military performed detailed forensic autopsies on the bodies located in the basement of the anatomical institute, and thirty-two of sixty-nine remains were identified. Twelve belonged to Soviet citizens and twenty to Poles. In thirty cases death was due to starvation, while twenty-eight persons had been hanged. The bodies were interred at the regional cemetery of the Red Army number 3 in Greifswald.[124]

After the first phase of identifying NS victims, anatomical departments were only sporadically approached on this topic until the public debate on the history of medicine in the Third Reich in the 1980s. Following the inquiry by the *Kultusministerkonferenz* in 1989, all institutes had to react, effectively starting a second phase of investigations. The University of Tübingen formed an external commission to investigate the specimen collections of all university institutes for the remains of NS victims, thereby launching the first systematic study of an anatomical collection. The commission was headed by the distinguished legal scholar Albin Eser, then director of the MPI for Foreign and International Criminal Law in Freiburg. The report of the commission,[125] which included a self-study by the anatomical institute, revealed that many specimens potentially hailing from the bodies of NS victims had been removed from the collections during the first years after the war. However, several microscopic slides and macroscopic specimens were still present at the anatomical institute in 1990 and some other departments. The committee recommended the removal of all identifiable anatomical specimens and all remains with dates from 1933 to 1945 and their burial in a grave of honor. Interestingly, the commission noted that it discerned a change in medical ethics during the second half of the twentieth century, and requested anatomists reflect on their research ethics concerning body procurement and the development of guidelines in this area.

In Heidelberg, an evaluation of the anatomical collection in 1989 revealed several embedded pieces of tissue for microscopy originating from NS victims as well as a skull, which were interred in the grave of honor at the *Bergfriedhof*. At the anatomical institute of the Charité in Berlin no specimens from NS victims were found in 1989. However, it was noted that several years before, an unnamed director had removed histological specimens and slides, which might have held tissues from NS victims.[126] In 1990 a student initiative in Hamburg led to the investigation of anatomical body procurement from 1933 to 1945 based on archival documents, including those from the concentration camp Fuhlsbüttel. Evidence was found for the delivery of hundreds of bodies of NS victims from the concentration camp and execution sites to the anatomical institutes, but there was no report on identification efforts.[127] In addition, the anatomical collection was investigated[128] and scrutinized for specimens from the NS period. Seven wet specimens of unclear provenance were removed and buried, according to Professor Emeritus Adolf-Friedrich Holstein.[129]

In Munich several inquiries from local, national, and international organizations took place from the 1970s to the early 1990s, and specimens of unclear provenance were removed during these undocumented investigations.[130] Anatomical specimens with possible NS connection were also removed in Erlangen and Göttingen. No specimens from NS victims were reported from Munich, Würzburg, Marburg, Rostock, Greifswald, Bonn, Cologne, Münster, Leipzig, Halle, Kiel, Hamburg, and Jena.[131] Freiburg, Giessen, and Frankfurt referred to the destruction of their anatomy buildings, assuming that remains from NS victims could not have existed any longer after the war.[132] Later investigations in Jena and Giessen revealed that this supposition was incorrect, as specimens from the NS period were indeed present for many years after the war.[133]

In a parallel development, Austrian universities—especially the anatomical institute of Vienna—came under international scrutiny with the realization that images from Eduard Pernkopf's atlas might be based on the bodies of NS victims.[134] The Vienna Senate Project investigated the anatomists and the body-procurement process at the anatomical institute during the NS period from 1938 to 1945, collecting the number of bodies received as well as the number of bodies of executed persons.[135] They noted the names of the executed, the reason for the death verdict, and information from all other bodies from traditional sources, including birth and death dates.[136] Authorities scrutinized other institutes and found specimens from NS victims, among them ninety-eight histological slides at the department of histology and embryology. They were all buried in a grave of honor provided by the city of Vienna.[137]

A special situation existed at the Museum of Natural History in Vienna, where investigations into the remains of NS victims in the anthropological collections had started in 1991, several years before the Senate Project, and four years after Aly's report on the connection between Posen and the Viennese museum.[138] Skulls and death masks of Jewish persons from the anatomical institute in Posen as well as skeletons removed from the Vienna Jewish Währinger Cemetery were returned to the Jewish community organization in Vienna in 1991. The skulls of Polish citizens acquired from Posen remained at the museum until 1999, when they were repatriated to Poland.[139] Other specimens from large anthropological studies, including, for example, photos, masks, and hair samples, as well as documentations of expert witness statements on heredity, were discovered in the museum's storage areas.[140] Further detailed investigations of the history of these specimens followed, as well as discussions on the ethics of keeping them within the museum collections or returning them to their places of origin. Several victims were identified.[141]

While established German anatomists still discouraged their colleagues from investigating NS history in the 1990s, the situation changed after the turn of the century in a third phase of studies. Possibly due to the internationally well-received efforts in Tübingen and Vienna, but probably also because of the advent of a new generation of German chairmen of anatomical departments who held no special loyalties to anatomists active during the NS period, systematic studies of anatomical institutes and individual anatomists became more frequent. In addition to the investigations mentioned above, more or less comprehensive studies are now available for many universities (see appendix, table 3), and more are under way.

The goal of most of these studies clearly lies in the search for remaining anatomical specimens from NS victims and their subsequent removal. The elucidation of the fate of individual victims has not been a priority so far. Generally, body registers contain the names of persons whose bodies were delivered and can be used for identification of NS victims, but these books are often missing or incomplete. Full identification is easiest for the group of executed victims because of the extensive administrative documentation surrounding these cases.[142] Their fates can be traced through various archival sources that include, apart from body registers, court documents, municipal records, and federal and university archives.[143] Thus Forsbach published a list with the names of all executed persons whose bodies were used at the anatomical institute in Bonn,[144] which is an exception among all departmental studies. In Jena and Vienna, the names of executed victims were found but not published at the time for privacy reasons.[145] Julia Buddecke went further in her study of the anatomical institute in Kiel, which includes not only the profiles of anatomists, but also reconstructions of biographies of three victims: twenty-nine-year-old Polish forced laborer Anezka Klofácová, who was executed on 24 March 1943; twenty-three-year-old Czech forced laborer Wenzel Vaska, who was executed in May 1943; and forty-seven-year-old German-Czech Frank Wollitzer, who was executed on 18 October 1943. Klofácová and Vaska had been accused of theft, and Wollitzer of black marketeering. Their bodies were delivered to the anatomical institute in Kiel without notification of the families. Buddecke's study of these three biographies is one of the first examples of a new approach to the memorialization of NS victims.[146]

New Approaches: Victim Identification

A new era of comprehensive investigation and documentation of the identities of NS victims in anatomy has begun with the work of Hans-Joachim Lang. He moved beyond the traditional anonymity of bodies in anatomical body procurement and started with the systematic reconstruction of names and biographies of NS victims from often fragmentary information from a wide variety of sources, including newly accessed archives. After noticing that the historical accounts of medicine in National Socialism left the victims in the background and nameless, he investigated the biographies of the eighty-six persons who became victims of Hirt's research plans and gave them back their names. The French authorities had declared the remains of these Jewish prisoners, found at the Strasbourg institute, to be unidentifiable, but in Mitscherlich and Mielke's report it was noted that Henry Henrypierre had recorded the numbers of the prisoners' tattoos from Auschwitz when they arrived at Strasbourg.[147] Lang found a copy of Henrypierre's note in the archive of the US Holocaust Memorial Museum, and from this initial set of data he launched a laborious quest for information on the lives of Hirt's victims, who had been selected by anthropologists at Auschwitz and then sent to the Natzweiler-Struthof concentration camp for their murder.[148] Hirt's victims came from all corners of Europe and ranged in age from the young teens to persons in their sixties; twenty-nine were women and fifty-seven were men. Their biographies are now accessible through an interactive website that promotes the further collection of data.[149]

Among the stories is Frank Sachnowitz's. He was born in Larvik, Norway, the youngest of seven siblings. His father had planted an apple tree to honor the birth of each of his children, and one in memory of their deceased mother. Although the Germans occupied Norway in spring 1940, the 2,000 Norwegian Jews were not persecuted until January 1942 when they had to mark their clothes with the letter "J." Life gradually worsened for them. In October 1942 all male Jews were taken prisoner, including the men of the Sachnowitz family. Ultimately all of them, including the women, were put on a transportation ship to Germany in late November 1942 and on arrival in Germany transferred to trains that brought them to Auschwitz on 1 December. There Frank's father and all but one of his siblings were murdered. His brother Hermann survived and immigrated to the United States. Many years later Hermann Sachnowitz still remembered vividly Frank's longing for their apple trees at home during their conversations in Auschwitz. Hermann saw his brother for the last time on 15 May 1943. Frank was eighteen years old when he was murdered on 17 or 19 August 1943.[150] With his work, Hans-Joachim Lang literally turned "numbers back into people".[151]

Lang's approach of systematic de-anonymization of victims in NS anatomy was also utilized in a study on those victims whose bodies were used by Stieve for his research.[152] Much has been written about Stieve, but little about the people whose tissues he used, despite the fact that the women on Stieve's list were known by name since he created a list in 1946. It contains the names of 182 persons executed at the Berlin Plötzensee, and documentation on persons executed in Berlin is collected at the *Gedenkstätte Deutscher Widerstand* (Memorial Center for the German Resistance). Given that women were in the minority among those executed following verdicts by civilian and military courts,[153] there are only a few studies on this particular group of prisoners. Women made up about 10 percent of the victims from Plötzensee, with the same ratio in Halle; only 4 percent were women in Kiel, and 2 percent in Bonn.[154] Any fates described so far were often those of prominent dissidents,[155] but it is just as important to learn about all other victims. Previous attempts to identify Stieve's victims from his publications did not lead to satisfactory results because of similarities in biographies.[156] Historian Ernst Klee was convinced he had recognized Cato Bontjes van Beek's body as a source of "material" in Stieve's 1946 publication.[157] However, apart from the fact that it seems tasteless and un-ethical to publicly connect names of NS victims with distinct histological images in Stieve's articles, there were at least two other twenty-two-year-old women on Stieve's list who had been imprisoned for the same period as van Beek and whom Klee did not consider.

In 1992 Brigitte Oleschinski was the first to report on a list of names of NS victims that Stieve had prepared for the Soviet military authorities.[158] After the German reunification in 1989, Oleschinski had been able to access archival materials of the German Central Justice Administration in the Soviet Occupying Zone, which are held at the Federal Archives in Potsdam. She found a postwar note by Harald Poelchau, a Protestant minister, who had given pastoral care to the prisoners in Plötzensee during the war and worked as an official for the administration searching for information on executed victims in 1946.[159] Poelchau related a conversation with Stieve from 30 November 1946, during which Stieve had handed him a typed list of names of persons whose bodies he had used for his research, mentioning that the original documents, in which he had noted all names that were known to him, had been burned. Stieve probably reconstructed this new list after the war from his scientific notes, as the body registers of the anatomical department, which would have been the only complete documentation of names, had vanished in 1945. This list is only an incomplete account of the total number of bodies that he used, as it contains mostly names of women, but he also used the bodies of male NS victims for his research, and he mentioned in his writings the names of additional women that were not on the list.[160] In a paper from 1946 Stieve reported results from 421 healthy adults, 188 women and 233 men.[161] While some of these persons may have been victims of bombing raids, it is most likely that the majority were executed NS victims. Several copies of the list exist in the Federal Archives.[162] The original is typewritten and amended with handwritten additions by an unidentified author. Stieve listed the names, ages, and dates of birth and execution of women and men under continuous numbers from one to 182. The names were sometimes misspelled or in two cases exchanged for other names, and some dates were wrong. The corrected data are shown in table 4 of the appendix.

In addition to the bodies, Stieve was interested in collecting biological data from these women, such as status of menstruation at time of death, length of amenorrhea, and number of children.[163] Biographical data were mentioned only insofar as they related to marriage, psychological background, duration of imprisonment, and time of announcement of execution date. He never noted the nationality or "race" of a person. All of this information was potentially available through court and prison records, which were accessible to him.[164] When Stieve was questioned after the war about the sources of his information, he explained that he had been indirectly—through his technician Pachaly—in communication with the staff and prison physician at the women's prison in Barnimstrasse concerning data on the menstrual cycle of the women.[165] Most women were held at Barnimstrasse before their final destination, the execution site at Plötzensee prison.[166] Some of them were transported to Barnimstrasse after longer incarceration in other countries, others were held there for all of their prison time. The prisoner cards from Plötzensee documented in the individual victim files at the *Gedenkstätte Deutscher Widerstand*[167] contain notes made by personnel on individual prisoners at Barnimstrasse,[168] with observations on prisoner appearance and state of mind when faced with a death verdict. For example, twenty-two-year-old Herta Lindner was described as "A delicate blonde, modest, warm-hearted, intelligent, helpful … convinced … brave … in spite of heartbreak,"[169] while an older woman, also convicted for treason, was described as "old, apparently primitive woman. Not responsive due to overwhelming fear. Clung to the prison guard. Was like an animal. Only case of its kind".[170] Stieve used this information in some of his more detailed case histories.[171] Also, the senior prosecutor provided additional documentation on executed persons Stieve was specifically interested in.[172]

Stieve's list contained information on 174 women and eight men. The male victims were numbers two, three, six, eight, ten, eleven, 156, and 168. No information other than names and dates could be found for four of the women, numbers thirty-two, seventy-eight, eighty-two, and ninety-nine. All of their bodies had been transported from Plötzensee to the anatomical institute between 1935 and 1944, with one each in 1935 and 1936, two in 1937, five in 1938, two in 1939, one in 1940, three in 1941, twenty-three in 1942, and most of them in 1943 and 1944, with seventy-two bodies in each year. Data from the Memorial Site Plötzensee show that the majority of all executions took place between 1942 and 1944;[173] the same is true for executions at the Vienna assize courts and the military shooting range in Vienna.[174] In 1945 Stieve reported to have used the bodies of 269 women altogether during NS times,[175] possibly including women whose bodies he used for teaching. Thus Stieve accepted a majority of the estimated 300-

plus executed women[176] for anatomical purposes. Among those executed at Plötzensee were married couples, but only in one case were husband and wife listed by Stieve: Georg and Anna Schwitzer, numbers eight and nine. In all other cases Stieve recorded exclusively the wife's body, such as Veronika Augustinak (number sixteen), Elise Hampel (number sixty-two) and all the married women of the Rote Kapelle and Baum-Gruppe.

The age of the persons on the list ranged from eighteen to sixty-eight years, with a preponderance of women of childbearing age, eighteen to forty-five years. A majority of eighty-seven persons had received death verdicts because of treason and espionage, followed by forty-five convicted for theft, black marketeering, fraud, or looting; twenty-one for subversion of the military or aiding the enemy; fifteen for murder or attempted murder; eight for arson, one for refusal of military oath, and one for providing abortions. While in 1942 and 1943 treason and espionage were the most common reasons for conviction, with fourteen of twenty-three and forty-seven of seventy-two respectively, this changed in 1944 when women were more commonly convicted for property crimes, with twenty-eight of seventy-two, and verdicts for subversion of the military as well as aiding the enemy became nearly as frequent as treason and espionage, with eighteen out of seventy-two. Reasons for death verdicts were similar for persons executed in Vienna and delivered to the anatomical department there, with treason most common between 1942 and 1944, followed by property crimes.[177] Also in Dresden, half of all executions at the Münchner Platz site followed death verdicts for political "crimes."[178] Of the persons on Stieve's list, 115 were German, fifteen French, six Belgian, fourteen Czech, five Austrian, twenty Polish, one American, and one Soviet. These numbers roughly reflect the overall population of persons executed at Plötzensee, half of which were German, with further strong contingents of Czech, Polish, and French citizens.[179]

As was to be expected in a population of women prisoners of childbearing age, some of them were pregnant when they were taken captive. This was significant for the judicial process, as pregnant women were prohibited from being executed according to regulations by the secretary of justice from 19 February 1939.[180] Stieve's list held the names of five women who were known to be pregnant or possibly pregnant. From the evidence in these five cases it seems that the NS judicial system reacted inconsistently in its decisions concerning pregnant prisoners. For both, thirty-four-year-old Hilde Coppi and twenty-year-old Liane Berkowitz, members of the Rote Kapelle, the executions were postponed until after the delivery of their children and some time for breastfeeding.[181] The decision was very different in the case of the first woman on Stieve's list. Twenty-four-year-old Charlotte Jünemann was executed on 26 August 1935 for murder/manslaughter, despite the fact that she was pregnant. Her story had gained some notoriety, because her three children had died due to neglect by Jünemann, while her husband was a patient in a psychiatric hospital. There was speculation in the US press whether the execution was going to be postponed because of Jünemann's pregnancy.[182]

The situation was similar for twenty-year-old Lucienne Tassin, number 106 on Stieve's list. This French forced laborer had been sentenced to death by *Sondergericht VI Berlin* on 24 September 1943 for looting. The notes from prison stated that she had claimed to be seven months pregnant. There was no documentation of a medical confirmation. The observer continued by characterizing her as "childlike, not criminal. Seems to have been a victim of circumstances (highly insufficient salary and provisions). Was, apparently because of pregnancy, immediately executed."[183] So in this case it seems to have been the fact of the pregnancy itself that led to a swift execution on 13 October 1943.

Other women must have believed themselves to be pregnant simply on the basis of being amenorrhoic. After all, it had not been common knowledge at the time that women stopped menstruating under chronic physical and psychological stress—in fact, Stieve's work helped to publicize this insight—and for most women a pregnancy was the only known explanation for a disruption of their monthly bleeding. This may have happened in the case of Elfriede Henkel, number sixty-four on Stieve's list. The forty-one-year-old woman had received a death verdict for repeated theft by *Sondergericht VII Berlin* on 26 May 1943. On 27 May the prosecutor informed officials at Barnimstrasse that Henkel believed herself to be pregnant but that the physician who examined her could not confirm this. However, should a pregnancy indeed exist, he recommended a termination on "hygienic" grounds, as he considered Henkel to be a "useless" person whose progeny was not expected to contribute to society.[184] In her petition for clemency from 28 May, Henkel mentioned her (unconfirmed) pregnancy as a reason for a conversion of the verdict, but she was executed on 25 June 1943.

The only other published case of the execution of a pregnant woman following an NS court verdict is that of Hildegard Trusch. The twenty-four-year-old had been sentenced to death because of looting and was killed at the execution site *Roter Ochse* in Halle. The NS authorities' reaction to the news of her pregnancy was similar to that in Elfriede Henkel's case. A physician recommended the termination of the pregnancy as the child was not viewed as a "valuable asset" to the German people. The official documents registered a miscarriage, but the historian Michael Viebig assumes that an abortion was performed and the term "miscarriage" was used to "keep the books clean".[185] Viebig also has evidence of two other pregnant women whose executions were planned at Halle, one of whom was in the early stages of her pregnancy and may have had an abortion before she was executed, and the other who delivered her child before her execution.[186] Another unnamed and not otherwise documented case of a pregnant woman on death row in Berlin was mentioned in Charlotte Pommer's writings.[187]

The question remains why the NS authorities acted inconsistently in cases of pregnant women on death row. While their age range was the same, their socioeconomic backgrounds and reasons for the death verdict were different. One possibility is that, based on NS racial hygienic discriminatory thinking, the authorities differentiated between obviously well-educated German women convicted of treason and uneducated German or foreign women accused of murder and theft. Roland Freisler, senior judge at the *Volksgerichtshof* during the war, had demanded in his previous position as secretary at the Reich Ministry of Defense in 1936 that a differentiation between groups of prisoners should be made based on their likelihood of reform or permanent asocial status. As a consequence, from 1937 on a criminal-biological service was introduced to help with this differentiation using scientific insights from racial hygiene.[188] The authorities may also have discriminated between the women's offspring. Thus the death verdict was carried over into the next generation in women who together with their children were deemed to be of "inferior value" or "useless" for society. An alternative explanation is offered by Viebig, who believes that the NS authorities' decisions were mostly driven by practical matters: whether they could get away with a quick execution of a pregnant woman without a public outcry, and how much trouble it was to house pregnant women or young mothers on death row.[189]

Most persons on Stieve's list had been convicted by either the Berlin *Volksgerichtshof* or *Sondergerichte* in various cities. Only a few had gone through military trials. Among the political prisoners had been twenty women of the international group *Rote Kapelle*. There were German, Belgian, and French women among these victims.[190] A further distinct group of German political prisoners were seven women of the *Baum-Gruppe*, a resistance group named after Herbert Baum, which had a majority of Jewish members, and the *Steinbrink-Gruppe*, named after Werner Steinbrink.[191] Six women of the latter group were of so-called Jewish descent, as was one of a group of thirteen Czech resistance fighters.[192] Polish women comprised a larger contingent of political prisoners. Fourteen of them worked in the organized Polish resistance,[193] while four others helped Polish and Russian prisoners.[194] In addition, four women were members of the French resistance.[195] Other political dissidents acted outside any organization, among them Otto and Elise Hampel, whose story was fictionalized by German author Hans Fallada in his last novel in 1947.[196] Elise's name is number sixty-two on Stieve's list.

Helene Delacher was born in 1904 in Burgfrieden/Lienz to an Austrian farmer.[197] She and her South-Tyrolian partner Alois Hochrainer joined the Jehovah's Witnesses in 1938. After their imprisonment for religious activity in 1940, Hochrainer had to return to South Tyrol, Italy, while Delacher remained in Austria. She continued to smuggle forbidden religious pamphlets over the border, an act which was considered to be "undermining the military". Delacher was intercepted on one of these clandestine trips on 14 June 1943 and imprisoned. Her trial was staged at the *Volksgerichtshof* in Berlin on 4 October 1943 and she was sentenced to death. Delacher's execution took place on 12 November 1943, and she became number fifty-four on Stieve's list. On 8 September 1999, fifty-six years after her murder, Helene Delacher was officially rehabilitated by the state court of Vienna.[198]

Vera Obolensky, known as Vicky, was active for the French resistance. She was born in Moscow in 1911 and immigrated with her family to France,[199] where she worked as a model and secretary and married another Russian emigrant, Prince Nikolai Obolensky. During the war she was active for the *Organisation Civile et Militaire*, a resistance group that gathered intelligence and tried to free prisoners of war. She was arrested on 17 December 1943 and sentenced to death for treason by a military court in Arras/France in May 1944. Obolensky was executed on 4 August 1944, and appeared as number 153 on Stieve's list.

Irene Wosikowski was also part of the French resistance. She was born in Danzig in 1910, lived in Kiel and Hamburg, and became a member of a communist youth organization. As a secretary for the Communist party, she had to flee Germany in 1934. After time in Moscow and the Czech Republic, she moved to Paris and worked as a newspaper correspondent, were she assisted French resistance groups. In 1940 she was interned with other German nationals by the French authorities in the concentration camp Gurs, from which she escaped and fled to Marseille to continue her political work. She was betrayed by a German informer and taken into custody on 26 July 1943. Despite severe and continued torture by the GeStapo in Marseille and later in Hamburg, she did not give up the names of her comrades. Wosikowski was sentenced to death by the *Volksgerichtshof* in Berlin on 13 September 1944, and executed on 27 October. She appeared as number 179 on Stieve's list.[200]

Among the Polish citizens executed at Plötzensee were the forty-two-year-old Veronika Augustyniak (Stieve list number sixteen) and the twenty-five-year-old Leokadia Zbierska (Stieve list number forty-one). Together with Augustyniak's husband Jozef and six other men and women, they had sheltered the Soviet prisoner of war Fyodor Asarow after his escape from a camp. All of them were put on trial for "aiding and abetting the enemy," sentenced to death, and executed in August and September 1942.[201] Only Augustyniak and Zbierska appeared on Stieve's list, even though the court papers on the Augustyniaks' case clearly stated that Jozef's body, too, had been transported to the anatomical institute.[202] The Augustyniaks were peasant farmers and not affiliated with any resistance organization. They left a nine-year-old daughter behind.[203]

Among the Polish women directly involved with Polish underground organizations were the thirty-five-year-old Miroslawa Kocowa and the twenty-one-year-old Halina Konieczna. Both were trained accountants and worked with the Berlin-Tegel intelligence center of the patriotic Polish National Armed Forces, a unit that was anti–National Socialist as well as anti-Communist in its outlook.[204] Kocowa acted as a courier between Berlin and the central command in Warsaw. The Berlin group was betrayed by collaborators and its members were arrested on 18 February 1942. Twenty-four-year old Wanda Węgierska, a clerical worker, was based in the central command in Warsaw, but was also active as a courier to areas in Germany with large populations of Polish forced workers. She was imprisoned in Lodz in March 1942 and subsequently tortured. Her father and brother died in concentration camps.[205] Henryka Veith, a twenty-eight-year-old actress and member of the National Armed Forces in the unit "Zadoch," was arrested in 1942. Her prison records describe her as a "very delicate, blond, blue-eyed, mature" woman of "warm character," who acted out of conviction.[206] All four resistance fighters were put on trial for high treason and received death penalties. Kocowa and Konieczna were executed on 16 February 1943 and appeared as numbers seventy-two and seventy-three on Stieve's list. Wanda Węgierska and Henryka Veith were killed on 25 June 1943 and received the numbers 111 and 109.

Wilhelmine Günther, executed on 9 June 1944 at the age of twenty-six, was among those who became politically active due to events that directly influenced her private life.[207] She lived with her German family in Posen and worked as a secretary. In 1941 she fell in love with the Polish citizen Antonin Jagly, thereby defying NS law. In December 1941 Jagly was arrested, possibly after a denunciation by Günther's father or neighbors, and imprisoned in a concentration camp. He perished there in September 1942, a month after Günther gave birth to their daughter Maria, who died as an infant. Already during her pregnancy Günther contacted friends involved in Polish underground army intelligence networks and started providing them with information about her German employers. Her group was betrayed in 1943 and Günther was imprisoned with her mother in November 1943. They were put on trial at the *Volksgerichtshof* in Berlin on

26 April 1944. Her mother was acquitted, while Günther was sentenced to death and executed. She appeared as number 128 on Stieve's list.

The close connection between Germany and Poland was often demonstrated by families like Günther's, who had roots in both countries. Marianne Gaszczak belonged to such a family.[208] She was born to Polish parents on 28 December 1914 in the Ruhr region of Germany. In 1929 she obtained a scholarship to attend a teacher seminar in Leszno, Poland, and remained in the country after receiving her certificate as an elementary school teacher. In 1935 she accepted a position in Glomsk, West Pomerania, where she became known for her generosity, patriotism, and love for the Polish culture. Under NS occupation all Polish schools were closed in late fall 1939 and Gaszczak was ordered to leave Glomsk. She went to Berlin were she became involved with Polish patriotic groups and was temporarily arrested. Because of her language skills she was employed by the Reich mail office, essentially a spying and censoring facility, where she had to read letters, mostly from forced laborers and their families. She was supposed to report on political and anti-NS remarks and create lists of the writers; however, she destroyed these lists. She was arrested in June 1942 and suffered long interrogations and torture until her trial by the *Volksgerichtshof* on 6 August 1943. Gaszczak was executed on 28 September 1943 and appeared as number sixty-one on Stieve's list.

Pelagia Scheffczyk's biography was very similar.[209] She, too, was born to Polish parents in the Ruhr region, on 8 March 1915. Her family moved to Poland in 1921, where she went to school and became a clerical worker. In 1940 she began work in a German company for office supplies in Kattowitz. Her co-worker Sigmund Witzcak was active for a Polish underground organization and gathered intelligence on the German occupiers. He recruited Scheffczyk as a courier for this organization, but she was caught on 1 December 1942. She was put on trial for espionage by the *Volksgerichtshof* and sentenced to death on 20 August 1943. The verdict was carried out on 5 October 1943 and Scheffczyk became number 100 on Stieve's list.

Members of the Czech resistance also appeared on Stieve's list. Twenty-three-year-old Herta Lindner was of German-Czech origin and lived in the border region of Teplitz-Schönau. Raised in a socialist family climate—her father was also a member of the Czech resistance—Lindner became involved with communist groups in this area. In 1940 she started working as a courier disguised as a member of a mountaineering club. She was arrested on 27 November 1941, only a day after her father's arrest. Despite being tortured she did not give up the names of her collaborators.[210] The trial took place on 23 November 1942 and she was sentenced to death for treason. Lindner was executed on 29 March 1943, appearing on Stieve's list as number seventy-nine.

Other members of her resistance group—forty-two-year-old Wilhelmine Rubal (list number ninety-five), twenty-three-year-old Eleonore Slach (list number ninety-seven), and forty-three-year-old Maria Krsnak (list number seventy-six)—suffered the same fate.[211] Rubal received her verdict from the *Volksgerichtshof* on 3 September 1942 and was executed together with Lindner. Slach and Krsnak, who were both described as active communists, were executed on the same day, 12 November 1942, together with Herta Lindner's father.[212]

The "crime" of "undermining the military spirit" was attributed to fourteen women, who "said too much": they criticized the NS regime openly in conversations with neighbors and acquaintances, and were betrayed. Most of them were executed in 1944, like Wanda Kallenbach.[213] She was born in Poland in 1902, was a laborer, and lived with her husband and eleven-year-old daughter in Berlin-Friedrichshain. In August 1943 she visited her sister and complained to people in the village about the bombings in Berlin, criticizing the government for false promises of safety. She was denounced, taken into custody on 20 January 1944, and put on trial at the *Volksgerichtshof* for "undermining the military," "friendliness to Jews," and former union activity. Kallenbach received a death verdict and was executed on 18 August 1944. She appeared as number 136 on Stieve's list. To honor her memory the city of Berlin named a new street after her in 2007.

Elfriede Scholz's story was very similar.[214] She was born in 1903 in Osnabrück, worked throughout her life as a seamstress, was twice unhappily married, and lost an infant daughter to a heart condition in 1923. During the last years of her life she lived and worked in Dresden, where in the summer of 1943 she was denounced by neighbors after remarking that Hitler was responsible for the death of the German soldiers killed in the war and that she would willingly shoot Hitler herself, given the opportunity.[215] Scholz was put on trial at the *Volksgerichtshof* on 29 September 1943. Senior judge Roland Freisler was possibly even more biased against Scholz than other women accused of the same "crime," as he stated during the trial that "your brother unfortunately escaped us, but the same will not happen with you".[216] Scholz's brother was Erich Maria Remarque, the pacifist author of *All Quiet on the Western Front*, who had emigrated from Germany in 1933. She was found guilty and executed on 16 December 1943. Her name appeared as number 105 on Stieve's list. Remarque only learned about his sister's death in 1946, and even later about the fate of her body through press reports on Stieve's work.[217]

Ehrengard Frank-Schultz was another woman who "said too much."[218] The fifty-nine-year-old widowed nurse worked as a Protestant deaconess. Sometime after the 20 July 1944 attack against Hitler she told a Red Cross colleague that she regretted the outcome of the plot and thought that a few years under foreign government would be better than the current violent regime. Her trial by the *Volksgerichtshof* on 6 November 1944 ended in a death sentence, which was carried out on 8 December 1944. Frank-Schultz became number 125 on Stieve's list.

There is also the fate of forty-one-year-old Marianne Latoschinski, mother of three children, who came from a family of communist convictions in Bernburg/Saale.[219] In 1944 she was denounced, possibly by neighbors, for having voiced the opinion that Hitler and others would be called into account for their activities and be "shot dead" at some point.[220] She was sentenced to death by the *Volksgerichtshof* on 1 August 1944 and executed on 29 September 1944. She appeared as number 143 on Stieve's list.

All other women on Stieve's list were petty criminals or had been convicted for manslaughter, arson, murder, and providing abortion. There were German citizens among them and ten forced laborers (eight women and two men). All of them were in their twenties or younger and of French, Polish, and Czech nationality. Very little is known about most of them, but the example of Bronisława Czubakowska shows that research by those interested—here, a Polish-German student group—can elucidate these fates even as late as sixty years after the events.[221] The twenty-six-year-old Czubakowska appeared as number twenty on Stieve's list and had been born in Zbiersk, Poland, in 1916. In April 1940 the German occupiers rounded up young people in her hometown and deported them to Brandenburg in Germany. She was put to work in a textile-production company, which employed 130 Polish forced laborers in addition to its several hundred German employees. The women lived in a camp on company grounds and had to adhere to strict rules, always under the threat of severe punishment. On 12 July 1941 a minor fire in one of the German women's restrooms was discovered, and two days later Czubakowska was interrogated by the GeStapo and arrested under the suspicion of arson and sabotage. On 10 September 1941 she was sentenced to seven years in prison, but a revision by the senior prosecutor of Potsdam led to a death verdict on 13 May 1942. She was executed on 15 August 1942.[222]

Even intensive research into the handling of the ashes of victims dissected at the Berlin anatomical department could not clarify their exact location. One of the likely last resting places is the mass grave for victims executed at Plötzensee at the cemetery in Alt-Glienicke. To honor Czubakowska's last wish, an urn with earth from this site was interred in her mother's grave in Zbiersk in 2005. Her family had never before heard anything about her final destiny in Germany.[223] Among those whose fates still need more exploration is the forced laborer Madeleine Parmentier, a twenty-year-old French citizen who worked in Halbe/Teltow outside Berlin in a furniture factory. Together with Jacques Polard and Jean Robin, possibly compatriots, she stole luggage from Berlin train stations during nighttime blackouts. They were caught and put on trial. Parmentier received a death sentence, which was carried out on 2 March 1944. In her prison notes she claims that her salary had been insufficient and that she had not owned even the most essential items. Her observer described her as "child-like."[224] Parmentier became number 155 on Stieve's list.

Denunciations by neighbors were the downfall of many of these women, among them Emma Bethge, who was born in 1891 and worked as a metal worker in Berlin. In the winter of 1943–44 she was observed picking up items of clothing from a bombing site by a woman whose window over-looked the site. The witness reported Bethge to the authorities, and she was taken into custody after a search of her apartment revealed several looted items: "10 pairs of men's socks, 2 shirts, 1 nightgown and 3 women's dresses, some of them very worn".[225] On 11 February 1944 Bethge was sentenced to death by a Berlin *Sondergericht*. Even though she was quite deaf, she was considered fit to stand trial. After her execution on 9 March 1944 she ended on Stieve's list as number 116.

Linking documents from Plötzensee with data gathered from Stieve proved productive in restoring the biographies of the women on his list, so this approach was also used to identify NS victims mentioned in anatomical publications from the University of Leipzig in combination with data from the execution site in Dresden.[226] Historian Birgit Sack and her colleagues at the memorial center *Münchner Platz* in Dresden have created a database for all persons who were executed there between 1933 and 1945.[227] Over 1,400 prisoners were executed in Dresden, and more than 800 of them were Czech, about 350 German, 110 Polish, and the rest from other countries. Half of them were convicted of political "crimes," a quarter for lesser offenses like theft, and the rest for "severe crimes" such as murder and rape. Some of these prisoners' bodies could be traced to the anatomical department in Leipzig, the only university to which Dresden delivered bodies, where they were used by chairman Clara and his colleagues for educational and research purposes.

Between 1933 and 1945 the anatomical department of the University of Leipzig produced thirty-eight publications based on "material" from the bodies of the executed. This was by far the highest number of such articles among all German anatomical departments.[228] Included in the Leipzig publications were seven studies by Clara and his doctoral students in which specimens procured from the bodies of the executed were identified by the first letters of the person's family name, as well as by age and time of imprisonment.[229] Investigations by Clara's collaborators Schiller and Heckel made use of unusually high numbers of bodies of the executed. Some bodies were "shared" among the researchers.[230] Altogether, seventy-five different abbreviations and ages appeared in the seven publications, indicating the use of "material" from seventy-five NS victims. Six of them were women, and the ages ranged from eighteen to seventy-three years. These persons probably represent only a fraction of the total number of bodies of the executed delivered and used in Leipzig. In collating the information of the abbreviated names and ages with data from the execution site, it was possible to discover the identity of fifty-seven NS victims (appendix, table 5). Information on the nationality of the victim was available in fifty-five cases, and on the reason for the death penalty in forty-two cases. Twenty-three persons were of German origin, twenty-six Czech, five Polish and one Austrian. Eighteen received the death penalty for minor crimes like "theft during blackout," "crime during air raid," or fraud, whereas eleven were convicted for violent crimes including rape, and thirteen for murder.

Only one man among this group of prisoners was convicted for political reasons; he was also the only one for whom additional biographical information could be found. This was thirty-three-year-old Ludwig Cyranek, a leader of the German Jehovah's Witnesses. He was sentenced to death by *Sondergericht I* in Dresden "on a charge of demoralization of the armed forces in concurrence of involvement with an anti-military association" on 4 July 1941.[231] Cyranek's execution was reported in the US daily press at the time without further comment.[232] It appears that Stieve used many more bodies of political dissidents than Clara and his group. One explanation might be that political dissent was a more common reason for the death verdict in women than in men. On the other hand, it is conceivable that those responsible in Dresden for the transport of bodies to Leipzig consciously avoided sending bodies of political prisoners, and instead sent so-called "criminals." Also, the so-called "crimes," including murder and attempted murder, could have had political motives. More archival work is necessary to properly document the lives of these victims.

While studies of the judicial NS system did not often focus on the connection between victims and anatomical institutes, this relationship comes up when referring to burial sites. Gunnar Richter's detailed investigation of the *Arbeitserziehungslager* Breitenau in North Hessen (an educational work camp, another form of penitentiary camp)[233] includes the stories of twenty-four forced laborers, most of them Polish, whose bodies were delivered to the anatomical department of the University of Marburg between 1941 and 1943 (appendix, table 6). The transports were recorded in the Marburg body register in thirteen cases and cited in other documents in eleven instances. The names may have been omitted from the register because only bodies that were eventually used for dissection and research were listed, while those with advanced postmortem changes were not.[234]

The twenty-four men were between eighteen and forty-four years old, and the cause of death was execution in eighteen cases, accident in one, suicide in one, seizures in one, and unknown in three. The victims of executions had received the death sentence for various reasons, but most commonly for "racial defilement," Many of these men worked on rural farms and were often friendly with the farmer and his staff and family, so that the development of amorous attachments was natural, though forbidden by NS authorities. Relationships became public either by the woman's pregnancy or by denunciations, which resulted in arrests for both partners. Execution was mandatory for Polish men, and the women were sent to concentration camps.[235] However, the couples were usually given some alternatives: Polish men could ask for a "racial" evaluation, and if they passed they were not executed. German women could declare that they were raped, thereby condemning their partners to death but saving themselves from incarceration. "Racial" evaluation for the men rarely resulted in a positive outcome. Another common reason for incarceration in Breitenau was "attempted murder," often in situations where forced laborers resisted violently against orders from their employers. Finally, six Polish citizens had been inmates of the Breitenau camp for various "crimes" and were executed by the GeStapo in retaliation for the murder of a policeman committed by another Polish man.

The victims' stories collected by Richter are often heartbreaking love stories, and the retelling of all of them would go beyond the scope of this book. One example shall stand for the whole group, the fate of Josef Jurkiewicz, who was executed on 26 January 1942, and his lover Anna S.[236] Jurkiewicz, born in 1909 in the province of Posen, was a former Polish prisoner of war and forced to perform labor on a farm near Hersfeld in Hessen since the spring of 1940. He was a skilled worker and became acquainted with his employer and his family. The farmer's sister Anna S. fell in love with him, but their relationship was discovered when Anna S. became pregnant. They were denounced by a village official in April 1941, and Jurkiewicz was imprisoned immediately, first in Kassel and then in Breitenau. He could not apply for a "racial" evaluation because he was already married in Poland and therefore could not offer marriage to Anna S. He was sentenced to death and executed nine months after his imprisonment. His body was delivered to the anatomical department in Marburg. Anna S. was interrogated several times before giving birth to a daughter in late June 1941. Shortly after the delivery she was incarcerated, first in Kassel, then in Breitenau, and finally in the concentration camp Ravensbrück, were she stayed until the end of the war. Her daughter Anneliese died from a gastrointestinal infection in a hospital within the first weeks of her life. It is not known if her death was due to intentional neglect by those caring for her, a common practice employed with children who were considered of "mixed race" and "unworthy of living."[237] After the war Anna S. was repeatedly denied any reparation payments which were available to other NS victims, with the explanation that neither political nor "racial" reasons had led to her imprisonment.[238]

Gunnar Richter's study of one of the smaller concentration camps exemplifies the opportunities that lie in this local research approach, which explores not only official documents but also oral testimonies from surviving witnesses. Such investigations offer the chance to learn more about those victims for whom information is otherwise hard to come by. While Anke Schulz's investigation of forced laborer camps in Hamburg revealed that victims of a mass execution were delivered to the anatomical institute, probably in 1943, an identification of individual victims was not included.[239] As of yet, no other studies have been found with similar information.

Last Wishes: Death and the Dead Body as Part of a Person's Biography

The women on Stieve's list were often incarcerated for long periods and had time to think about their impending death. Some felt that they were aging rapidly due to the mental and physical torture they had to endure, a sentiment exemplified in Libertas Schulze-Boysen's poem, "Oh Grace, to Ripen While Young in Body."[240] Stieve did indeed observe histological changes of premature aging of the reproductive organs in some women and interpreted them as the results of psychological stress.[241] While he was so meticulous in his objective observations on the histological samples, he never remarked on the signs of violent punishment that must have been visible on so many of the tortured victims.

Despite the severity of NS legislation against all prisoners, those on death row were allowed to write a last letter. While most of these notes contained as a main topic thoughts for their loved ones, some of the women also had clear wishes concerning the last resting place of their bodies: many wanted to be with their mothers. Bronisława Czubakowska wrote, "Make sure that my mortal remains are buried with my mother's,"[242] and Herta Lindner put her last wish in these words: "I want her [Lindner's mother], if possible, to take me home, so that I can at least be with her in death."[243] Instead, their wishes went unfulfilled, and they became subjects of anatomical research. Twenty-nine-year-old Libertas Schulze-Boysen's final message to her mother reveals the desire she shared with Czubakowska and Lindner: "Don't fret about things that perhaps could have been done, one way or another—fate has claimed my death. I wanted it this way.... As a last wish I have asked that my 'material substance' be left to you. If possible, bury me in a beautiful place amidst sunny nature. ... Now, my darling, the bell tolls for me."[244] However, NS legislation from 1942 decreed that families of executed persons would no longer be informed about the execution date, so that they could not claim their loved ones' bodies.[245] Schulze-Boysen had been sentenced to death for treason by the Reich military court and was executed on 22 December 1944. Her body arrived—against her wishes—on a dissection table of the anatomical department of Berlin fifteen minutes after her decapitation. She became number thirty-seven on Stieve's list.

Taking into account the clearly expressed wishes of the prisoners in their last letters, the use of the women's bodies for anatomical purposes and the anatomists' apparent indifference to the personhood of their research subjects appears doubly troubling. If one considers the human body and what happens to it after death as part of a person's biography,[246] then Stieve contributed to the violent abbreviation of the life stories of NS victims.[247] The women and men whose bodies were used by Stieve, Clara, and others came from all walks of life—they were domestic and industrial workers, homemakers, teachers, and academics; some were politically interested, others not. None of them volunteered to become research subjects; on the contrary, many wanted their remains to rest with their families. One of the failures of anatomists in the Third Reich lay in their refusal to care about the personal history of their research subjects, who had been victimized by a violent regime.

Commemoration of the Victims

Dignified and suitable solutions have been found to remember victims of the National Socialist system over the last few decades. Ultimately all resolutions must be based on a detailed study of the true facts of this history, which alone provide the foundation for an honest and worthwhile commemoration. The research must include the identification of both victims and perpetrators. Anatomists and scientists in related disciplines also have to address the question of the possible continued use of tissues derived from victims, which have to be identified and properly buried, if possible in a named grave. The search for remaining specimens from NS victims is still ongoing, even though the likelihood of their existence diminishes with each passing year.[248] While many anatomical institutes have investigated their collections, they have not yet addressed the microscopic slides given to students during their studies in WWII.[249] In Berlin alone, each student in Stieve's dissection course, which sometimes included more than a thousand persons per semester, received a set of 160 histological slides, which they were allowed to keep.[250]

An example of a particularly dignified memorial is the burial for neuroanatomical specimens that had been taken from children at the *Spiegelgrund* hospital in Vienna. The memorial site now represents each child individually.[251] Other ways to remember include research on individuals and publications of their biographies, including information on both perpetrators and victims. Several organizations have also put together memorial statements, which list names and biographical information. These memorials have varying formats, ranging from review articles[252] to books like Lang's report on Hirt's victims[253] and general statements of responsibility by several groups of physicians.[254]

A more interactive and extensive approach to commemoration is possible through online databanks. Paul Weindling and his colleagues are currently establishing such a data collection for all victims of human experiments and coercive research during National Socialism.[255] An interactive database for the NS victims in anatomy could contain all personal information of a victim, including name, age, gender, nationality, and profession, as well as facts of imprisonment or hospitalization and other conditions that led to their delivery to an anatomical institute. Information on the fate of the victim's body could be noted, including its final resting place. The advantage of a database lies in the ease of adding information as it becomes known, ready availability for analysis of group data, and the possibility of ultimately making it public, similar to the website that Lang has created for Hirt's victims.[256] In the long term, many biographies could be publicized and made available for a general audience, specifically for families who are still looking for information about their loved ones. An online database of individuals has been created by the University of Vienna for the victims among its employees. It is called a "memorial book" to emphasize the importance of the victims' remaining "part of the collective memory of today's University".[257] Similarly, the victims of anatomy could become part of the collective memory of German medicine.

In addition to all of these ways of commemoration, an annual memorial service of remembrance for the victims of NS medicine, not just in Germany but also worldwide, has been discussed by several authors.[258] A burial alone might be problematic, as neurologist Jürgen Peiffer pointed out: "There is a dangerous possibility that we may bury our bad consciences together with these tissue remains, thereby avoiding the necessity of remembering the past at least once every semester, together with the students".[259] Such a ceremony would remind the global community of health workers about the legacy of this history. It could facilitate, in Peiffer's words, "facing ourselves, our own history, and asking the question whether science, students, and academic teachers are prepared to learn from the past in order to influence the future".[260] However, keeping the victims' lives in our midst by remembering them has to be at the center of all efforts of memorialization, because "forgetting them would be the victims' final annihilation".[261]

Notes

1. Czech political prisoner on German death row, Fuchik 1949, 45.
2. Seidelman 2014.
3. Fangerau and Krischel 2011.
4. Weindling 2015.
5. Fuchs et al. 2007.
6. Weindling 2010; Susanne Zimmermann 2007.
7. See historian Astrid Ley's work, quoted in Blog T4 2014.
8. Bundesarchivgesetz 1998, §5.
9. Oleschinski 1992.
10. Susanne Zimmermann 2007.
11. Forsbach 2006.
12. Lang 2007.
13. Hildebrandt 2013a
14. Bazelon 2013; Lenzen-Schulte 2013; Gill 2013; Bochsler 2013; Der Standard 2013; Pringle 2013; Elmostrador.mundo 2013.
15. Oehler-Klein et al. 2012; Weindling 2010.
16. Ude-Koeller et al. 2012; Angetter 1998.
17. Schönhagen 1992, 84.
18. Ibid., 71; Peiffer 1991, 128.
19. Schönhagen 1992, 71–79.
20. Blessing et al. 2012.
21. Richter 2009.
22. Weindling 1992.
23. Essential literature on NS "euthanasia": Platen-Hallermund 1948; Klee 1985; Müller-Hill 1984; Roelcke 2010; Fuchs et al. 2007; Hohendorf and Rotzoll 2014.
24. Schönhagen 1992; Grundmann and Aumüller 1996, "Anatomen in der NS-Zeit"; Aumüller und Grundmann 2002, "Anatomy During the Third Reich"; Redies et al. 2005b, "Origin of the Corpses Received by the Anatomical Institute"; Redies et al. 2012, "Dead Bodies for the Anatomical Institute"; Eckart 2012, Medizin in der NS-Diktatur.
25. Redies et al. 2012.
26. Bussche 1989, 156; Beushausen et al. 1998, 233; Noack 2008, 14–15.
27. Schultka and Viebig 2012.
28. Blessing et al. 2012.
29. Ibid.
30. Redies et al., 2012.
31. Lichtblau 2013.
32. Kogon 2006.
33. Rothmaler 1990; Rothmaler 1991.
34. Hildebrandt 2008.
35. Grundmann and Aumüller 1996; Aumüller and Grundmann, 2002; Redies et al. 2005b; Czech 2015.
36. Goeschel 2009, 217.
37. Kwiet and Eschwege 1984, Selbstbehauptung und Widerstand.

38. Kater 1999.
39. Grundmann and Aumüller 1996; Richter 2009, 282–83.
40. Seeger 1998; Dr. Birgit Sack, *Gedenkstätte Landgericht Dresden, Münchner Platz,* personal communication.
41. Winkelmann 2008; Hildebrandt 2013a.
42. Berenbaum 1990; USHMM 2014.
43. Feig 1990.
44. Hildebrandt 2013b.
45. Seidelman 2012.
46. Lang 2007; Lang 2013b; Steegmann 2005a.
47. Aly 1987; Aly 1994; Heimann-Jelinek 1999.
48. Hildebrandt 2013a.
49. Forsbach 2006, 554.
50. Angetter 2000.
51. Berenbaum 1990; Kidder 2012.
52. Kater 1969; Garbe 2008; Hesse 2001
53. King 1990, 188.
54. Garbe 2008, 667, note 177.
55. Hildebrandt 2013a.
56. Toledano 2013.
57. Hartmann and Hartmann 1986.
58. Hebestreit 2009.
59. Halbrainer 1998.
60. Ibid.
61. Hildebrandt 2013a.
62. Evans 1996.
63. Plato et al. 2008.
64. Wagner 2010.
65. Korte 1991; Form and Schiller 2005; Frewer et al. 2009.
66. Schwarze 1997; Zimmermann V. 2007.
67. Rosmus 1993.
68. Speck 2002, 244.
69. Ude-Koeller et al. 2012.
70. Schönhagen 1992.
71. Hildebrandt 2013a.
72. Richter 2009.
73. Hildebrandt 2013b.
74. Aly 1987, 1994; Neander 2006; Tomkiewicz and Semków 2013; Alvermann 2014; Lang 2007.
75. Universität Tübingen 1990, 19.
76. Ibid.., 22.
77. Ibid.., 21.
78. Ibid..
79. Angetter 1998; Czech 2015, 148.
80. Neugebauer 1998.
81. Alt-Hiezinger 2013; on Lederer and Scholz: Österreichisches Biobliographisches Lexikon 2013.
82. Neugebauer 1998.
83. Kurier 1962.
84. Neugebauer 1998.
85. Czech 2015, 158.
86. Ibid., 162.

87. Ibid., 164–65.
88. Ibid., 176–79.
89. Forsbach 2006, 541.
90. Noack 2012.
91. DKP 1998; Pfoertner 2003.
92. Noack 2012, 289.
93. Ibid., 290.
94. Ibid.
95. Eckart et al. 2006, 662.
96. Kultusministerkonferenz 1994.
97. Letter from Eduard von Bargen to Benninghoff, April 25 1948, estate papers Benninghoff.
98. Nałkowska 2000; Neander 2006; Tomkiewicz and Semków 2013.
99. Neander 2006, 66.
100. Nałkowska 2000, 3–4.
101. Neander 2006, 77–78.
102. Jaroszewski 1993, 57–63.
103. Tomkiewicz and Semków 2013, 68.
104. Ibid., 68–69.
105. Ibid., 69–70.
106. Neander 2006, 66–67.
107. Nałkowska 2000, 4–5.
108. Neander 2006, 65.
109. Nałkowska 2000, xiii.
110. Neander 2006; Tomkiewicz and Semków 2013.
111. Noack 2008, 20.
112. Ibid., 21.
113. Buddecke 2011.
114. Ibid.
115. Ratschko 2013, 420.
116. Grundmann and Aumüller 1996, 329; Kultusministerkonferenz 1994.
117. Roth 1984.
118. The author is indebted to Dr. Raphael Toledano for information on the French investigation report of January 1946. Dr. Toledano is the foremost authority on the anatomical department in Strasbourg 1933–45.
119. Lang 2007, 2013b.
120. Champy and Risler 1945.
121. Bever 2015.
122. Aly 1994, 135–36.
123. Majewski 2013, 142–45; Szołdrska 1948, 40.
124. Vsevolodov 2014; Holtz 2014; Alvermann 2014.
125. Universität Tübingen 1990.
126. Kultusministerkonferenz 1994.
127. Rothmaler 1990.
128. Rothmaler 1991.
129. Professor A. F. Holstein in a letter to the author, 27 May 2013.
130. Hildebrandt et al. 2014.
131. Kultusministerkonferenz 1994; Redies et al. 2012.
132. Kultusministerkonferenz 1994.
133. Redies et al. 2012; Oehler-Klein et al. 2012.
134. Williams 1988; Hildebrandt 2006.
135. Akademischer Senat der Universität Wien 1998; Malina and Spann 1999.

136. Angetter 1998, and personal communication by Dr. Angetter.
137. Angetter 2000.
138. Aly 1987, 1994.
139. Heimann-Jelinek 1999; Seidelman 2012.
140. Teschler-Nicola and Berner 1998.
141. Heimann-Jelinek 1999; Pawlowsky 2005; Spring 2005; Lange 2011; Berner 2005; Berner 2011.
142. Hildebrandt and Redies 2012.
143. Viebig 2012.
144. Forsbach 2006.
145. Redies et al. 2005; Angetter 1998; personal communications from Dr. Redies and Dr. Angetter.
146. Buddecke 2011.
147. Mitscherlich and Mielke 1947, 105.
148. Lang 2007, 2013b.
149. Lang 2013a.
150. Lang 2007, 91–99; Sachnowitz 2002.
151. Snyder 2010, 408.
152. Hildebrandt 2013a.
153. Winkelmann 2008; Scherrieble 2008.
154. Halle: Scherrieble 2008; Kiel: Seeger 1998; Bonn: Forsbach 2006.
155. E.g., Oleschinski 1992, Zimmermann 2007.
156. Hildebrandt 2013a, 14.
157. Klee 2004, 108.
158. Oleschinski 1992.
159. Ibid.: archival signature DJV JV A 1663/46, 4.12.1946, BA Potsdam, P-1/Nr. 2.
160. Oleschinski 1992.
161. Stieve 1946.
162. Hildebrandt 2013a.
163. E.g., Stieve 1952.
164. Winkelmann 2008.
165. Oleschinski 1992; Schagen 2005; Winkelmann 2008.
166. Gélieu 1994.
167. Filenames abbreviated GDW/P followed by prisoner's name.
168. Compare Gélieu 1994, 197–99.
169. GDW/P: Lindner, Herta.
170. Gélieu 1994, 197.
171. E.g., Stieve 1952a, 49.
172. Schagen 2005.
173. Gedenkstätte Plötzensee 2012a.
174. Angetter 1998.
175. Winkelmann 2008.
176. Gélieu 1994; Frauengefängnis Barnimstrasse 2012.
177. Angetter 1998.
178. Personal communication from Dr. Birgit Sack.
179. Gedenkstätte Plötzensee 2012b.
180. Massnahmen aus Anlass von Todesurteilen, Reichsminister der Justiz 447-III.a 4 318.39, 19. Februar 1939; the author is indebted to historian Michael Viebig for this information.
181. Gedenkstätte Deutscher Widerstand 2012c; Gedenkstätte Deutscher Widerstand 2012d.
182. *Reading Eagle* 1935; *Time* 1935.

183. GDW/P: Tassin, Lucienne; prison card.
184. GDW/P: Henkel, Elfriede; Staatsanwalt/Leiter der Anklage bei dem Sondergericht am 27. Mai 1943 an den Vorstand des Frauengefängnisses.
185. Viebig in Halleforum.de 2007.
186. Viebig 2012.
187. Orth 2013.
188. Knauer 2002.
189. Michael Viebig, personal communication.
190. Literature on the *Rote Kapelle,* e.g.: Schilde 1993; Trepper 1995; Brysac 2000; Vinke 2003; Nelson 2009; Conze et al. 2010.
191. E.g., Kwiet and Eschwege 1984; Wippermann 2001; Scheer 2004.
192. See also Oleschinski 1997, 29.
193. E.g., Tomaszewski 2006.
194. E.g., GDW/P: Augustyniak, Veronika.
195. E.g., Perrault 1998.
196. Kuhnke 2001; Fallada 2011.
197. A Letter to the Stars 2012.
198. Moos 1999.
199. Perrault 1998.
200. Kraushaar 1970a; Röhl 2004; Hälker 2012; Bake 2012.
201. Gostomski and Loch 1993.
202. GDW/P: Augustyniak, Veronika, Der Oberreichsanwalt beim Volksgerichthof 11 J 21/42, 10 August 1942—Verfügung betr. Vollstreckung von Todesurteilen.
203. Plötzensee Memorial Center 2012.
204. National Armed Forces of Poland 2012; GDW/P: Koniszna, Halina and Kocowa, Miroslawa.
205. Tomaszewski 2006.
206. GDW/P: Veith, Henryka.
207. Bojarski 2010.
208. Glomsk 2012; GDW/P: Gaszczak, Marianne.
209. Bottrop 2012.
210. Step21 2008; Wunderlich 2011.
211. Gostomski and Loch 1993.
212. GDW/P: Slach, Eleonore, Krsnak, Maria, Rubal, Wilhelmine.
213. Peters 2006; GDW/P: Kallenbach, Wanda; StA. Chbg. Reg. Nr. 2917/44.
214. GDW/P: Scholz, Elfriede; Glunz and Schneider 1997.
215. Hochhuth 2004a; Hochhuth 2004b.
216. Scholz 2007, 3.
217. Hochhuth 2004a, 2004b.
218. Gerechte der Pflege 2012.
219. Steinborn 2009.
220. Wagner 1974, 129–30.
221. GDW/P: Czubakowska, Bronisława ; Leutner 2007; Berg 2005; Püschel 2002.
222. Leutner 2007.
223. Berg 2005; Leutner 2007.
224. GDW/P: Parmentier, Madeleine.
225. GDW/P: Bethge, Emma, Urteil Sondergericht 11. Februar 1944.
226. Hildebrandt, 2013b.
227. Haase and Sack 2001; personal communication by Dr. Birgit Sack.
228. Hildebrandt 2013c.
229. Clara 1937; Michaelis 1938; Schulze 1938; Kretschmar 1940; Schiller 1942; Heckel 1942; Müller 1942.

230. Hildebrandt 2013b, 526, 527.
231. Garbe 2008, 334.
232. *Long Island Daily Press* 1941.
233. Richter 2009.
234. Ibid., 346.
235. Richter 2009, 321.
236. Pseudonym given by Richter for privacy protection.
237. Rosmus 1993, 34.
238. Richter 2009.
239. Schulz 2010, 45.
240. Gollwitzer et al. 1955.
241. Stieve 1942, 1702.
242. Püschel 2002.
243. Step1, 2008.
244. Gollwitzer et al. 1955.
245. Noack 2012.
246. Winkelmann and Schagen 2009.
247. Wischmann 2011.
248. E.g., Hildebrandt et al. 2014.
249. Hoffmann 1951.
250. Romeis 1953, 406.
251. Czech 2002; Weindling 2013.
252. E.g., Hildebrandt and Aumüller 2012.
253. Lang 2007.
254. E.g., Kolb et al. 2012; Schneider 2011s.
255. Weindling and Villiers 2014.
256. Lang 2013a.
257. University of Vienna 2013.
258. Peiffer 1991; Seidelman 1996; Cohen 2009.
259. Peiffer 1991, 126.
260. Ibid., 125.
261. Lang 2007, 13.

Archival Materials

Gedenkstätte Deutscher Widerstand
  – File names abbreviated GDW/P followed by prisoner's name
Universitätsarchiv Marburg
  – Estate papers Benninghoff

Bibliography

A Letter to the Stars. 2012. "Helene Delacher." Accessed 10 March 2015. http://www
  .lettertothestars.at/himmelsbriefe.php?s=1&opfer__id=62066.
Akademischer Senat der Universität Wien. 1998. *Senatsprojekt der Universität Wien: Untersuchungen zur anatomischenWissenschaft in Wien 1938–1945.* Unpublished manuscript.
Alt-Hiezinger. 2013. "Gestorben für Österreich. Jacob Kastelic." Accessed 29 April 2013.
  http://www.alt-hietzinger.at/archiv/personen/jakobkastelic.shtml.

Alvermann, Dirk. 2015. "'Praktisch begraben"—NS-Opfer in der Greifswalder Anatomie 1935–1947." In: "… die letzten Schranken fallen lassen": Studien zur Universität Greifswald im Nationalsozialismus, ed. Dirk Alvermann, 311–51. Köln Weimar Wien: Böhlau Verlag.

Aly, Götz. 1987. "Das Posener Tagebuch des Anatomen Hermann Voss." In: Biedermann und Schreibtischtäter: Materialien zur deutschen Täter-Biographie, ed. Götz Aly, Peter Chroust, and Christian Pross, 15–66. Berlin: Rotbuch Verlag.

———. 1994. "The Posen Diaries of the Anatomist Hermann Voss." In: Cleansing the Fatherland: Nazi Medicine and Racial Hygiene, ed. Götz Aly, Peter Chroust, and Christian Pross, 99–155. Baltimore: Johns Hopkins University Press.

Angetter, Daniela. 1998. "Erfassung der von der NS-Justiz in Wien in der Zeit von 1938–1945 Hingerichteten, die als Studienleichen dem anatomischen Institut der Universität Wien zugewiesen wurden." In: Senatsprojekt der Universität Wien: Untersuchungen zur anatomischenWissenschaft in Wien 1938–1945, ed. Akademischer Senat der Universität Wien, 81–92. Unpublished manuscript.

———. 2000. "Anatomical Science at University of Vienna 1938–45." Lancet 355: 1445–57.

Aumüller, Gerhard, and Kornelia Grundmann. 2002. "Anatomy during the Third Reich: The Institute of Anatomy at the University of Marburg, as an Example." Annals of Anatomy 184: 295–303.

Bake, Rita. 2012. "Irene Wosikowski." Landeszentrale für politische Bildung Hamburg. Accessed 10 March 2015. http://www.hamburg.de/clp/frauenbiografien-schlagwortregister/clp1/hamburgde/onepage.php?BIOID=3116.

Bazelon, Emily. 2013. "The Nazi Anatomists." Slate. Accessed 28 August 2014. http://www.slate.com/articles/life/history/2013/11/nazi_anatomy_history_the_origins_of_conservatives_anti_abortion_claims_that.html.

Berenbaum, Michael. 1990. A Mosaic of Victims: Non-Jews Persecuted and Murdered by the Nazis. New York: New York University Press.

Berg, Guido. 2005. "Und heute geschah es." Potsdamer neueste Nachrichten 13 September 2005. Accessed 30 April 2012. http://www.pnn.de/potsdam/115423/.

Berner, Margit. 2005. "'Judentypologisierungen' in der Anthropologie am Beispiel der Bestände des Naturhistorischen Museums, Wien." Zeitgeschichte 2(32): 111–16.

———. 2011. "'Die haben uns behandelt wie Gegenstände': Anthropologische Untersuchungen an jüdischen Häftlingen im Wiener Stadion während des Nationalsozialismus." In: Sensible Sammlungen. Aus dem anthropologischen Depot, ed. Margit Berner, Anette Hoffmann, and Britta Lange, Britta, 147–67. Hamburg: Philo Fine Arts.

Beushausen, Ulrich, Hans-Joachim Dahms, Thomas Koch, Almut Massing, and Konrad Obermann. 1998. "Die medizinische Fakultät im Dritten Reich." In: Die Universität Göttingen unter dem Nationalsozialismus, ed. Heinrich Becker, Hand Joachim Dahms, and Cornelia Wegeler, 183–286. München: K. G. Saur.

Bever, Lindsey. 2015. "Remains of Holocaust experiment victims found at French forensic institute." The Washington Post July 22 2015. http://www.washingtonpost.com/news/morning-mix/wp/2015/07/22/remains-of-holocaust-victims-used-as-guinea-pigs-found-at-french-forensic-institute/. Accessed 12 August 2015.

Blessing, Tim, Anna Wegener, Hermann Koepsell, and Michael Stolberg. 2012. "The Würzburg Anatomical Institute and Its Supply of Corpses (1933–1945)." Annals of Anatomy 194 (2012): 281–85.

Blog T4. 2014. "Blog Gedenkort T4." Accessed 9 March 2015. http://blog.gedenkort-t4.eu/2014/08/05/28-8-2014-topographie-des-terrors-den-namenlosen-einen-namen-geben-die-namensnennung-von-euthanasie-opfern-aus-juristischer-sicht/.

Bochsler, Katharina. 2013. "Die Frauen auf Dr. Stieves Liste." SRF, Schweizer Radio und Fernsehen. 15 February 2013. Accessed 30 September 2013. http://www.srf.ch/wissen/mensch/die-frauen-auf-dr-stieves-liste.

Bojarski, Piotr. 2010. "Wilka, Niemka, która walczyła w AK." *Poznan Gazeta*, 6 November 2010. Accessed 19 August 2015. http://poznan.wyborcza.pl/poznan/1,36001,79 99019,Wilka__Niemka__ktora_walczyla_w_AK.html

Bottrop. 2012. "Pelagia Scheffczyk: Ernst-Ender-Straße 19. Patenschaft für den Stolperstein: Familie Fröhlich-Landig." Accessed 15 July 2012. http://www.bottrop.de/stadtle ben/kultur/stolpersteine/060814_stolper_Scheffczyk.php.

Brysac, Shareen Blair. 2000. *Resisting Hitler: Mildred Harnack and the Red Orchestra*. Oxford: Oxford University Press.

Buddecke, Julia. 2011. *Endstation Anatomie: Die Opfer nationalsozialistischer Vernichtungsjustiz in Schleswig-Holstein*. Hildesheim, Germany: Georg-Olms Verlag.

Bundesarchivgesetz. 1998. "Gesetz über die Sicherung und Nutzung von Archivgut des Bundes." Accessed 19 March 2014. http://www.gesetze-im-internet.de/bundesrecht/ barchg/gesamt.pdf.

Bussche, Hendrik van den, ed. 1989a. *Medizinische Wissenschaft im "Dritten Reich." Kontinuität, Anpassung und Opposition an der Hamburger Medizinischen Fakultät*. Berlin/Hamburg: Dietrich Reimer Verlag.

Champy, Christian, and Dr. Risler. 1945. "Sur une Série de Préparations Histologiques Trouvées das la Laboratoire d'un Professeur Allemand. Expériences Faites sur L'Homme au Camp de Struthof. Communication Seance du 1 Mai 1945." *Bulletin de l'Académie de Médicine* 129: 263–65.

Clara, Max. 1937. "Über das Vorkommen von Atraktosomen in den Schleimzellen der menschlichen Drüsen." *Zeitschrift für Zellforschung und mikroskopische Anatomie* 25(5): 655–93.

Cohen, M. Michael, Jr. 2009. "Overview of German, Nazi and Holocaust Medicine." *American Journal of Medical Genetics*, Part A: 687–707.

Conze Eckart, Norbert Frei, Peter Hayes, and Moshe Zimmermann. 2010. *Das Amt und die Vergangenheit: Deutsche Diplomaten im Dritten Reich und in der Bundesrepublik*. München: Blessing.

Czech, Herwig. 2002. "Der Fall Heinrich Gross. Die wissenschaftliche Verwertung der 'Spiegelgrund'-Opfer in Wien." *Context XXI*. Accessed 12 March 2015. http://www .contextxxi.at/context/content/view/164/93/.

———. 2015. "Von der Richtstätte auf den Seziertisch: Zur anatomischen Verwertung von NS-Opfern in Wien, Innsbruck und Graz." *Jahrbuch des Dokumentationsarchiv des österreichischen Widerstandes*, 2015:141–90.

Der Standard. 2013. "Die 182 Leichen des Dr. Stieve." Accessed 28 August 2014. http:// derstandard.at/1358304935403/Die-182-Leichen-des-Dr-Stieve.

DKP München. 1998. *Die wiedergefundene Liste*. München: Selbstverlag.

Eckart, Wolfgang Uwe. 2012. *Medizin in der NS-Diktatur. Ideologie, Praxis, Folgen*. Wien: Böhlau Verlag.

Eckart, Wolfgang Uwe, Volker Sellin, and Eike Wolgast. 2006. *Die Universität Heidelberg im Nationalsozialismus*. Heidelberg: Springer.

Elmostrador.mundo. 2013. "Las Víctimas Olvidadas de la Anatomía Nazi." Accessed 28 August 2014. http://www.elmostrador.cl/mundo/2013/02/04/las-victimas-olvidadas-de- la-anatomia-nazi/.

Evans, Richard J. 1996. *Rituals of Retribution: Capital Punishment in Germany 1600–1987*. Oxford: Oxford University Press.

Fallada, Hans. 2011. *Jeder stirbt für sich allein*. Berlin: Aufbau-Verlag. First published in 1947.

Fangerau, Heiner, and Mathis Krischel. 2011. "Der Wert des Lebens und das Schweigen der Opfer: Zum Umgang mit den Opfern nationalsozialistischer Verfolgung in der Medizinhistoriographie." In: *NS-"Euthanasie" und Erinnerung. Vergangenheitsaufarbeitung-Gedenkformen-Betroffenenperspektiven*, ed. Stephanie Westermann, Richard Kühl, and Tim Ohnhäuser, 19–28. Berlin: Lit Verlag.

Feig, Konnilyn. 1990. "Non-Jewish Victims in the Concentration Camps." In: *A Mosaic of Victims: Non-Jews Persecuted and Murdered by the Nazis*, ed. Michael Berenbaum, 161–78. New York: New York University Press.

Form, Wolfgang, and Theo Schiller. 2005. *Politische Justiz in Hessen. Die Verfahren des Volksgerichtshofs, der politischen Senate der Oberlandesgerichte Darmstadt und Kassel 1933–1945 sowie Sondergerichtsprozesse in Darmstadt und Frankfurt/M. (1933/34)*. Marburg: Elwert Verlag.

Forsbach, Ralf. 2006. *Die Medizinische Fakultät der Universität Bonn im "Dritten Reich."* München: R. Oldenbourg Verlag.

Frauengefängnis Barnimstrasse. 2012. "Insassinnen von 1933 bis 1945." Accessed 26 July 2012. http://www.ml-architekten.de/barnim/html/1933_1945.html.

Frewer, Andreas, Bernhard Bremberger, Günther Siedbürger, eds. 2009. *Der "Ausländereinsatz" im Gesundheitswesen (1939–1945): Historische und ethische Probleme der NS-Medizin*. Stuttgart: Franz-Steiner Verlag.

Fuchik, Julius 1949. *Notes from the Gallows*. Salt Lake City, UT: Peregrine Smith Books.

Fuchs, Petra, Maike Rotzoll, Ulrich Müller, Paul Richter, and Gerrit Hohendorf. 2007. *"Das Vergessen der Vernichtung ist Teil der Vernichtung selbst." Lebensgeschichten von Opfern der nationalsozialistischen "Euthanasie."* Göttingen: Wallstein Verlag.

Garbe, Detlef. 2008. *Between Resistance and Martyrdom. Jehovah's Witnesses in the Third Reich*. Madison: University of Wisconsin Press.

Gedenkstätte Plötzensee. 2012a. "Hinrichtungen in Plötzensee zwischen 1933 und 1945." Accessed 10 March 2015. http://www.gedenkstaette-ploetzensee.de/zoom/02_1_dt.html.

———. 2012b. "Hinrichtungen in Plötzensee zwischen 1933 und 1945." Accessed 10 March 2015. http://www.gedenkstaette-ploetzensee.de/zoom/02_2_dt.html.

Gedenkstätte Deutscher Widerstand. 2012c. "Liane Berkowitz." Accessed 29 July 2012. http://www.gdw-berlin.de/en/recess/biographies/biographie/view-bio/berkowitz/.

———. 2012d. "Hilde Coppi." Accessed 29 July 2012. http://www.gdw-berlin.de/en/recess/biographies/.

Gélieu, Claudia von. 1994. *Frauen in Haft. Gefängnis Barnimstrasse. Eine Justizgeschichte*. Berlin: Elefanten Press.

Gerechte der Pflege. 2012. "Ehrengard Frank-Schultz." Accesed 7 September 2014. http://www.gerechte-der-pflege.net/wiki/index.php/Ehrengard_Frank-Schultz.

Gill, Victoria. 2013. "Victims of Nazi Anatomist Named." *BBC Health News*, 27 January 2013. Accessed 30 September 2013. http://www.bbc.co.uk/news/health-21086388?print=true.

Glomsk. 2012. "Maria Gaszczak." Accessed 16 May 2012. http://www.glomsk.toowo.pl/strony/gaszczak.html.

Glunz, Claudia, and Thomas Schneider. 1997. *Elfriede Scholz, geb. Remark. Im Namen des deutschen Volkes. Dokumente einer justiziellen Ermordung*. Osnabrück: Universitätsverlag Rasch.

Goeschel, Christian. 2009. *Suicide in Nazi Germany*. Oxford: Oxford University Press.

Gollwitzer, Helmut, Käthe Kuhn, and Reinhold Schneider, R, eds. 1955. *Du hast mich heimgesucht in der Nacht. Abschiedsbriefe und Aufzeichnungen des Widerstandes 1933–1945. Zweite Auflage*. München: Chr. Kaiser Verlag.

Gostomski, Victor von, and Walter Loch. 1993. *Der Tod von Plötzensee. Erinnerungen, Ereignisse, Dokumente*. Frankfurt: Bloch Verlag.

Grundmann, Kornelia, and Gerhard Aumüller. 1996. "Anatomen in der NS-Zeit: Parteigenossen oder Karteigenossen? Das Marburger anatomische Institut im Dritten Reich." *Medizinhistorisches Journal* 31 (3–4): 322–57.

Haase, Norbert, and Birgit Sack. 2001. *Münchner Platz. Die Strafjustiz der Diktaturen und der historische Ort*. Leipzig: Gustav Kiepenheuer Verlag.

Hälker, Kurt. 2012. "La Femme Allemande." Accessed 10 March 2015. http://drafd.org/?archiv,DrafdInfo200105_Wosikowsk.

Halbrainer, Heimo. 1998. "Widerstand und Verfolgung in der Steiermark während der Zeit des Nationalsozialismus. 'Er hat dem Teufel den Eid nie geschworen' Franz Dollnig—Der steirische Franz Jaegerstätter." Accessed 18 March 2014. http://korso.at/korso/DStmk/eid.html.

Halleforum.de. 2007. "Vom Lazarett zum Mordplatz Halle," *Halleforum.de* 24.11.2007. Accessed 29 August 2014. http://www.chefduzen.de/index.php?topic=12889.0;wap2.

Hartmann, Albrecht, and Heidi Hartmann. 1986. *Kriegsdienstverweigerung im Dritten Reich*. Frankfurt: Haart+Herchen Verlag.

Hebestreit, Steffen. 2009. "'Kriegsverräter' rehabilitieren. Einsicht nach 64 Jahren." *Frankfurter Rundschau* 1 July 2009. Accessed 9 March 2015. http://www.fr-online.de/politik/-kriegsverraeter—rehabilitieren-einsicht-nach-64-jahren,1472596,3362122.html.

Heckel, Lothar. 1942. "Untersuchungen über das Vorkommen von Vitamin C in der Nebenniere des Menschen." *Zeitschrift für Mikroskopisch-Anatomische Forschung* 52: 393–417.

Heimann-Jelinek, Felicitas. 1999. "Zur Geschichte einer Ausstellung. Masken. Versuch über die Schoa." In: *"Beseitung des Jüdischen Einflusses…" Antisemitische Forschung, Eliten und Karrieren im Nationalsozialismus*, ed. Fritz Bauer Institut, 131–45. Frankfurt/New York: Campus Verlag.

Hesse, Hans. 2001. *Persecution and Resistance of Jehovah's Witnesses during the Nazi Regime: 1933–1945*. Bremen: Edition Temmen.

Hildebrandt, Sabine. 2006. "How the Pernkopf Controversy Facilitated a Historical and Ethical Analysis of the Anatomical Sciences in Austria and Germany: A Recommendation for the Continued Use of the Pernkopf Atlas." *Clinical Anatomy* 19: 91–100.

———. "Capital Punishment and Anatomy: History and Ethics of an Ongoing Association." *Clinical Anatomy* 21: 5–14.

———. 2013a. "The Women on Stieve's List: Victims of National Socislism Whose Bodies Were Used for Anatomical Research." *Clinical Anatomy* 26: 3–21.

———. 2013b. "Current Status of Identification of Victims of the National Socialist Regime Whose Bodies Were Used for Anatomical Purposes." *Clinical Anatomy* 27: 514–36.

———. 2013b. "Research on Bodies of the Executed in German Anatomy: An Accepted Method that Changed during the Third Reich. Study of Anatomical Journals from 1924 to 1951." *Clinical Anatomy* 26: 304–26.

Hildebrandt, Sabine, Reinhard Putz, Mathias Schütz, Florian Steger, Jens Waschke. 2014. "Evaluation of the Anatomical Collection at the Anatomical Department I of LMU Munich Concerning Specimens Deriving from Victims of National Socialism." *Website Anatomische Gesellschaft*. Accessed 25 July 2014. http://anatomische-gesellschaft.de/images/data/1_Evaluation_of_the_Anatomical_Collection_LMU_Munich.pdf.

Hildebrandt, Sabine, and Christoph Redies. 2012. "Anatomy in the Third Reich." *Annals of Anatomy* 194: 225–27.

Hochhuth, Rolf. 2004a. "Remarque in Plötzensee." In: *Nietzsches Spazierstock. Gedichte, Tragikomödie "Heil Hitler!" Prosa*, by Rolf Hochhuth, 19–34. Reinbek bei Hamburg: Rowohlt.

———. 2004b. "Elfriede Scholz, geb. Remark." In: *Nietzsches Spazierstock. Gedichte, Tragikomödie "Heil Hitler!" Prosa*, by Rolf Hochhuth, 40–41. Reinbek bei Hamburg: Rowohlt.

Hoffmann, Auguste. 1951. "Hermann Stieve zu seinem 65. Geburtstag am 22. Mai 1951." *Zeitschrift für Mikroskopisch-Anatomische Forschung* 57: 117–28.

Hohendorf, Gerrit, and Maike Rotzoll. 2014. "Medical Research and National Socialist Euthanasia: Carl Schneider and the Heidelberg Research Children from 1942 until 1945." In: *Human Subjects Research after the Holocaust*, ed. Sheldon Rubenfeld and Susan Benedict, 127–38. Cham: Springer.

Holtz, Britta. 2015. "Protokoll der gerichtsmedizinischen Untersuchung der im Keller des Anatomischen Instituts der Universität Greifswald vorgefundenen Leichen, 13.–15. November 1947." In: *"… die letzten Schranken fallen lassen": Studien zur Universität*

Greifswald im Nationalsozialismus, ed. Dirk Alvermann, 280–90. Köln Weimar Wien: Böhlau Verlag.

Kater, Michael H. 1969. "Die Ernsten Bibelforscher im Dritten Reich." *Vierteljahreshefte für Zeitgeschichte* 17(2): 181–218.

———. 1999. "Das Böse in der Medizin: Nazi-Ärzte als Handlanger des Holocaust." In: *"Beseitigung des Jüdischen Einflusses...." Antisemitische Forschung, Eliten und Karrieren im Nationalsozialismus*, ed. Fritz Bauer Institut, 219–39. Frankfurt: Campus Verlag.

Kidder, Annemarie S. 2012. *Ultimate Price: Testimonies of Christians Who Resisted the Third Reich*. New York: Orbis Books.

King, Christine. 1990. "Jehovah's Witnesses under Nazism." In: *A Mosaic of Victims: Non-Jews Persecuted and Murdered by the Nazis*, ed. Michael Berenbaum. New York: New York University Press, 188–93.

Klee, Ernst. 1985. *Dokumente zur Euthanasie*. 1997 ed. Frankfurt am Main: Fischer Taschenbuchverlag GmbH.

———. 2004. *Auschwitz, die NS-Medizin und ihre Opfer*. 3rd ed. Frankfurt am Main: Fischer Verlag.

Knauer, Wilfried. 2002. "NS-Strafvollzug: 'Bessern' oder 'vernichten.'" In: *Justiz im Nationalsozialismus. Über Verbrechen im Namen des Deutschen Volkes*, ed. Niedersächsische Landeszentrale für politische Bildung, 36–48. Baden-Baden: Nomos-Verlagsgesellschaft.

Kogon, Eugen. 2006. *Der SS-Staat. Das System der deutschen Konzentrationslager*. 43rd ed. München: Wilhelm Heine Verlag. First published in 1946.

Kolb, Stephan, Paul Weindling, Volker Roelcke, and Horst Seithe. "Apologizing for Nazi Medicine: A Constructive Starting Point." *Lancet* 380: 722–23.

Korte, Detlef. 1991. *"Erziehung ins Massengrab." Die Geschichte des "Arbeitserziehungslagers Normark" Kiel Russee 1944–1945*. Kiel: Neuer Malik Verlag.

Kraushaar, Luise.1970a. *Deutsche Widerstandkämpfer 1933–1945. Band 1*. Berlin: Dietz Verlag

Kretschmar, Hans-Joachim. 1940. "Über das regelmäßige Vorkommen einer Flimmerepitheltasche im Bereich der Tonsilla palatina beim Menschen." *Zeitschrift für Mikroskopisch-Anatomische Forschung* 47: 137–50.

Kuhnke Manfred. 2001. *Falladas letzter Roman: Die wahre Geschichte*. Friedland: Steffen Verlag.

Kultusministerkonferenz. 1994. *Abschlussbericht "Präparate von Opfern des Nationalsozialismus in anatomischen und pathologischen Sammlungen deutscher Ausbildungs- und Forschungseinrichtungen."* Bonn, den 15.01.1994. Copy from the William Seidelman Collection of Papers.

Kurier. 1962. "Geköpfte lagen 15 Jahre im anatomischen Institut." *Kurier* 17 January 1962.

Kwiet, Konrad, and Helmut Eschwege. 1984. *Selbstbehauptung und Widerstand: Deutsche Juden im Kampf um Existenz und Menschenwürde 1933–1945*. Hamburg: Christians.

Lang, Hans-Joachim. 2007. *Die Namen der Nummern: Wie es gelang, die 86 Opfer eines NS-Verbrechens zu identifizieren*. Überarbeitete Ausgabe. Frankfurt am Main: S. Fischer Verlag.

———. 2013a. "Die Namen der Nummern." Accessed 2 May 2013. http://www.die-namen-der-nummern.de/index.html.

———. 2013b. "August Hirt and 'Extraordinary Opportunities for Cadaver Delivery' to Anatomical Institutes in National Socialism: A Murderous Change in Paradigm." *Annals of Anatomy* 195: 373–80.

Lange, Britta. 2005. "Sensible Sammlungen." In: *Sensible Sammlungen. Aus dem anthropologischen Depot*, ed. Margit Berner, Anette Hoffmann, and Britta Lange, 15–40. Hamburg: Philo Fine Arts.

Lenzen-Schulte, Martina. 2013. "Anatomen ohne Gewissen und Hitlers Henker. Die Nutzung von Opfern des Nationalsozialismus zu fragwürdigen Forschungszwecken." *Frankfurter Allgemeine Zeitung* 31 Juli 2013.

Leutner, Klaus. 2007. "Im Namen des deutschen Volkes? Auf der Grabsuche nach Bronislawa Czubowska." *Gedenkstättenrundbrief* 135: 27–40. Accessed 10 March 2015. http://www.gedenkstaettenforum.de/nc/gedenkstaetten-rundbrief/rundbrief/news/im_namen_des_deutschen_volkes/.

Lichtblau, Eric. 2013. "The Holocaust Just Got More Shocking." *New York Times*, Sunday 3 March 2013, 3.

Long Island Daily Press. 1941. "Nazis Execute One Objector." *Long Island Daily Press*, 21 March 1941. Accessed 12 March 2015. http://fultonhistory.com/Newspaper%2014/Jamaica%20NY%20Long%20Island%20Daily%20Press/Jamaica%20NY%20Long%20Island%20Daily%20Press%201941/Jamaica%20NY%20Long%20Island%20Daily%20Press%201941%20-%200408.pdf .

Majewski, Olaf Edward. 2013. "Medizin an der Reichsuniversität Posen (1941–1945) und der polnischen Untergrunduniversität der westlichen Gebiete U. Z. Z. (1942–1945)." Dissertation Medical Faculty Universität Heidelberg.

Malina, Peter, and Gustav Spann. 1999. "Das Senatsprojekt der Universität Wien 'Untersuchungen zur Anatomischen Wissenschaft in Wien 1938–1945.'" *Wiener Klinische Wochenschrift* 111(18): 743–53.

Michaelis, Werner. 1938. "Variationsstatistische Untersuchungen über die Kerngrössen und das Verhältnis von ein- und zweikernigen Zellen in der menschlichen Leber." *Zeitschrift für Mikroskopisch-Anatomische Forschung* 43: 567–80.

Mitscherlich, Alexander, and Fred Mielke. 1947. *Das Diktat der Menschenverachtung.* Heidelberg: Verlag Lambert Schneider.

Moos, Reinhard. 1999. "Die Rehabilitierung von Kriegsdienstverweigerern am Beispiel der Zeugen Jehovas." *Rundbrief. Verein zur Erforschung nationalsozialistischer Gewaltverbrechen und ihrer Aufarbeitung. Verein zur Förderung justizgeschichtlicher Forschungen*, Nr. 2, Dec 1999. Accessed 10 March 2015. http://old.doew.at/thema/rehabil/moos.html.

Müller, Rolf. 1942. "Untersuchungen über das Vorkommen von Vitamin C im Hoden des Menschen." *Zeitschrift für Mikroskopisch-Anatomische Forschung* 52: 440–54.

Müller-Hill, Benno. 1984. *Tödliche Wissenschaft.* Reinbek bei Hamburg: Rowohlt Taschenbuch Verlag.

Nałkowska, Zofia. 2000. *Medallions.* Evanston, IL: Northwestern University Press. Originally published in Polish in 1946.

National Armed Forces of Poland. 2012. "Polish underground soldiers 1944–1963: The Untold Story." Accessed 11 September 2014. http://www.nationalarmedforces.com/.

Neander, Joachim. 2006. "The Danzig Soap Case: Facts and Legends around 'Professor Spanner' and the Danzig Anatomic Institute 1944–1945." *German Studies Review* 29(1): 63–86.

Nelson, Anne. 2009. *Red Orchestra.* New York: Random House.

Neugebauer, Wolfgang. 1998. "Zum Umgang mit sterblichen Resten von NS-Opfern nach 1945." In: *Senatsprojekt der Universität Wien: Untersuchungen zur anatomischenWissenschaft in Wien 1938–1945*, ed. Akademischer Senat der Universität Wien, 459–65. Unpublished manuscript.

Noack, Thorsten. 2008. "Begehrte Leichen. Der Berliner Anatom Hermann Stieve (1886–1952) und die medizinische Verwertung Hingerichteter im Natinoalsozialismus." *Medizin, Gesellschaft und Geschichte. Jahrbuch des Instituts für Geschichte der Medizin der Robert Bosch Stiftung* 26: 9–35.

———. 2012. "Anatomical Departments in Bavaria and the Corpses of Executed Victims of National Socialism." *Annals of Anatomy* 194: 286–92.

Oehler-Klein, Sigrid, Dirk Preuss, and Volker Roelcke. 2012. "The Use of Executed Nazi Victims in Anatomy: Findings from the Institute of Anatomy at Giessen University, Pre- and Post-1945." *Annals of Anatomy* 194: 293–97.

Österreichisch Akademie der Wissenschaften, ed. 1959. *Österreichisches Biographisches Lexikon 1815–1950*. Vol. 2. Wien: Verlag der Österreichischen Akademie der Wissenschaften.

Oleschinski, Brigitte. 1992. "Der 'Anatom der Gynäkologen': Hermann Stieve und seine Erkenntnisse über die Todesangst und weiblichen Zyklus." *Modelle für ein deutsches Europa- Ökonomie und Herrschaft im Grosswirtschaftsraum*, ed. Horst Kahrs, Ahlrich Meyer and MG Esch, 211–18. Berlin: Rotbuch Verlag.

———. 1997. *Gedenkstätte Plötzensee*. Berlin: Möller Druck und Verlag.

Orth, Barbara. 2013. *Gestapo im OP: Bericht der Krankenhausärztin Charlotte Pommer*. Berlin: Lukas-Verlag.

Pawlowsky, Verena. 2005. "Erweiterung der Bestände. Die Anthropologische Abteilung des Naturhistorischen Museums 1938–1945." *Zeitgeschichte* 2(32): 69–90.

Peiffer, Jürgen. 1991. "Neuropathology in the Third Reich: Memorial to Those Victims of National-Socialist Atrocities in Germany Who Were Used by Medical Science." *Brain Pathology* 1: 125–31.

Perrault, Gilles. 1998. *La longue traque*. Paris: Librairie Arthème Fayard.

Pfoertner, Helga. 2003. *Mit der Geschichte leben: Mahnmale, Gedenkstätten, Erinnerungsorte für die Opfer des Nationalsozialismus in München 1933–1945. Band 2, I-P.* München: Literareon im Herbert Utz Verlag.

Platen-Hallermund, Alice. 1948. *Die Tötung Geisteskranker*. Heidelberg: Verlag der Frankfurter Hefte.

Plato, Alexander von, Almut Leh, Christoph Thonfeld, eds. 2008. *Hitlers Sklaven. Lebensgeschichtliche Analysen zur Zwangsarbeit im internationalen Vergleich*. Wien: Böhlau Verlag.

Plötzensee Memorial Center. 2012. "Josef and Veronika Augustyniak." Accessed 13 August 2012. http://www.gedenkstaette-ploetzensee.de/04_e.html.

Pringle, Heather. 2013. "The Sad Fate of Libertas Schulze-Boysen." Accessed 28 August 2014. http://www.lastwordonnothing.com/2013/02/18/the-sad-fate-of-libertas-schultze-boysen/.

Püschel, Almuth. 2002. *Verwehte Spuren. Zwangsarbeit in Potsdam. Fremdarbeiter und Kriegsgefangene*. Wilhelmshorst: Märkischer Verlag.

Ratschko, Karl-Werner. 2013. "Kieler Hochschulmediziner in der Zeit des Nationalsozialismus. Die Medizinische Fakultät der Christian-Albrechts-Universität im 'Dritten Reich.'" Dissertation Philosophische Fakultät Christian-Albrechts-Universität Kiel.

Reading Eagle. 1935. "Unborn Child May Save German Woman from Ax. April 1, 1935." Accessed 13 July 2012. http://news.google.com/newspapers?nid=1955&dat=19350401&id=BmIhAAAAIBAJ&sjid=tYcFAAAAIBAJ&pg=5567,22896.

Redies, Christoph, Michael Viebig, Susanne Zimmermann, and Rosemarie Fröber. 2005. "Origin of the Corpses Received by the Anatomical Institute at the University of Jena during the Nazi Regime." *Anatomical Record* 285 (Part B: New Anat.): 6–10.

Redies, Christoph, Rosemarie Fröber, Michael Viebig, and Susanne Zimmermann. 2012. "Dead Bodies for the Anatomical Institute in the Third Reich: An Investigation at the University of Jena." *Annals of Anatomy* 194: 298–303.

Richter, Gunnar. 2009. *Das Arbeitserziehungslager Breitenau (1940–1945). Ein Beitrag zum nationalsozialistischen Lagersystem*. Kassel: Verlag Winfried Junior. Accessed 11 March 2013. https://kobra.bibliothek.uni-kassel.de/bitstream/urn:nbn:de:hebis:34-201 1120539885/1/RichterArbeitserziehungslagerBreitenau.pdf.

Röhl, John Charles Gerald. 2002. *Kaiser, Hof und Staat: Wilhelm II. und die deutsche Politik*. München: C. H. Beck.

Roelcke, Volker. 2010. "Psychiatrie im Nationalsozialismus. Historische Kenntnisse, Implikationen für aktuelle ethische Debatten." *Nervenarzt* 11: 1317–25.

Romeis, Benno. 1953. "Hermann Stieve +." *Anatomischer Anzeiger* 99: 401–40.

Rosmus, Anna. 1993. *Wintergrün. Verdrängte Morde*. Konstanz: Labhard.

Roth, Karl Heinz. 1984a. "Grosshungern und Gehorchen. Das Universitätsklinikum Eppendorf." In: *Heilen und Vernichten im Mustergau Hamburg. Bevölkerungs- und Gesundheitspolitik im Dritten Reich*, ed. Angelika Ebbinghaus, and Karl-Heinz Roth, 109–135. Hamburg: Konkret Literatur Verlag.

Rothmaler, Christiane. 1990. "Gutachten und Dokumentation über das Anatomische Institut des Universitätskrankenhauses Eppendorf der Universität Hamburg 1933–1945." "1999." *Zeitschrift für Sozialgeschichte des 20. u 21. Jahrhunderts* 2: 78–95.

———. 1991. "Die Sammlung des Anatomischen Instituts der Hansischen Universität in Hamburg: Didaktisches Konzept und Aufbau 1919 bis 1945." In: *Ideologie der Objekte, Objekte der Ideologie; Naturwissenschaft, Medizin und Technik in Museen des 20. Jahrhunderts*. ed. Deutsche Gesellschaft für Geschichte der Medizin, Naturwissenschaft und Technik (73. Jahrestagung), 55–63. Kassel: Georg Wenderoth Verlag..

Sachnowitz, Herman. 2002. *The Story of "Herman der Norweger" Auschwitz prisoner #79235*. Lanham, MD: University Press of America.

Schagen, Udo. 2005. "Die Forschung an menschlichen Organen nach 'plötzlichem Tod' und der Anatom Hermann Stieve (1886–1952)." In: *Die Berliner Universität in der NS-Zeit. Band II: Fachbereiche und Fakultäten*, ed. Rüdiger von Bruch and Rebecca Schaarschmidt, 35–54. Stuttgart: Franz Steiner Verlag Wiesbaden GmbH.

Scheer, Regina. 2004. *Im Schatten der Sterne. Eine Jüdische Widerstandsgruppe*. Berlin: Aufbau-Verlag.

Scherrieble, Joachim. 2008. *Der Rote Ochse Halle (Saale). Politische Justiz 1933–1945, 1945–1989*. Berlin: Christoph Links Verlag.

Schilde, Kurt. 1993. *Eva-Maria Buch und die "Rote Kapelle." Erinnerungen an den Widerstand gegen den Nationalsozialismus*. Berlin: Overall Verlag.

Schiller, Erich. 1942. "Über den Fettgehalt der Leber beim gesunden Menschen." *Zeitschrift für Mikroskopisch-Anatomische Forschung* 51: 309–21.

Schneider, Frank. 2011. *Psychiatrie im Nationalsozialismus. Erinnerung und Verantwortung*. Berlin: Springer.

Schönhagen, Benigna. 1992. "Das Gräberfeld X auf dem Tübinger Stadtfriedhof. Die verdrängte 'Normalität' nationalsozialistischer Vernichtungspolitik." In: *Menschenverachtung und Opportunismus, Tübingen: Zur Medizin im Dritten Reich*, ed. Jürgen Peiffer, 69–92. Tübingen: Attempto.

Scholz, Harald. 2007. "Angeklagt wegen Wehrkraftzersetzung—eine Frau vor dem Volksgerichthof." Accessed 10 March 2015. http://www.kultur-im-olg.de/assets/templates/kuk/downloads/termin07/2007-08-15vortreoeremark.pdf.

Schultka, Rüdiger, and Michael Viebig. 2012. "The Fate of Bodies of Executed Persons in the Anatomical Institute in Halle between 1933 and 1945." *Annals of Anatomy* 194: 274–80.

Schulz, Anke. 2010. *Hamburger Zwangsarbeiterlager in der Lederstrasse 1939–1945*. Aachen: Shaker Verlag.

Schulze, Heta. 1938. "Über den Fettgehalt in dem Epithel des Gallengangssystems bei verschiedenen Säugetieren." *Zeitschrift für Mikroskopisch-Anatomische Forschung* 44: 489.

Schwarze, Gisela. 1997. *Kinder, die nicht zählten. Ostarbeiterinnen und ihre Kinder im zweiten Weltkrieg*. Essen: Klartext.

Seeger, Andreas. 1998. ""Gegen Schwerstverbrecher ist in Kriegszeiten die zugelassene Todesstrafe grundsätzlich die gebotene." Todesurteile des Sondergerichtes Altona/Kiel 1933–1945." In: *"Standgericht der inneren Front" Das Sondergericht Altona/Kiel 1932–1945*, ed. Robert Bohn and Uwe Danker, 166–89. Hamburg: Ergebnisse Verlag.

Seidelman, William E. 1996. "Nuremberg Doctors' Trial: Nuremberg Lamentation: For the Forgotten Victims of Medical Science." *British Medical Journal* 313: 1463–67.

———. 2012. "Dissecting the History of Anatomy in the Third Reich: 1989–2010: A Personal Account." *Annals of Anatomy* 194: 228–36.

———. 2014. "Requiescat sine Pace. Recollections and Reflections on the World Medical Association, the Case of Prof. Dr. Hans Joachim Sewering and the Murder of Babette Fröwis." In: *Silence, Scapegoats, Self-Reflection: The Shadow of Nazi Medical Crimes on Medicine and Bioethics*, ed. Volker Roelcke, Sascha Topp and Etienne Lepicard, 281–300. Göttingen: V&R unipress.

Snyder, Timothy. 2010. *Bloodlands: Europe between Hitler and Stalin*. New York: Basic Books.

Speck, Dieter. 2002. "Universitätskliniken und Zwangsarbeit: Das Beispiel Freiburg." In: *Medizin und Zwangsarbeit im Nationalsozialismus. Einsatz und Behandlung von "Ausländern" im Gesundheitswesen*, ed. Andreas Frewer and Günther Siedbürger, 231–52. Frankfurt/New York: Campus Verlag.

Spring, Claudia. 2005. "Vermessen, deklassiert und deportiert. Dokumentation zur anthropologischen Untersuchung an 440 Juden im Wiener Stadion im September 1939 unter der Leitung von Josef Wastl vom Naturhistorischen Museum." *Zeitgeschichte* 2(32): 91–110.

Steegmann, Robert. 2005a. "La Faculté de Médecine de la Reichsuniversität de Strasbourg et las expérimentations médicales au KL-Natzweiler." In: *Les Reichsuniversitäten des Strasbourg et de Poznan et les resistances universitaires 1941–1944*, ed. Christian Bächler, François Igersheim, and Pierre Racine, 143–58. Strasbourg: Presse Universitaires des Strasbourg.

Steinborn, Carsten. 2009. "Ein Foto von Marianne Latoschinski ist aufgetaucht." *Mitteldeutsche Zeitung* 28.1.0.09. Accessed 10 March 2015. http://www.mz-web.de/bernburg/ein-foto-von-marianne-latoschinski-ist-aufgetaucht,20640898,17952118.html.

Step21 [Weisse Flecken]. 2008. "Herta Lindner: ein Leben im deutsch-tschechischen Widerstand." *Step21*. Accessed 1 May 2012. http://www.step21.de/uploads/tx_templa voila/080123_WF2_zeitung_Komplett_FINAL.pdf.

Stieve, Hermann. 1942. "Der Einfluss von Angst und psychischer Erregung auf Bau und Funktion der weiblichen Geschlechtsorgane." *Zentralblatt für Gynäkologie* 66: 1698–1708.

———. 1946. "Über Wechselbeziehungen zwischen Keimdrüsen und Nebennierenrinde." *Das deutsche Gesundheitswesen* 18(1): 537–45.

———. 1952. *Der Einfluss des Nervensystems auf Bau und Tätigkeit der Geschlechtsorgane des Menschen*. Stuttgart: Thieme.

Szołdrska, Halska. 1948. *Walka z Kultura Polska*. Poznan: Odbito w drukarni Uniwersytetu Poznanskiego.

Teschler-Nicola, Maria, and Margit Berner. 1998. "Die Anthropologische Abteilung des Naturhistorischen Museums in der NS-Zeit: Berichte und Dokumentation von Forschungs- und Sammlungsaktivitäten 1938–1945." In: *Senatsprojekt der der Universität Wien*, ed. Akademischer Senat der Universität Wien, 333–58. Accessed 27 September 2013. http://www.nhm-wien.ac.at/jart/prj3/nhm/data/uploads/mitarbeiter_dokumente/berner/Senatsber.pdf.

Time. 1935. "Germany: Meanest mother." *Time*, 9 September 1935. Accessed 13 July 2012. http://www.time.com/time/magazine/article/0,9171,748953,00.html.

Toledano, Raphael. 2013. "Deliveries of Dead Bodies at the Anatomical Institute of the Reichsuniversität Strassburg between 1941 and 1944." *Pulse-Project*. Accessed 13 March 2015. http://www.pulse-project.org/node/588.

Tomaszewski, Irene, ed. 2006. *Inside a Gestapo Prison: The Letters of Krystyna Wituska, 1942–1944*. Detroit: Wayne State University Press.

Tomkiewicz, Monika, and Piotr Semków. 2013. *Soap from Human Fat: The Case of Professor Spanner*. Gdynia, Poland: Wydawnictwo Róża Wiatrów.

Trepper, Leopold. 1995. *Die Wahrheit. Autobiographie des "Grand Chef" der Roten Kapelle*. Freiburg: Ahriman-Verlag.

Ude-Koeller, Susanne, Wilfried Knauer, and Christoph Viebahn. 2012. "Anatomical Practice at Göttingen University since the Age of Enlightenment and the Fate of Victims from Wolfenbüttel Prison under Nazi Rule." *Annals of Anatomy* 194: 304–13.

Universität Tübingen. 1990. "Ergebnisbericht. Überprüfung der Sammlungen des anatomischen Institutes auf das Vorhandensein von Präparaten von NS-Opfern. 3.4.1989." In: *Berichte der Kommission zur Überprüfung der Präparatesammlungen in den medizinischen Einrichtungen der Universität Tübingen im Hinblick auf Opfer des Nationalsozialismus,* by Universität Tübingen. Manuscript, collection Dr. William Seidelman.

University of Vienna. 2013. *Memorial Book for the Victims of National Socialism at the University of Vienna in 1938.* Accessed 9 May 2013. http://gedenkbuch.univie.ac.at/index .php?id=435&no_cache=1&L=2&id435=&no%28underscore%29cache1=&L2=.

USHMM. 2014a. "A Mosaic of Victims: In Depth." *United States Holocaust Memorial Museum.* Accessed March 27, 2014. http://www.ushmm.org/wlc/en/article.php?Mod uleId=10007329.

Viebig, Michael. 2012. "'…The Cadaver Can Be Placed at Your Disposition Here': Legal, Administrative Basis of the Transfer of Cadavers in the Third Reich, Its Traces in Archival Sources." *Annals of Anatomy* 194: 267–73.

Vinke, Hermann. 2003. *Cato Bontjes van Beek: "Ich habe nicht um mein Leben gebettelt": Ein Porträt.* Zürich-Hamburg: Arche Verlag AG.

Vsevolodov, Vladimir. 2015. "Unterlagen in russischen Archiven zur Untersuchung der sowjetischen Militärkommission im Anatomischen Institut der Universität Greifswald 1947." In: *"… die letzten Schranken fallen lassen": Studien zur Universität Greifswald im Nationalsozialismus,* ed. Dirk Alvermann, 291–310. Köln, Weimar, Wien: Böhlau Verlag.

Wagner, Walter. 1974. *Der Volksgerichtshof im nationalsozialistischen Staat. Quellen und Darstellungen zur Zeitgeschichte. Band 16/III. Die Deutsche Justiz im Nationalsozialismus.* Stuttgart: Deutsche Verlagsanstalt.

Wagner, Jens-Christian. 2010. "Zwangsarbeit im Nationalsozialismus: ein Überblick." In: *Zwangsarbeit. Die Deutschen, die Zwangsarbeiter und der Krieg,* ed. Volkhard Knigge, Rikola-Gunnar Lüttgenau, Jans-Christian Wagner, 180–93. Weimar: Druckhaus Gera GmbH.

Weindling, Paul J. 1992b. "Psychiatry and the Holocaust." *Psychological Medicine* 22: 1–3.

———. 2010. "Psychiatrische Opfer von Humanexperimenten im Nationalsozialismus. Jeder Mensch hat einen Namen." *Psychiatrie* 7(4): 255–60.

———. 2013. "From Scientific Object to Commemorated Victim: The Children of the Spiegelgrund." *History and Philosophy of the Life Sciences* 35: 415–30.

———. 2015. *Victims and Survivors of Nazi Human Experiments: Science and Suffering in the Holocaust.* London: Bloomsbury Academic.

Weindling, Paul J, and Anna von Villiers. 2014. *Victims of Human Experiments and Coercive Research under National Socialism.* Accessed 14 August 2014. http://www.history .brookes.ac.uk/research/centres/hms/vhens/.

Williams, David J. 1988. "The History of Eduard Pernkopf's Topographische Anatomie des Menschen." *Journal of Biocommunication* 15(2): 2–12.

Winkelmann, Andreas. 2008. "Wann darf menschliches Material verwendet werden? Der Anatom Hermann Stieve und die Forschung an Leichen Hingerichteter." In: *Die Charité im Dritten Reich. Zur Dienstbarkeit medizinischer Wissenschaft im Nationalsozialismus,* ed. Sabine Schleiermacher and Udo Schagen, 105–20. Paderborn: Ferdinand Schöningh.

Winkelmann, Andreas, and Udo Schagen. 2009. "Hermann Stieve's Clinical-Anatomical Research on Executed Women during the 'Third Reich.'" *Clinical Anatomy* 22(2): 163–71.

Wippermann, Wolfgang. 2001. *Die Berliner Gruppe Baum und der jüdische Widerstand. Beiträge zum Thema Widerstand 19.* Accessed 12 September 2014. http://www.gdw-berlin .de/fileadmin/bilder/publ/beitraege/B19.pdf.

Wischmann, Tewes. 2011. "'Paracyclische Ovulationen' und 'Schreckblutungen'—eine Fortsetzung." In: *Nichts ist unmöglich!? Frauenheilkunde in Grenzbereichen. Beiträge der 39. Jahrestagung der Deutschen Gesellschaft für Psychosomatische Frauenheilkunde und Ge-*

burtshilfe (DGPFG e. V.), ed. Susanne Ditz, Brigitte Schlehofer, Friederike Siedentopf, Christof Sohn, Wolfgang Herzog, and Martina Rauchfuss, 147–52. Frankfurt am Main: Mabuse-Verlag.

Wunderlich, Jens. 2011. "Meine Brigade trug ihren Namen." *Rotfuchs/115: Tribüne für Kommunisten und Sozialisten* Nr. 161, Juni 2011. Accessed 12 September 2014. http://www.rotfuchs.net/zeitung/archiv/2011/RF-161-06-11.pdf.

Zimmermann, Susanne. 2007. ""… er lebt weiter in seinen Arbeiten, die als unverrückbare Steine in das Gebäude der Wissenschaft eingefügt sind"—Zum Umgang mit den Arbeiten des Anatomen Hermann Stieve (1886–1952) in der Nachkriegszeit." In: *Täterschaft- Strafverfolgung- Schuldentlastung. Ärztebiographien zwischen nationaler Gewaltherrschaft und deutscher Nachkriegsgeschichte*, ed. Boris Böhm, and Norbert Haase, 29–40. Leipzig: Leipziger Universitätsverlag.

Zimmermann, Volker. 2007. "Die medizinische Fakultät der Göttinger Georgia Augusta während der NS-Diktatur." In: *Leiden verwehrt vergessen. Zwangsarbeiter in Göttingen und ihre medizinische Versorgung in den Universitätskliniken*, ed. Volker Zimmermann, 17–54. Göttingen: Wallstein Verlag.

Chapter 8

THE SCIENCE OF ANATOMY IN
NATIONAL SOCIALIST GERMANY

> It would be comforting to believe, of course, that good science
> tends to travel with good ethics, but the sad truth seems to be
> that cruelty can coexist fairly easily with "good science."
> Robert Proctor, 2000[1]

Anatomists, who forever had to cope with shortages of "material" for their teaching and research purposes, realized very quickly the new situation arising through the changes in body supply under the NS regime. Not only did most of them have no qualms about using the bodies of NS victims, but they also saw it as their professional duty and unique career opportunity to optimize the use of the valuable new "asset." The result of this attitude was that after the change in traditional body supply during the Third Reich, the science itself also started to change.

Science, not Pseudoscience

Scientific investigations performed during National Socialism, including studies in racial hygiene, human experiments, and anatomical research, have often been called "pseudoscientific." The term "pseudoscience" describes a system of theories, assumptions, and methods that are purportedly scientific, but not rooted in the scientific method. The word is usually employed in an effort to distance the speaker from this type of scientific endeavor, thereby allowing the dismissal of any further reflection on the time period and its science as of no relevance for the present and the future. After WWII the medical community denounced much of the controversial scientific work performed during the NS period, especially in racial hygiene and human experiments, as "pseudoscience". Physicians declared not to have been part of such unscientific activities. This defensive argument was widely accepted by the public[2] and helped establish the prevailing silence on the involvement of physicians in harmful research and ideological support for the NS regime for decades after WWII. The fact was generally suppressed that scientists working in NS Germany were often competent and sometimes produced exciting new results, even in the face of US interest in German science after the war.[3] Also, racial hygiene was an internationally accepted field of science in the early twentieth century, even if it had its critics both inside and outside of Germany.[4] The concept of "pseudoscience" and the inadequacy of scientific results during National Socialism belong to the many myths surrounding the perception of relations between the NS regime and medicine.[5] The truth is much more complicated. Medical atrocities were not committed by only a few fervent NS physicians and researchers, but by many members of the general medical scientific community, who were intricately woven into and collaborated with the NS system.[6]

Anatomists were among these scientists. They were professionally adequate, if not necessarily inspired in their research. German anatomy was also not quite as isolated from international contacts as previously claimed, as relationships with Sweden were still quite active[7] and anatomists attended international meetings in Italy and Hungary.[8] The teaching and science of anatomy in NS Germany was not a field isolated in time and space, but continued along lines of thought and method reaching back into the nineteenth century, and included the use of bodies of the executed.[9] Indeed, it continued into the postwar period in Germany. Some of the work was new, other work was still focused on concepts in need of innovation, but most of the teaching and research was based on the scientific method and did not fit the term "pseudoscience" at all. The biomedical sciences were certainly not destroyed by the National Socialists, as researchers liked to pronounce after the war, but certain areas were suppressed whereas others were encouraged.[10] For example, the dynamic concept of racial hygiene by Saller and Weidenreich was repressed by the NS regime and these scientists found their careers disrupted, but Fischer's static construct of race led to his promotion. Similarly, NS legislation produced new "opportunities" for anatomists through the availability of a plentiful body supply. The NS regime left an imprint on the science of anatomy, mostly so in the eager and unquestioning acceptance of bodies of NS victims for dissection and research, and in the "opportunity" of moving toward human experimentation.

Use of Bodies of the Executed in Anatomy

Bodies of executed persons were historically the first legal source for anatomical dissection and are still used in parts of the world.[11] Nobel Laureate Werner Forssmann began his medical studies at the University of Berlin in 1922 and described the routine of tissue acquisition: "[The anatomical institute] sent an assistant and a senior technician to each of the executions, which had now become a rarity, to conserve organs and tissues freshly for microscopic examination."[12] As numbers of executions soared in Germany after 1933, bodies of executed persons were used widely for medical education and research. The bodies of the executed were also used in the creation of scientific teaching materials, among them collections and books. At the anatomical institute of the University of Hamburg, Johannes Brodersen created exhibits from these bodies for the teaching collection,[13] and Pernkopf used them for his atlas. Anatomist Gerhard Aumüller and historian Cornelia Grundmann[14] were the first to point out that some anatomists reported the use of "material" of such bodies in publications from the war period. To find out whether research on the executed and its explicit mention in publications was unique to the time of the Third Reich and if it was more common in Germany than elsewhere, a 2013 study analyzed publications in German and Anglo-American anatomical journals from 1924 to 1951. The aim was to assess the explicit mention of "material" from the bodies of the executed.[15] The majority of German-language papers originated in Germany and Austria, and most of the English-language papers in the United Kingdom and the United States.

The articles on human anatomy frequently omitted any explanation of the provenance of the tissues used. "Materials" explicitly listed originated from various sources, including routine anatomical dissection in medical education, autopsies in departments of pathology, surgical specimens, embryological specimens from gynecological departments and autopsies, historical collections, radiographs from living persons, anthropological studies on living and dead persons, volunteer living persons (e.g., biopsy material), and executed persons. Of the 7,438 German-language anatomical papers, 166 explicitly mentioned the use of "material" from bodies of the executed (2.2 percent), whereby three of the 166 were studies from departments of pathology. An additional seventeen articles originated in other European countries, among them seven from Switzerland, four from Austria before 1938, two from Estonia, two from Hungary, and two from Sweden. Before 1933, thirty-three of a total of 3,734 papers (1 percent) listed the use of the executed. In the early NS period from 1933 to 1938 the number rose to forty-six out of 2,265 (2 percent), and increased again during the war years from 1939 to 1945 to seventy-three out of 984 papers (7 percent).

After the war, from 1946 to 1951, the number was fifteen out of 486 articles (3 percent). In comparison, the British *Journal of Anatomy* presented 1,154 papers from 1924 to 1951, with no report of the use of the executed. Similarly, the American journals *Anatomical Record* and *American Journal of Anatomy* published a total of 2,711 and 837 papers respectively, with only one paper in each mentioning the use of material from executed persons. A study from South Africa reported the use of "material" from bodies of executed "Bantu males"[16] and a US anatomist studied the foot of a "young (22 years) executed criminal".[17] The journal with the highest output of studies using "material" from the executed was the *Zeitschrift für mikroskopisch-anatomische Forschung*, which was founded in 1924 under the editorship of Hermann Stieve. Twenty-three of the thirty-one German anatomical departments published such articles, with six of them accounting for three-quarters of the total. Clara's group in Leipzig was the most prolific with thirty-eight publications, followed by Blotevogel's in Breslau with fourteen, Heiss' in Königsberg with thirteen, Stieve's in Berlin with twelve, Stöhr's in Bonn with nine and Petersen's and Elze's group in Würzburg with nine. Clara authored ten papers, von Hayek nine, Josef Wallraff (Breslau) eight, Bargmann six, Schiller six, and Stieve five.

Historically, bodies of the executed were preferred "material" for histological studies because the time of death was known, so the removal of tissues could be planned in advance and the "freshness" of the tissues could be ensured.[18] The latter was particularly important in those tissues that deteriorated quickly after death, such as adrenal glands, mucosa and certain nervous tissues. Obtaining tissues from the bodies of the executed was difficult during the Weimar Republic due to the decline in execution rates. Therefore, this type of "material" became very valuable. Anatomists complained that procurement of "normal cases, for example executed persons, is very difficult, nearly impossible"[19] and helped each other by providing "ideal material" for research on sensitive tissues.[20] For German anatomists the use of tissues from the executed had become a "gold standard" for the quality of histological studies even before 1933. With the rising numbers of these bodies after 1933, anatomists discovered new qualities in these "life-like" tissues. They used bodies of the executed as "control material" on a regular basis, and applied new techniques directly in the execution chambers, such as posing of bodies in certain positions and new embalming approaches. Often large supplies of such bodies were available, including those of women beginning in 1935.[21] Even after the war, papers were published that most likely used "material" from the NS period, based on the identity of the authors and the sources listed.[22] The Berlin and Leipzig anatomical departments are documented to have accepted bodies of the executed in the first years after the war, as did the anatomical institute of Graz.[23]

It is unclear why anatomists who published in English-language journals never mentioned the use of "material" from bodies of the executed. It seems doubtful that the many studies on large numbers of unclaimed bodies from the anatomical dissection laboratories of medical schools in the United States did not include the executed, as many states at the time allowed capital punishment. Also, while the use of bodies of the executed for anatomical purposes was expressly forbidden in the UK since the Warburton Act in 1832, not all countries of the British Commonwealth had passed similar legislation by the 1930s.[24] Thus it is quite possible that executed persons were among the number of unclaimed bodies used for anatomical dissection. This lack of an emphasis on reporting bodies of the executed as a special source of tissue may also indicate that anatomists who published in English-language journals apparently did not rely on the "opportunities" provided by "material" from the bodies of the executed. Even though some British anatomists apparently envied their German colleagues for their access to "such valuable sources of material,"[25] they still performed similar investigations, often with routine postmortem and surgical material or with animal models.[26] Thus it is at least debatable whether there was a true scientific need for this standard.

Just as authors in the United States and the United Kingdom did not mention the use of "material" from the executed, neither did anatomists who worked in the Soviet Union. Until 1933 German-language journals frequently published papers by Soviet authors, many of them studies with large numbers of bodies. A typical example was the research by A. Chanamirjan from the North Ukrainian State University in Rostow-on-Don, who investigated 260 bodies of fetuses, neonates, and adults.[27] It has to be assumed that the bodies in these large-scale studies were unclaimed and, given Soviet politics at the time,[28] may have included those of executed persons. E. Stankiewitsch from the Belarusian State University in Minsk addressed the problem of obtaining fresh human material by studying how long it took neuronal cells in dogs to deteriorate after death. He found that the cells remained structurally intact until twenty-four hours after death and thus saw the use of "older" postmortem human tissues as justified.[29]

What are the possible explanations for the difference in attitude toward working with "material" from the executed between German and Anglo-American anatomy? Anatomists from both backgrounds had certainly been used to working with bodies from the executed throughout several centuries of scientific history,[30] and historical anatomical collections in the United Kingdom, the United States, and Germany still hold specimens and models from the bodies of executed persons obtained in the eighteenths, nineteenth and twentieth century.[31] Maybe it was simply a tradition of German-language papers to expressly mention executed persons. However, this would not account for the German anatomists' view of these tissues as a "gold standard." At this point there is not enough information to explain why anatomists trained in the German tradition felt the need to refer to this standard before 1933, and an examination of the important French literature in anatomy is still missing. However, German anatomists' familiarity with the use of the executed for published research before 1933 may explain why they so easily and often gladly accepted the "opportunities" offered to them through the execution practices of the NS regime.

Human Experiments in Anatomy

Experimentation in anatomy traditionally belongs to the postmortem domain. Anatomists remove unaltered tissues, sometimes organs injected before removal, and at most, pose the joints or the whole body to study body mechanics.[32] However, by 1942 anatomists began to experiment with live humans. Winkelmann and Noack[33] were the first to point to a human experiment performed by Clara, who wrote in a 1942 study of vitamin C distribution in the human nervous system: "The material evaluated in the current study stems from 5 apparently healthy adult individuals of different ages, who without exception all died of a sudden death after varying periods of imprisonment; a 33¾ year old male individual received 1 pill of Cebion (Merck) four times daily for the last 5 days before his death".[34] Clara had recognized that he had a unique "opportunity": access to "material" from prisoners of the NS regime not only after the victim's death, but also before. He shared the organs of the man mentioned in the 1942 study with his pupils Lothar Heckel, Rolf Müller, and Erich Schiller.[35] Clara, who often contacted the authorities directly with his professional needs,[36] had apparently received permission to administer vitamin C tablets to this prisoner. While the vitamin C in itself was harmless, the coercive nature of this experiment on a death-row prisoner is evident.[37] Clara's experiment signifies the decisive transgression from anatomists' work with the dead to their work with the still living but "future dead." The prisoner's death became part of Clara's anatomical research plan, which resembled the medical experiments performed by psychiatrist Carl Schneider on children within the "euthanasia" program in Heidelberg.[38]

It seems that the war years, with their increasing brutalization of normal life, radicalized some anatomists to cross the traditional boundary separating living and dead subjects. Kremer's life and work has been described in chapter 6. Like Clara, this anatomist used the "opportunities" provided by the NS regime for research, expanding his previous series of animal studies to humans in Auschwitz, and contributing to the death of many prisoners. However offensive Kremer's activities were, there is no evidence that he had a concrete research plan when he arrived at Auschwitz. In contrast, Hirt had a clearly outlined research design for his studies on prisoners from concentration camps. Early on he became the true master and theoretician of anatomical work with the "future dead" when he realized the potential new opportunities for research offered by the murderous reality of the NS regime.[39] By the late 1930s, he had shifted his scientific investigations from basic anatomy and microscopic technology to several new fields, among them the exploration of racial anthropology and the effect of warfare agents on the human body. In 1942 Hirt performed coercive medical experiments with mustard gas on prisoners in the concentration camp Natzweiler-Struthof, which led to the suffering and death of many victims.[40] This research continued Hirt's work on animal models he had performed in 1941.[41] While the mustard gas experiments were outside the realm of traditional anatomical study, his plans for a so-called "racial" skeleton collection belonged to the anatomical field of physical anthropology. Hirt saw this plan as an extension of the famous craniological collection created by Gustav Schwalbe at the anatomical department in Strasbourg.[42] In June 1943, at Hirt's behest, SS anthropologists Bruno Beger and Hans Fleischhacker selected 115 prisoners from Auschwitz. Among them were 109 Jews and so-called "Asiatic types," in whom Beger had a specific scientific interest.[43] Eighty-six Jewish prisoners were then transported by train to Natzweiler-Struthof and murdered there in a gas chamber in August 1943. Their bodies were transferred to the anatomical department in Strasbourg.[44] Hirt himself had handed the Zyklon B gas capsules with instructions for the murder of these prisoners to Josef Kramer, the director of the camp.[45] The bodies of the victims were still in the anatomical department when the Allied Forces liberated Strasbourg in November 1944. Due to a lack of documentation it is still unclear whether this group of victims was also used as the source for studies on male reproductive organs performed by Hirt's assistant Kiesselbach.[46]

Clara, Kremer, and Hirt are the only anatomists known to have performed human experiments. Results of these studies were published by Clara, but there is no documentation that Kremer and Hirt actually did research on the "material" they had gathered. Clara, on the other hand, had plans to expand his human experiments after his recruitment to the University of Munich in 1942. In a 1943 letter to the prison authorities in Munich Stadelheim, he asked that synthetic vitamin C be secretly added to the food of prisoners on death row.[47] Apparently Clara's request was granted, as he reported on at least three more prisoners with premortem vitamin C ingestion in his postwar publications.[48]

Other Medical Disciplines Using the Bodies of NS Victims

Many other medical disciplines made use of the bodies of victims of the NS regime for research. In fact, it is hard to find a medical field that did not use tissues from these victims. Some of them have been investigated in detail, such as forensic medicine and neuropathology, while systematic studies are still missing for others, like pathology. A complete overview of currently available documentation on individual medical specialties would go beyond the scope of this book, but examples pertinent with respect to their association with anatomy will be discussed here. The fields closest to anatomy, in terms of the epistemological premise of acquiring knowledge through the study of bodies of dead persons, are forensic medicine, pathology, and neuropathology. The practitioners of these disciplines developed a pattern of ethical transgressions similar to that in anatomy: individual forensic specialists and pathologists crossed the boundary from working with the dead to experimenting with the "future dead." Similarly, those within clinical specialties, who traditionally studied the living, now also started to integrate the planned death of their subjects into their research designs and transgressed ethical lines in a complimentary but reverse pattern to anatomy. Under the conditions set by the NS regime all manner of ethical offenses became possible. Access to prisoner-subjects of human experiments was granted through the SS and Himmler's *Ahnenerbe.*[49]

Forensic medicine was deeply affected by NS policies, not only in its relationship with the government and in the delivery of bodies of NS victims, but also in its scientific research and practice.[50] Concepts of racial biology and inheritance became prominent, and forensic medical specialists contributed to a change in the legal system by helping to shift the focus of criminal law from the crime itself to the perpetrators and their biology. A prominent representative of this new school of criminal biology was Georg Dahm, rector of the University of Kiel. He claimed that a thief was not defined as a person who steals a thing, but rather as a person who is a thief by nature,[51] thereby declaring criminality to be a heritable trait that was unchangeable and independent of environmental factors. This argument left many criminal offenders vulnerable to the application of the sterilization laws. As forensic specialists also contributed evaluations for the Hereditary Health Courts, they became doubly responsible for the victims' sterilization: first by providing the biologistic argument for the conviction, and then by writing biological evaluations of the individual "criminals." Forensic pathologists were also called as expert witnesses in cases of self-mutilation in the military. Specialists at the institute of forensic medicine at the University of Vienna helped with the development of more efficient methods of mass murder, among them the murder of victims with automobile exhaust fumes.[52]

At the Institute for Forensic Medicine of the University of Halle, which was responsible for the investigation of unnatural deaths, human experiments were performed. Director Gerhard Schrader supervised the studies of doctoral student Siegfried Krefft on forced laborers who were executed by hanging.[53] Krefft's medical dissertation, titled "On the Genesis of Bleeding into the Cervical Musculature during Death by Hanging," was based on the close observations of the deaths of fifteen prisoners. According to forensic pathologist Herber, Krefft's dissertation was of little or no scientific value.[54] Nevertheless, Krefft had a long postwar professional career.[55]

Krefft's observational research borders life and death, and as such is a transgression of the traditional boundary in forensic medicine. There is at least one example of a forensic pathologist who crossed this boundary completely by performing experiments on living human beings.[56] Gerhart Panning became director of the institute for forensic medicine at the University of Berlin in 1942. He was also a 6th Army staff doctor and served as a forensic adviser at the Eastern Front during the war.[57] His research interest was in ballistics, and he performed human experiments in this field on Jewish Soviet prisoners of war in August 1941, in collaboration with the chief of the *Sonderkommando*, SK4a Paul Blobel.[58] He sought to scientifically confirm the Soviets' use of explosive infantry ammunition, which was internationally banned according to the Geneva Convention, by recording the morphology of wounds inflicted by these bullets. The experiment was performed in a field close to Zhytomyr in the Ukraine with the assistance of SS personnel. Panning pointed to the parts of the victims that were to be injured, and had the SS men repeatedly shoot at least six or seven prisoners with the Soviet ammunition. The victims suffered greatly before their death. Immediately thereafter Panning performed autopsies in a nearby farmhouse.[59] He published his results in a professional journal in 1942, without reference to the true source of his data.[60]

Human experiments in forensic pathology were also carried out at the concentration camp Buchenwald. SS physician Erich Wagner pursued a study on tattoos there under the mentorship of Friedrich Timm, forensic pathologist at the University of Jena,[61] in which he collected tattoos from 800 prisoners, whom he had selected from the camp population and killed soon thereafter. Their skin was also processed at the pathology department together with the skin from other prisoners, which was used for items of daily living for SS officers, such as lampshades.[62] Wagner received his doctoral title in 1940. He was imprisoned in 1958 and escaped trial by committing suicide in 1959.[63] The processing of skin from deceased prisoners, especially those with tattoos, was also reported from the concentration camps Mauthausen-Gusen and Dachau.[64] While some of these observations were greatly overblown in postwar press reports, they do have real incidents of abuse at their core.[65] Wagner's study and the historical facts of skin processing are activities far removed from true scientific inquiry. However, the story of the abuse of "Jewish bodies" including their skin became such an important topos of the postwar Holocaust narrative, that it is essential to include this core of facts here.[66]

Pathologists also became intimately involved in activities specific to the NS period. Military science was increasingly integrated into teaching and research, and bodies of NS victims served for research purposes.[67] Many of the concentration camps had pathology departments, which were run by SS physicians and prisoner physicians.[68] Also, pathology departments outside the camps collaborated with camp physicians. Recently discovered documents confirm the suspicion long held by historians that Sigmund Rascher, an SS officer who performed deadly hypothermia and high-altitude experiments on prisoners in Dachau, sent five bodies of his victims to the pathological department at the Schwabing Hospital in Munich for autopsies.[69] At the department of pathology of the University of Munich a special "investigative unit" was created for the armed forces, which performed autopsies on members of the military and prisoners of war.[70]

SS officer Robert Neumann, assistant at the pathology department of the University of Berlin, performed human experiments in the concentration camps Oranienburg, Auschwitz, and Buchenwald in 1939 and 1940. After testing a new apparatus for liver biopsies on bodies of dead prisoners he started taking liver tissue samples from living victims, most of whom died after the operation. In October 1940 Neumann became the director of the pathological institute at Tongji University in Shanghai, where he pursued studies in racial hygiene.[71] He evaded legal investigations after the war and returned to Germany in 1954, where he found employment as a pharmacological consultant.[72]

Those who studied in the field of neuropathology profited greatly from the increased availability of human "material," specifically as a result of the NS "euthanasia" program. During WWII, 250,000 to 300,000 persons were murdered within the framework of the "euthanasia" program.[73] The brains of these victims were used for research by neuropathologists. Julius Hallervorden of the KWI for Brain Research in Berlin was one of the most prominent in the field. He was also the *Prosektor* and medical examiner at the psychiatric hospital and later "euthanasia" center in Brandenburg-Görden and several other Berlin institutions. He performed histopathological studies on a total of 697 brains.[74] His colleagues Hans Joachim Scherer (Breslau) and Berthold Ostertag (Berlin) also used brains from children killed in the "euthanasia" program.[75] In June 1941 Heinrich Gross, attending physician at the Vienna psychiatric hospital *Am Spiegelgrund*, visited Görden for a training session to learn the techniques of "euthanasia." Subsequently, 336 children died under his care in Vienna until he was called to military service in 1943. He again worked at *Spiegelgrund* in 1944. Tissues from the children's bodies were still stored for decades after the war, and Gross started publishing results of his histopathological work based on these specimens in 1952.[76] *Spiegelgrund* also became a venue for experiments on the "future dead." Pediatrician Elmar Türk tested a new tuberculosis vaccine on disabled children who were later murdered and autopsied. In three cases the patient's records included a "wish list" sent by Türk to the neuropathologist, detailing clinically pertinent facts to be considered during the autopsy. For Türk the murder of the children became part of his research plan,[77] similar to work by his colleague Carl Schneider at the University of Heidelberg.[78] While the MPI for Brain Research and the *Spiegelgrund* facility purged their collections of specimens from the NS period and buried them several decades after the war, other collections are still in the process of documenting and evaluating their history and the provenance of tissues, such as the neuropathological collection of the C. and O. Vogt Archives at the University of Düsseldorf.[79]

Otmar von Verschuer, who succeeded Fischer as director of the KWI Anthropology in 1942, was internationally recognized for his twin research. Twins were considered to be the ideal subjects for studies on the influence of genetics and environment on the development of the physical and psychological characteristics of human beings. Necessarily, this was research on the living, as the simultaneous death of twins was rare. This changed under the conditions of the NS regime. In 1943–44 von Verschuer employed his former doctoral student Josef Mengele as an assistant at the KWI Anthropology. During this time Mengele also worked as an SS physician at the concentration camp Auschwitz, where he experimented on human subjects, especially children and twins. Mengele provided blood samples for von Verschuer's institute from diverse "racial" groups, eyes and internal organs from murdered children, two skeletons from Jews, and sera from twins, who he had infected with typhoid fever. Von Verschuer and his colleagues used 200 blood samples from Auschwitz to determine the heritability of certain proteins. They also studied eyes sent from Auschwitz to determine inheritance characteristics of eye color. Both projects were funded by the German Research Foundation.[80]

Gynecologists, who were intimately involved in the implementation of NS policies through the nearly 400,000 forced sterilizations performed between 1933 and 1945,[81] also made use of the newly abundant "material" from the executed. In 1941, Hans Fuchs, director of the gynecological hospital in Posen, wrote to the general attorney asking for "fresh material from testes of criminals executed in Posen".[82] Fuchs's colleague Boris Belonoschkin planned on using this "material" for his work on the influence of stress on fertility in men. The request was apparently granted, as Belonoschkin published the results of an investigation on spermatozoa from the testes and epididymis of fourteen executed prisoners in a gynecological journal in 1943.[83] The author discussed in his article the specific difficulties of obtaining fresh human testes for these studies. Like Stieve, Belonoschkin was able to obtain information on the duration of imprisonment before a victim's execution, which ranged from seventeen days to three years in the men who were between twenty-one and forty-four years old. The tissues were processed between fifty minutes and five hours after the execution.[84] Contrary to Stieve's results, which were known to him, Belonoschkin's concluded that the length of imprisonment and mental stress had only a minor effect on the function of reproductive organs in men.[85]

While Belonoschkin, like Stieve, still worked with the traditional tissues from the dead, gynecologist Karl Ehrhardt, director of the department of gynecology at the University of Graz, worked right at the border of life and death. Ehrhardt not only performed unnecessary and mutilating surgical procedures, forced sterilizations, and abortions on Russian forced-laborer women, but he also experimented on their fetuses, which were studied in turn by Alfred Pischinger, chairman of the anatomical department. Gabriele Czarnowski has explored Ehrhardt's activities in detail.[86] Ehrhardt's 1939 recruitment from Frankfurt had apparently been a political one—he was member of NSDAP and SS since 1933. His research focused on "fetography", i.e., intrauterine imaging techniques of the fetus and placenta with various contrast materials. In 1937 he was granted funding by the German Research Organization for experiments on the intrauterine biological actions of the fetus and published an article in which he described drinking movements in a fetus. The subjects of his investigation were a twenty-one-year old woman, who had been evaluated as "feeble minded," and her fetus, who was to be aborted for eugenic reasons. The radiological contrast medium Thorotrast was injected into the amniotic cavity fifteen hours before the fetus was removed via an abdominal incision. During the same surgery the woman was sterilized. Ehrhardt described how he was able to retrieve the fetus ("egg," "fruit" in the nomenclature of the time) unharmed in its membranes. Ehrhardt observed the movements of the fe-

tus through the translucent membranes for five minutes and then proceeded to take radiographs.[87] Results from these experiments on three- to
six-month-old intrauterine fetuses were presented by the anatomist Wilhelm Pfuhl in Ehrhardt's name at the annual meeting of the *Anatomische
Gesellschaft.*[88]

Ehrhardt again collaborated with an anatomist, this time with embryologist Alfed Pischinger. In 1941 they published side-by-side papers on the
results of their investigations of Ehrhardt's intra-amnial Thorotrast injections. Ehrhardt compared the contrast media Thorotrast and Umbrathor
and noticed again that they were well tolerated by the living fetus.[89] In this
paper he mentioned that the indication for abortion was a eugenic one in
some cases.[90] Pischinger focused on whether the enrichment of contrast
medium in the fetal lungs was due to physiological breathing actions or an
artifact, and concluded that his data were not compatible with physiological intrauterine fetal breathing. To prove his point, he suggested an additional experiment with ink, a method that had been used in similar animal
experiments. Ehrhardt performed this test on four fetuses and the results
supported Pischinger's conclusions.[91] Czarnowski has pointed out that this
change of methods may have been the first application of intra-amnial
ink injections in humans, and the collaboration between Ehrhardt and
Pischinger led to a new kind of human experiment.[92]

In a paper from February 1945 Ehrhardt described research results on
two near-term fetuses and referred to the method of formalin-induced
abortion, which he had developed earlier, as a particularly "gentle" one,
whereby the "fruit" died within twelve to forty-eight hours of intra-amnial
injection and was "expelled" after "hours, days, weeks".[93] This spare scientific language has to be translated: Ehrhardt injected a formalin preparation into the abdomen of a heavily pregnant woman shortly before delivery,
watched the death of the fetus via radiographs, and then left the woman—
who may or may not have known what he had done, who may or may not
have understood the language—to give birth to a dead child hours, days,
or weeks later. Ehrhardt never went on trial for these experiments; the
women, whose names and fates are not yet known, have never been com-
pensated—if such compensation were ever possible. Ehrhardt's method
of formalin-induced abortion was quoted in a 1967 book on "artificial
abortion" not as a crime, but as a "conservative method."[94] After the war
Ehrhardt worked in private practice in Frankfurt am Main.[95] Pischinger, a
member of NSDAP and SA, was dismissed from his position after the war
but was reinstated shortly thereafter, and ended his career as chairman of
histology and embryology at the University of Vienna.[96] Pischinger's collaborative work with Ehrhardt is missing in the bibliography in Pischinger's
1984 obituary.[97]

The transgressions and lapses in medical ethics committed by Ehrhardt and Pischinger in their observations of the dying and dead fetuses were not considered unethical at the time, as this "method" of experimenting with freshly aborted fetuses was actually accepted in contemporary medicine and pursued in the United States.[98] This approach seems unconscionable from a modern point of view, which is based on a different set of ethics and social perceptions of fetuses and research performed on them. The NS regime officially sanctioned the abuse of vulnerable women and their children, women who were forced into the termination of their pregnancies and sterilization against their will and often without a real understanding of the consequences.

Many other academic disciplines profited from easy access to the "future dead." Zoologist Gotthilft von Studtnitz, chairman of the department of zoology at the University of Halle, had received funding from the navy for a study on the improvement of night vision. To this purpose he not only investigated the heads of decapitated NS victims from the execution site in Halle, but he also experimented on twenty-one prisoners on 19 June 1944. Von Studtnitz arranged that the prisoners were given a substance to drink 6 hours before their deaths. The prisoners were told that the difficult to swallow oily liquid was a sedative. Immediately after the execution the eyes were removed for investigation.[99] Otto Reche, director of the department of racial hygiene and ethnology at the University of Leipzig, wanted to take anthropometric measurements and study skin samples from prisoners after their execution. However, Dresden was destroyed during bombing raids on 13 and 14 February 1945, and the investigations were likely never performed.[100]

Exploring the use of the bodies of NS victims for scientific purposes confirms the wide range of ethical transgressions committed by medical researchers and fostered by the NS regime. Other than anatomy, many academic disciplines were involved in the use of these victims' remains, and ethical boundaries were crossed in a manifold fashion. Clinicians, who traditionally worked exclusively with living human beings, started to include the planned death of their research subjects in their study designs, and anatomists and pathologists crossed the boundary from work with the dead to work with the "future dead." The science of anatomy itself changed during National Socialism through abuses of the human body in life and death for scientific profit.

Notes

1. Proctor 2000, 335.
2. Ibid., 336; Schleiermacher and Schagen 2012.
3. Proctor 2000; Schleiermacher and Schagen 2008; Jacobsen 2014.

4. Lipphardt 2008.
5. Proctor 2000; Roelcke 2010.
6. Roelcke 2010.
7. Hansson and Hildebrandt 2014.
8. Winkelmann 2012.
9. Noack 2008.
10. Proctor 1992; Walker 2003.
11. Hildebrandt 2008.
12. Forssmann 1972, 62.
13. Rothmaler 1991.
14. Aumüller, and Grundmann 2002
15. See Hildebrandt 2013b, 305.
16. Gillman 1934, 211
17. Jones 1941, 7.
18. Hildebrandt 2008.
19. Gagel and Bodechtel 1929, 132.
20. Bargmann 1931, 85; Volkmann 1933, 219.
21. Hildebrandt 2013a.
22. Schiller 1949a; Schiller 1949b; Schiller 1950; Stöhr 1943, 1948; Hayek 1950, referring to Hayek 1941; Herrlinger 1947; Herrlinger 1949.
23. Winkelmann 2007a; Feja and Riha 2014; Czech 2015.
24. Hildebrandt 2008.
25. Lewinsky and Stewart 1936, 99, quoted from Winkelmann and Schagen 2009, 169.
26. E.g., Shanklin 1951; Rasmussen 1928.
27. Chanamirjan 1929.
28. Snyder 2010.
29. Stankiewitsch 1934.
30. Hildebrandt 2008.
31. Grundmann and Aumüller 2012; Worden 2003; Chaplin 2005.
32. E.g., Hayek 1935; Hewel 1941.
33. Winkelmann and Noack 2010.
34. Clara 1942, 362.
35. Heckel 1942; Müller 1942; Schiller 1942; see Hildebrandt 2013c.
36. Noack 2012.
37. Winkelmann and Noack 2010.
38. Roelcke et al. 2001.
39. Lang 2013.
40. Mitscherlich and Mielke 1997, 219.
41. Kasten 1991, 190.
42. Lang 2007.
43. Wojak 1999; Weindling 2015, 155.
44. Pringle 2006; Lang 2007, 2013.
45. Lachman 1977.
46. Klee 2004, 376.
47. Schütz et al. 2013, 300.
48. Schütz et al. 2015
49. Schleiermacher 1988.
50. For a comprehensive overview of forensic medicine in the Third Reich, see Herber 2002; also, Lignitz 2004; Arias 2009.
51. Buddecke 2011.
52. Arias 2009.
53. Scherrieble 2008, 216.

54. Herber 2002, 232–333.
55. Ibid., 503; Catalogus Professorum Halensis 2014; Roth 2006, 126.
56. Herber 2002, 276; Preuss and Madea 2009.
57. Lower 2005.
58. Special detachment: troupes selected to find and kill "enemies," see Lower 2005.
59. Herber 2002, 277.
60. Preuss and Madea 2009, 16, 17.
61. Herber 2002, 263.
62. Hackett 1995, 223–25; Kogon 2006, 180.
63. Herber 2002, 263.
64. E.g., Le Chêne 1971; Conot 1983, 288, 289.
65. Neander 2009.
66. Jacobson, 2010.
67. Lampert 1991; Prüll 2003; Babaryka 2008.
68. Nyiszli 1993; Kogon 2006; Freyeisen 2000; Mitscherlich and Mielke 1960, 2007.
69. U.S. Naval Technical Mission in Europe, technical report no. 331-45, 1945, Beecher papers.
70. Babaryka 2008.
71. Freyeisen 2000, 238.
72. Prüll 2003, 393; Klee 2004, 36; Freyeisen 2000, 233.
73. Faulstich 2000.
74. Müller-Hill 1984; Peiffer 1997; Peiffer 2004, 104, 147–51; Klee 2003; Aly 2005; Hughes 2007, 119.
75. Peiffer 1997.
76. Neugebauer 1997; Spann 1999; Czech 2002.
77. Dahl 2000.
78. Roelcke et al. 2001; Fuchs et al. 2007; Hohendorf and Rotzoll 2014.
79. Professor Katrin Amunts, personal communication via electronic mail, 19 November 2013.
80. Müller-Hill 1984, 24; Bergmann et al. 1989, 137; Kröner 1998, 53; Ehrenreich 2007, 66; Weiss 2010, 110–18.
81. Bock 1986; Czarnowski 2004, 227.
82. Waltenbacher, personal communication via electronic mail, 16 December 2008, based on German Federal Archives file BA / Alt R 22 / 1318 / Bl. 28 f.
83. Belonoschkin 1943.
84. Ibid., 358, 364.
85. Ibid., 367.
86. Czarnowski 2004; Czarnowski 2012; Czarnowski 2014.
87. Ehrhardt 1937.
88. Ehrhardt 1938.
89. Ehrhardt 1941.
90. Ibid., 114.
91. Pischinger 1941.
92. Czarnowski 2004, 257.
93. Ehrhardt 1945, 182.
94. Czarnowski 2012, 146.
95. Klee 2003.
96. Schwarzacher 1984; Scheiblechner 2002.
97. Schwarzacher 1984.
98. Wilson 2014.
99. Scherrieble 2008, 219–20; Waltenbacher 2008, 227–29.
100. Scherrieble 2008, 221–22.

Archival Sources

Francis A. Countway Library, Center for the History of Medicine
– Estate papers of Henry K. Beecher, Box 11, Folder 75: U.S. Naval Technical Mission in Europe, Technical report no. 331–45, 1945

Bibliography

Aly, Götz. 2005. *Macht-Geist-Wahn: Kontinuitäten deutschen Denkens*. Frankfurt am Main: Fischer Taschenbuch Verlag, 73–93.

Arias, Ingrid. 2009. *Die Wiener Gerichtsmedizin im Nationalsozialismus*. Wien: Verlagshaus der Ärzte.

Aumüller, Gerhard, and Kornelia Grundmann. 2002. "Anatomy during the Third Reich: The Institute of Anatomy at the University of Marburg, as an Example." *Annals of Anatomy* 184: 295–303.

Babaryka, Gregor. 2008. "Das pathologische Institut der Universität München in der Ära Max Borst von 1910 bis 1946." In: *Die Universität München im Dritten Reich. Aufsaetz. Teil II*, ed. Elisabeth Kraus, 63–132. München: Herbert Utz Verlag.

Bargmann, Wolfgang. 1931. "Über Struktur und Speicherungsvermögen des Nierenglomerulus." *Zeitschrift für Zellforschung* 14: 73–137.

Belonoschkin, Boris. 1943. "Biologie der Spermatozoen im menschlichen Hoden und Nebenhoden." *Archiv für Gynäkologie* 174(3): 357–68.

Bergmann, Anna, Gabriele Czarnowski, and Annegret Ehmann. 1989. "Menschen als Objekte humangenetischer Forschung und Politik im 20. Jahrhundert." In: *Der Wert des Menschen: Medizin in Deutschland 1918–1945*, ed. Christion Pross and Götz Aly, 121–42. Berlin: Edition Hentrich.

Bock, Gisela. 1986. *Zwangssterilisation im Nationalsozialismus: Studien zur Rassenpolitik und Frauenpolitik*. Opladen: Westdeutscher Verlag.

Buddecke, Julia. 2011. *Endstation Anatomie: Die Opfer nationalsozialistischer Vernichtungsjustiz in Schleswig-Holstein*. Hildesheim: Georg-Olms Verlag.

Catalogus professorum halensis. 2014. "Siegfried Krefft." Accessed 14 April 2014. http://www.catalogus-professorum-halensis.de/krefftsiegfried.html.

Chanamirjan, A. 1929. "Versuch eines systematischen Studiums der Variationen der Wirbelarterie." *Anatomischer Anzeiger* 68: 163–78.

Chaplin, Simon. 2005. "John Hunter and the Anatomy of the Museum." *History Today* 55(2). Accessed 28 February 2012. http://www.historytoday.com/simon-chaplin/john-hunter-and-anatomy-museum.

Clara, Max. 1942. "Beiträge zur Histotopochemie des Vitamin C im Nervensystem des Menschen." *Zeitschrift für miskroskopisch-anatomische Forschung* 52: 359–92.

Conot, Robert E. 1983. *Justice at Nuremberg*. New York: Harper & Row Publishers.

Czarnowski, Gabriele. 2004. "Vom "reichen Material … einer wissenschaftlichen Arbeitsstätte." Zum Problem missbräuchlicher medizinischer Praktiken an der Grazer Universitäts-Frauenklinik in der Zeit des Nationalsozialismus." In: *NS-Wissenschaft als Vernichtungsinstrument. Rassenhygiene, Zwangssterilisation, Menschenversuche und NS-Euthanasie in der Steiermark*, ed. Wolfgang Freidl, and Werner Sauer, 225–73. Wien: Facultas.

———. "Österreichs 'Anschluss' an Nazi-Deutschland und die österreichische Gynäkologie." In: *Herausforderungen. 100 Jahre Bayerische Gesellschaft für Geburtshilfe und Frauenheilkunde*, ed. Christoph Anthuber, Matthias W. Beckmann, Johannes Dietl, Fritz Dross, and Wolfgang Frobenius. Stuttgart: Georg Thieme Verlag. Accessed 13 March 2015. http://www.bggf.de/cms/assets/pdfs/9_Österreich.pdf.

———. 2014. "Involuntary Abortion and Coercive Research on Pregnant Forced Laborers in National Socialism." In: *Human Subjects Research after the Holocaust,* ed. Sheldon Rubenfeld and Susan Benedict, 99–108. Cham: Springer.

Czech, Herwig. 2002. "Der Fall Heinrich Gross. Die wissenschaftliche Verwertung der 'Spiegelgrund'-Opfer in Wien." *Context XXI.* Accessed 12 March 2015. http://www .contextxxi.at/context/content/view/164/93/.

———. "Von der Richtstätte auf den Seziertisch: Zur anatomischen Verwertung von NS-Opfern in Wien, Innsbruck und Graz." *Jahrbuch des Dokumentationsarchiv des österreichischen Widerstandes,* 2015: 141–90.

Dahl, Matthias. 2000. "Die Tötung behinderter Kinder in der Anstalt am Spiegelgrund 1940–1945." In : *NS-Euthanasie in Wien,* ed. Eberhard Gabriel and Wolfgang Neugebauer, 75–92. Wien: Böhlau.

Ehrenreich, Eric. 2007. "Otmar von Verschuer and the 'Scientific' Legitimization of the Nazi Anti-Jewish Policy." *Holocaust and Genocide Studies* 21(1): 55–72.

Ehrhardt, Karl. 1937. "Der trinkende Fötus: Eine röntgenologische Studie." *Münchner Medizinische Wochenschrift* 84: 1699–1700.

———. 1938. "Der trinkende und atmende Fetus (vorgezeigt und mitgeteilt von W. Pfuhl, Frankfurt a. M.)." *Verhandlungen der Anatomischen Gesellschaft* 87: 420–22

———. 1941. "Weitere Erfahrungen mit meiner Methode der intraamnialen Thoriuminjektion (Fetale Organographie)." *Zentralblatt für Gynäkologie* 65: 114–20.

———. "Atmet das Kind im Mutterleib?" *Medizinische Zeitschrift* 5: 182–83.

Faulstich, Heinz. 2000. "Die Zahl der "Euthanasie"-Opfer." In: *Die historischen Hintergründe medizinischer Ethik,* ed. Andreas Frewer and Clemens Eickhoff, 218–29. Frankfurt: Campus-Verlag.

Feja, Christine, and Ortrun Riha. 2014. *"Hier hilft der Tod dem Leben": Das Leipziger Institut für Anatomie und das Leichenwesen 1933–1989.* Aachen: Shaker Verlag.

Forssmann, Werner. 1972. *Erinnerungen eines Chirurgen.* Düsseldorf: Droste Verlag.

Freyeisen, Astrid. 2000. *Shanghai und die Politik des Dritten Reiches.* Würzburg: Konigshausen & Neumann.

Fuchs, Petra, Maike Rotzoll, Ulrich Müller, Paul Richter, and Gerrit Hohendorf. 2007. *"Das Vergessen der Vernichtung ist Teil der Vernichtung selbst." Lebensgeschichten von Opfern der nationalsozialistischen "Euthanasie."* Göttingen: Wallstein Verlag.

Gagel, O., and G. Bodechtel. 1929. "Die Topik und feinere Histologie der Ganglienzellgruppen in der Medulla oblongata und im Ponsgebiet mit einem kurzen Hinweis auf die Gliaverhältnisse und die Histopathologie." *Zeitschrift für Anatomie und Entwicklungsgeschichte* 91(1–3): 130–250.

Gillman, J. 1934. "The Cellular Cycle, the Golgi Apparatus and the Phenomenon of Reversal in the Human Thyroid Parenchyma." *Anatomical Record* 60(2): 209–30.

Grundmann, Kornelia, and Aumüller, Gerhard, eds. 2012. *Das Marburger Medizinhistorische Museum: Museum anatomicum.* Marburg/Lahn: Druckhaus Marburg.

Hackett, David A. 1995. *The Buchenwald Report.* Boulder: Westview Press.

Hansson, Nils, and Sabine Hildebrandt. 2014. "Swedish-German Contacts in Anatomy 1930–1950: The Example of Gösta Häggqvist and Hermann Stieve." *Annals of Anatomy* 196: 259–67.

Hayek, Heinrich. 1935. "Das Verhalten der Arterien bei Beugung der Gelenke." *Zeitschrift für Anatomie und Entwicklungsgeschichte* 105(1): 25–36.

———. 1941a. "Über die Verengung der Bronchi und Bronchioli durch die Muskulatur." *Wiener Klinische Wochenschrift* 54: 114–16.

———. 1950a. "Die Muskulatur im Lungenparenchym des Menschen." *Zeitschrift für Anatomie und Entwicklungsgeschichte* 115(1): 88–94.

Heckel, Lothar. 1942. "Untersuchungen über das Vorkommen von Vitamin C in der Nebenniere des Menschen." *Zeitschrift für Mikroskopisch-Anatomische Forschung* 52: 393–417.

Herber, Friedrich. 2002. *Gerichtsmedizin unterm Hakenkreuz.* Leipzig: Militzke Verlag.

Herrlinger, Robert. 1947. "Das Blut in der Milzvene des Menschen." *Anatomischer Anzeiger* 96: 226–35.

———. 1949. "Neue funktionell-histologische Untersuchungen an der menschlichen Milz." *Zeitschrift für Mikroskopisch-Anatomische Forschung* 114(4): 341–65.

Hewel, Julius. 1941. "Über die Beweglichkeit der menschlichen Leber." *Anatomischer Anzeiger* 90: 273–96.

Hildebrandt, Sabine. 2008. "Capital Punishment and Anatomy: History and Ethics of an Ongoing Association." *Clinical Anatomy* 21: 5–14.

———. 2013a. "The Women on Stieve's List: Victims of National Socislism Whose Bodies Were Used for Anatomical Research." *Clinical Anatomy* 26: 3–21.

———. 2013b. "Research on Bodies of the Executed in German Anatomy: An Accepted Method that Changed during the Third Reich. Study of Anatomical Journals from 1924 to 1951." *Clinical Anatomy* 26: 304–26.

———. 2013c. "Current Status of Identification of Victims of the National Socialist Regime whose Bodies were Used for Anatomical Purposes." *Clinical Anatomy* 27: 514–536.

Hohendorf, Gerrit, and Maike Rotzoll. 2014. "Medical Research and National Socialist Euthanasia: Carl Schneider and the Heidelberg Research Children from 1942 until 1945." In: *Human Subjects Research after the Holocaust*, ed. Sheldon Rubenfeld and Susan Benedict, 127–38. Cham: Springer.

Hughes, J. Trevor. 2007. "Neuropathology in Germany during World War II: Julius Hallervorden (1882–1965) and the Nazi Programme of 'Euthanasia.'" *Journal of Medical Biography* 15: 116–22.

Jacobson, Mark. 2010. "The Lampshade: A Holocaust Detective Story from Buchenwald to New Orleans. New York: Simon & Schuster.

Jones, Russell L. 1941. "The Human Foot: An Experimental Study of Its Mechanics, and the Role of Its Muscles and Ligaments in the Support of the Arch." *American Journal of Anatomy* 68(1): 1–39.

Kasten, Frederick H. 1991. "Unethical Nazi Medicine in Annexed Alsace-Lorraine: The Strange Case of Nazi Anatomist Professor Dr. August Hirt." In: *Essays in Modern German History and Archival Policy*, ed. George O. Kent, 173–208. Fairfax, VA: George Mason University Press.

Klee, Ernst. 2003. *Das Personenlexikon zum Dritten Reich: Wer war was vor und nach 1945?* Frankfurt am Main: S. Fischer.

———. 2004. *Auschwitz, die NS-Medizin und ihre Opfer. Third Edition.* Frankfurt am Main: Fischer Verlag.

Kogon, Eugen. 2006. *Der SS-Staat. Das System der deutschen Konzentrationslager.* 43rd ed. München: Wilhelm Heine Verlag. Originally published in 1946.

Kröner, Hans-Peter. 1998. *Von der Rassenhygiene zur Humangenetik.* Stuttgart: Gustav Fischer Verlag.

Lachman, Ernest. 1977. "Anatomist of Infamy: August Hirt." *Bulletin of the History of Medicine* 51: 594–602.

Lampert, Udo. 1991. "Zur Situation der Pathologischen Anatomie an den deutschen Hochschulen während des zweiten Weltkrieges." In: *Der Arzt als "Gesundheitsführer": Ärztliches Wirken zwischen Resourcenerschliessung und humanitärer Hilfe im Zweiten Weltkrieg*, ed. Sabine Fahrenbach, Achim Thom, Gerhard Baader, N. Decker, and Wolfgang Uwe Eckart, 143–50. Frankfurt: Mabuse-Verlag.

Lang, Hans-Joachim. 2007. *Die Namen der Nummern: Wie es gelang, die 86 Opfer eines NS-Verbrechens zu identifizieren. Überarbeitete Ausgabe.* Frankfurt am Main: S. Fischer Verlag.

———. 2013. "August Hirt and 'Extraordinary Opportunities for Cadaver Delivery' to Anatomical Institutes in National Socialism: A Murderous Change in Paradigm." *Annals of Anatomy* 195: 373–80.

Le Chêne, Evelyne. 1971. *Mauthausen: The History of a Death Camp*. London: Methuen.

Lewinsky, W., and D. Stewart. 1936. "The Innervation of the Periodontal Membrane." *Journal of Anatomy* 71: 98–103.

Lignitz, Eberhard. 2004. "The History of Forensic Medicine in Times of the Weimar Republic and National Socialism: An Approach." *Forensic Science International* 144: 113–24.

Lipphardt, Veronika. 2008. "Das 'schwarze Schaf' der Biowissenschaften: Marginalisierungen und Rehabilitierungen der Rassenbiologie im 20. Jahrhundert." In: *Pseudowissenschaft*, ed. Dirk Rupnow, Veronika Lipphardt, Jens Thiel, and Christina Wessely, 223–50. Frankfurt am Main: Suhrkamp.

Lower, Wendy. 2005. *Nazi Empire-Building and the Holocaust in the Ukraine*. Chapel Hill: University of North Carolina Press.

Mitscherlich, Alexander, and Fred Mielke. 1947. *Das Diktat der Menschenverachtung*. Heidelberg: Verlag Lambert Schneider.

Müller, Rolf. 1942. "Untersuchungen über das Vorkommen von Vitamin C im Hoden des Menschen." *Zeitschrift für Mikroskopisch-Anatomische Forschung* 52: 440–54.

Müller-Hill, Benno. 1984. *Tödliche Wissenschaft*. Reinbek bei Hamburg: Rowohlt Taschenbuch Verlag.

Neander, Joachim. 2009. "A Strange Witness to Dachau Human Skin Atrocities: Anton Pacholegg a.k.a. Anton Baron von Guttenberg a.k.a. Antoine Charles de Guttenberg." *theologie.geschichte.de Band 4*, accessed 29 May 2014. http://universaar.uni-saarland.de/journals/index.php/tg/article/viewArticle/472/511#83.

Neugebauer, Wolfgang. 1997. "Wiener Psychiatrie und NS-Verbrechen. Referat im Rahmen der Arbeitstagung 'Die Wiener Psychatrie im 20. Jahrhundert,' Wien, Institut für Wissenschaft und Kunst, 20./21. Juni, 1997." Accessed 11 September 2014. http://old.doew.at/frames.php?/thema/thema_alt/justiz/euthjustiz/euth.html .

Noack, Thorsten. 2008. "Begehrte Leichen. Der Berliner Anatom Hermann Stieve (1886–1952) und die medizinische Verwertung Hingerichteter im Natinoalsozialismus." *Medizin, Gesellschaft und Geschichte. Jahrbuch des Instituts für Geschichte der Medizin der Robert Bosch Stiftung* 26: 9–35.

———. 2012. "Anatomical Departments in Bavaria and the Corpses of Executed Victims of National Socialism." *Annals of Anatomy* 194: 286–92.

Nyiszli, Miklos. 1993. *Auschwitz: A Doctor's Eyewitness Account*. New York: Arcade Publishing. Originally written in 1946.

Peiffer, Jürgen. 1997. *Hirnforschung im Zwielicht: Beispiele verführbarer Wissenschaft aus der Zeit des Nationalsozialismus*. Husum: Mathiessen Verlag.

———. 2004. *Hirnforschung in Deutschland 1849–1974*. Berlin: Springer Verlag.

Pischinger, Alfred. 1941. "Über das Wesen kindlicher Atembewegungen vor der Geburt." *Zentralblatt für Gynäkologie* 65: 120–24.

Preuss, Johanna, and Burkhard Madea. 2009. "Gerhard Panning (1900–1944): A German Forensic Pathologist and His Involvement in Nazi Crimes during Second World War." *The American Journal of Forensic Medicine and Pathology* 30: 14–17.

Pringle, Heather. 2006. *The Master Plan: Hitler's Scholars and the Holocaust*. New York: Hyperion.

Proctor, Robert N. 1992. "Nazi Biomedical Policies." In: *When Medicine Went Mad: Bioethics and the Holocaust*, ed. Arthur Caplan, 23–42. Totowa: Humana Press.

———. 2000. "Nazi Science and Nazi Medical Ethics: Some Myths and Misconceptions." *Perspectives in Biology and Medicine* 43(3): 335–46.

Prüll, Cay-Rüdiger. 2003. *Medizin am Toten oder am Lebenden: Pathologie in Berlin und in London, 1900–1945*. Basel: Schwabe & Co AG Verlag.

Rasmussen, A. T. 1928. "The Weight of the Principal Components of the Normal Male Adult Human Hypophysis Cerebri." *American Journal of Anatomy* 42(1): 1–26.

Roelcke, Volker. 2010. "Medicine during the Nazi Period: Historical Facts and Some Implications for Teaching Medical Ethics and Professionalism." In: *Medicine after the Holocaust: From the Master Race to the Human Genome and Beyond*, ed. Sheldon Rubenfeld, 17–29. New York: Palgrave MacMillan.

Roelcke, Volker, Gerrit Hohendorf, and Maike Rotzoll. 2001. "Psychiatric Research and 'Euthanasia': The Case of the Psychiatric Department at the University of Heidelberg, 1941–1945." *Psychoanalytic Review* 88(2): 275–94.

Roth, Karl-Heinz. 2006. "Flying Bodies, Enforcing States: German Aviation Medical Research from 1925 to 1975 and the Deutsche Forschungsgemeinschaft." In: *Man, Medicine and the State*, ed. Wolfgang Uwe Eckart, 107–37. Stuttgart: Steiner.

Rothmaler, Christiane. 1991. "Die Sammlung des Anatomischen Instituts der Hansischen Universität in Hamburg: Didaktisches Konzept und Aufbau 1919 bis 1945." In: *Deutsche Gesellschaft für Geschichte der Medizin, Naturwissenschaft und Technik (73. Jahrestagung). Ideologie der Objekte, Objekte der Ideologie; Naturwissenschaft, Medizin und Technik in Museen des 20. Jahrhunderts*, 55–63. Kassel: Georg Wenderoth Verlag.

Scheiblechner, Petra. 2002. "*...Politisch ist er einwandfrei...*" *Kurzbiographien der an der Medizinischen Fakultät der Universität Graz in der Zeit von 1938 bis 1945 tätigen Wissenschaftlerinnen*. Graz: Akademische Druck-u.Verlagsanstalt.

Scherrieble, Joachim. 2008. *Der Rote Ochse Halle (Saale). Politische Justiz 1933–1945, 1945–1989*. Berlin: Christoph Links Verlag.

Schiller, Erich. 1942. "Über den Fettgehalt der Leber beim gesunden Menschen." *Zeitschrift für Mikroskopisch-Anatomische Forschung* 51: 309–21.

———. 1949a. "Variationsstatistische Untersuchungen über Kerneinschlüsse und – Kristalle der menschlichen Leber." *Zeitschrift für Zellforschung* 34(4): 337–55.

———. 1949b. "Kerneinschlüsse und Amitose." *Zeitschrift für Zellforschung* 34(4): 356–61.

———. 1950. "Die morphologische Manifestation von Kernstoffwechselstörungen." *Verhandlungen der Anatomischen Gesellschaft* 48: 239–40.

Schleiermacher, Sabine. 1988. "Die SS-Stiftung 'Ahnenerbe.' Menschen als Material für 'exakte' Wissenschaft." In: *Menschenversuche: Wahnsinn und Wirklichkeit*, ed. Rainer Osnowski, 70–87. Köln: Kölner Volksblatt Verlag.

Schleiermacher, Sabine, and Udo Schagen 2008. "Medizinische Forschung als Pseudowissenschaft: Selbstreinigungsrituale der Medizin nach den Nürnberger Ärzteprozess." In: *Pseudowissenschaft*, ed. Dirk Rupnow, Veronika Lipphardt, Jens Thiel, and Christina Wessely, 251–78. Frankfurt am Main: Suhrkamp.

———. 2012. "Semantik als Strategie: Die Klassifizierung der Medizinverbrechen in Konzentrationslagern als Pseudowissenschaft." In: *Geschlecht und "Rasse" in der NS-Medizin*, ed. Insa Eschebach and Astrid Ley, 157–77. Berlin: Metropol Verlag.

Schütz, Mathias, Jens Waschke, Georg Marckmann, and Florian Steger. 2013. "The Munich Anatomical Institute under National Socialism: First Results and Prospective Tasks of an Ongoing Research Project." *Annals of Anatomy* 195: 296–302.

Schütz, Mathias, Maximilian Schochow, Jens Waschke, Georg Marckmann, and Florian Steger. 2015. "Anatomische Vitamin C-Forschung im Nationalsozialismus und in der Nachkriegszeit: Max Claras Humanexperimente an der Anatomischen Anstalt München." *Medizinhistorisches Journal* 50: 330–355.

Schwarzacher, Hans-Georg. 1984. "Nachruf auf Alfred Pischinger." *Zeitschrift für Mikroskopisch- Anatomische Forschung* 98(6): 801–4.

Shanklin, William M. 1951. "The Histogenesis and Histology of an Integumentary Type of Epithelium in the Human Hypophysis." *Anatomical Record* 109(2): 217–31.

Snyder, Timothy. 2010. *Bloodlands: Europe between Hitler and Stalin*. New York: Basic Books.

Spann, Gustav. 1999. "Untersuchungen zur anatomischen Wissenschaft in Wien 1938–1945. Senatsprojekt der Universität Wien. Eine Zusammenfassung. Auszug aus: Dokumentationsarchiv des österreichischen Widerstandes. Jahrbuch 1999." Accessed 11 September 2014. http://www.doew.at/cms/download/3of0q/spann_jb_1999.pdf.

Stöhr, Philipp. 1943. "Studien zur normalen und pathologischen Histologie vegetativer Ganglien.I." *Zeitschrift für Zellforschung* 32(5): 587–635.

———. 1948. "Studien zur normalen und pathologischen Histologie vegetativer Ganglien. II." *Zeitschrift für Anatomie und Entwicklungsgeschichte* 114(1–2): 14–52.

Volkmann, Rüdiger von. 1933. "Morphologie, Entstehung und Vorkommen des Abnutzungspigmentes im Epithel des menschlichen Plexus choroideus." *Zeitschrift für Anatomie und Entwicklungsgeschichte* 102(2–3): 211–31.

Walker, Mark. 2003. "'Nazi Science'? Natural Science in National Socialism." In: *"Kämpferische Wissenschaft": Studien zur Universität Jena im Nationalsozialismus*, ed. Uwe Hossfeld, Jürgen John, Oliver Lemuth, Rüdiger Stutz, 993–1012. Köln: Böhlau Verlag.

Waltenbacher, Thomas. 2008. *Zentrale Hinrichtungsstätten. Der Vollzug der Todesstrafe in Deutschland von 1937–1945. Scharfrichter im Dritten Reich*. Berlin: Zwilling.

Weindling, Paul J. 2015. *Victims and Survivors of Nazi Human Experiments: Science and Suffering in the Holocaust*. London: Bloomsbury Academic.

Weiss, Sheila. 2010. *The Nazi Symbiosis: Human Genetics and Politics in the Third Reich*. Chicago and London: University of Chicago Press.

Wilson, Emily K. 2014. "Ex Utero: Live Human Fetal Research and the Films of Davenport Hooker." *Bulletin of the History of Medicine* 88: 132–60.

Winkelmann, Andreas. 2007a. "Die menschliche Leiche in der heutigen Anatomie." In: *Grenzen des Lebens. Beiträge aus dem Institut Mensch, Ethik und Wissenschaft*. Band 5, ed. Sigrid Graumann, and Katrin Grüber, 62–74. Berlin: LIT Verlag Dr. W. Hopf.

———. 2012. "The Anatomische Gesellschaft and National Socialism: A Preliminary Analysis Based on Society Proceedings." *Annals of Anatomy* 194: 243–50.

Winkelmann, Andreas, and Thorsten Noack. 2010. "The Clara Cell: A 'Third Reich Eponym'?" *European Respiratory Journal* 36: 722–27.

Wojak, Irmtrud. 1999. "Das 'irrende Gewissen' der NS-Verbrecher und die deutsche Rechtsprechung. Die 'Jüdische Skelettsammlung' am Anatomischen Institut der 'Reichsuniversität Strassburg'". In: *„Beseitigung des jüdischen Einflusses ...": Antisemitische Forschung, Eliten und Karrieren im Nationalsozialismus, Jahrbuch 1998/99 zur Geschichte und Wirkung des Holocaust*, ed. Andreas Hofmann and Irmtrud Wojak, 101–30. Frankfurt am Main: Campus.

Worden, Gretchen. 2002. *Mütter Museum of the College of Physicians of Philadelphia*. New York: Blast Books.

Chapter 9

AFTER THE WAR

All the agony and terror that came to pass in places of execution, in torture chambers, madhouses and operating theatres ... People would like to be allowed to forget much of this.

Rainer Maria Rilke, 1910[1]

After the war, people in Germany wanted to be allowed to forget. And while they may have been unable to do just that, they at least kept quiet about their experiences in the Third Reich. Controversies remained secret. A great silence ensued.

Directly after WWII Germany was a country in chaos. Large parts of the population were displaced, among them the liberated camp inmates and refugees from former German regions, especially in the east. The country itself was divided into four occupational zones with new military authorities and disparate legislation.[2] Communication and travel were disrupted by the destruction of the organizational and physical infrastructure. All of this reflected on the universities and their remaining personnel: faculties were disbanded, some universities temporarily closed, buildings lay in ruins. Dismissals from university positions due to involvement in NS activities were common, and handled differently from zone to zone, as were the denazification tribunals that every professional German had to undergo.[3]

German Anatomy, 1945–50

Immediately after the truce of May 1945, the leadership of the *Anatomische Gesellschaft* started gathering information on the status of the anatomical departments and their colleagues. The institutes in Strasbourg, Danzig,

Dorpat, Posen, Prague, Breslau, Vienna, Graz, and Innsbruck were not under German control any longer. A 1946 survey conducted by Stieve and Benninghoff revealed that of twenty-five institutes, only five retained their chairman, with Stieve in Berlin, Stöhr in Bonn, Wagenseil in Giessen, Blechschmidt in Göttingen, and Benninghoff in Marburg.[4] Anatomists who had been dismissed due to NS discriminatory legislation but remained in Germany were immediately reinstated at three institutes, Hoepke in Heidelberg, Veit in Cologne, and von Lanz in Munich. Several chairmen and assistants were imprisoned because of their NS activities: Nauck and August Wilhelm Brockmann in Freiburg, Klaus Niessing in Leipzig, Heinz Feneis in Marburg, Clara in Munich, Kremer in Münster, Wetzel in Tübingen, and Pernkopf in Vienna.

Most anatomical departments appointed interim chairmen. In some cases these were anatomists in junior positions who were the only faculty left after all others had been imprisoned or suspended, including Ludwig Keller in Freiburg and Franz Bauer in Vienna.[5] Many anatomists were at least temporarily suspended. During these periods of unemployment some finished studies begun during the war years, while others wrote the books they later became famous for, among them Bargmann, von Hayek, and Voss. Eventually the majority found their way back to academic life, including SS member Hellmut Becher in Münster. Most of the more fervent supporters of the NS regime, such as Pernkopf, Wetzel and Nauck, were barred from leadership positions at German or Austrian universities. Clara, whose return into his former position was strongly opposed by the Munich faculty, moved to the University of Istanbul, Turkey, as a guest professor.[6] Enno Freerksen became director of a private research institute.[7]

There was a severe lack of suitable candidates for chairman positions, especially in the Soviet occupational zone. The atmosphere at some universities was difficult due to distrust and denunciations. In Berlin, Stieve had to defend himself against accusations of having abused bodies of NS victims.[8] In Erlangen a deep conflict arose between assistant Otto Popp and suspended director Hasselwander. Already in 1944 Hasselwander had expressed his opinion that Popp's position at the anatomical institute was no longer supportable because of his unsatisfactory professional performance.[9] Popp had left the Universities of Greifswald and Berlin under a cloud of discord with his superiors Hirt, Peter, and Stieve.[10] After the war Hasselwander suspected Popp of spreading intrigue against him, including the accusation of murder of a Polish citizen. A major power struggle at Erlangen also erupted between dismissed histologist Johannes Hett and the new director of the institute Karl Bauer.[11] Hett insisted on his reinstitution into his old position at the institute after his denazification, but Bauer had made other dispositions of the duties within the department.[12]

Apart from personnel problems and inquiries by the military authorities and families who were searching for the bodies of NS victims, anatomical departments had to deal with the physical reconstruction of the destroyed buildings. Only the institutes in Erlangen, Halle, Marburg, and Tübingen remained intact, whereas Frankfurt, Freiburg, Giessen, Göttingen, Kiel, Leipzig, and Münster were completely in ruins, and the rest were more or less extensively damaged. Otto Veit took the rebuilding of his institute in Cologne into his own hands, as did some of his other colleagues.[13] He recruited forty-eight students who were willing to help with the construction work, some of whom were not yet accepted as medical students and hoped for preferred treatment. Veit visited the building site daily, but ran into conflict with those students who were not accepted as medical candidates and were greatly disappointed in him.[14]

Rebuilding efforts on a different level were necessary in order to reestablish scientific communication within Germany and internationally. The loss of interaction especially with Anglo-American anatomy had led to a deficit in technological development. During the 1930s and early 1940s electron microscopy had become an important tool in other countries, whereas in Germany the technique had not become routine, even though the technology was based on German ideas from the early 1930s.[15] Also, the effects of the "brain drain" through the forced emigration of important scholars in the 1930s became obvious, as most of these researchers had no intention of returning to Germany.

Finally, one of the greatest immediate problems in postwar German anatomy was the acute lack of bodies for dissection. Most anatomical departments, such as those in Munich, Giessen, and Vienna, still had bodies in storage from the NS period and continued to used them for teaching purposes for the first years after the war.[16] In Leipzig the Soviet authorities prohibited using the remaining nineteen bodies of executed persons, mostly Czech resistance fighters, despite the repeated petitions by new chairman Kurt Alverdes.[17] New bodies were hard to come by. One reason may have been that the discovery of NS victims' bodies in the anatomical departments, especially those of political dissidents, had become public knowledge and people did not trust the anatomists. Another problem was the disruption of the local physical and administrative infrastructure for body procurement, as the adherence to respective laws often depended on the goodwill of local administrators and on the official allocation of means for transportation. The situation in Marburg is documented in Benninghoff's letters—in late 1947 he asked Starck, chairman of anatomy in Frankfurt, for one body, as "it is now at the time when we are at the end of our material."[18] The situation remained difficult over the next several years. In 1951 Benninghoff acknowledged a lack of bodies to a colleague who had offered the delivery of children's bodies, which Benninghoff gladly accepted.[19] In 1952 Benninghoff had to decline a petition for eight bodies for a surgical dissection course because there had been no new body delivery in the previous five months, forcing up to twenty-four medical students to dissect a single body.[20] The use of children's bodies for the dissection course was an accepted practice in anatomical education as late as the 1960s, such as at the University of Innsbruck in Austria. The supply of bodies of stillborn children, which often went unclaimed by their parents, was plentiful, because the anatomical departments provided burials for them. Thus children's bodies were used for advanced dissection courses in order to save those of adults for other purposes.[21] The situation only changed decisively with the development of new societal sensibilities and legislation concerning stillborn children, and with the advent of functional body donation programs in the second half of the twentieth century.

By the 1950s the anatomical institutes were reestablished and most of the new anatomists were in fact those who had worked in Germany during the Third Reich. The reinstated anatomists who had been expelled from their positions during the NS period had to collaborate with those colleagues whose careers had prospered under the NS regime. Little is known about the interactions and potential tensions between these groups of colleagues, who had such different experiences. The personal continuity of academics working at the anatomical departments, and the authoritarianism, paternalism, and unquestioning loyalty still prevalent at German universities after the war, may have contributed to the fact that any reflection on the activities within the anatomical field between 1933 and 1945 was discouraged for many decades to come. The anatomists—like the rest of Germany—wanted to start anew and forget the Third Reich. This is obvious in most of the obituaries written by German anatomists for their colleagues in the 1950s and 1960s. The NS period simply did not exist in the lives of the eulogized, not in the case of anatomists who had supported the regime nor in the biographies of those who had been forced to emigrate.[22] Examples include Helmut Ferner's 1967 obituary of Clara, in which the author emphasized that Clara's recruitment to Leipzig was based on his professional merits and not, as "the jealous" would have it, on political reasons. He mentioned Clara's immigration to Istanbul without referring to any political problems, making the international move sound like a special honor.[23] And in von Hayek's 1957 obituary of Politzer, the author talked about the anatomist's work in India and the difficulties he encountered on his return to Vienna after the war, but without once addressing the fact that Politzer's emigration was anything other than voluntary.[24] This pattern of silence was only broken many years later.

The Medical Students: Surrounded by Violence, Death, and Silence

Those who were medical students in the Third Reich became part of the silence. They had learned to remain quiet during their medical education, when they saw their more courageous or less lucky fellow students murdered because of acts of outspoken political resistance against the NS regime. The motivations for student opposition were manifold, ranging from experiences on the war front to religious convictions.[25] Adverse experiences in dissection courses, such as encounters with bodies of NS victims, have never been mentioned as reasons for dissent in students, but might have contributed. Among the most prominent medical student dissidents were the members of the *Weisse Rose* in Munich: Hans Scholl and his sister Sophie, a biology student, Alexander Schmorell, Willi Graf, and Christoph Probst, who were all executed in 1943. Frederick Geussenhainer of the *Weisse Rose* in Hamburg perished in the concentration camp Mauthausen in 1945, while Gretha Rothe, also from Hamburg, died in the prison Leipzig-Mensdorf in April 1945. Frankfurt medical student Arnd von Wedekind was executed on 3 September 1943.[26] The surviving medical students remained silent for a long time. When some of them published memoirs decades later, they reported several of the typical encounters in the dissection room, but mostly they spoke of violence, death, and secrecy. There is little documentation on the impact of these experiences on the private and professional lives of the physicians of this generation. Having survived a medical training that had been suffused with the various manifestations of war and the brutality of a cruel regime, they kept busy rebuilding Germany and their lives after the war, a task that consumed all their energy. Reflections on the past were avoided or came much later, and then only for a few.

During the war all students had to perform civilian labor for the war effort, and male students could expect to be called up for active military service at any time. Some had just returned from the front when they embarked on their medical education. The confidentiality learned in their medical training transformed into secrecy and quiet acceptance of authoritative voices. Stieve strictly enforced the prosecutor's command of secrecy concerning the provenance of the bodies during the dissection course. He even called the police when students brought unauthorized visitors to the dissection labs, citing the potential danger of enemy propaganda.[27] During one of his skirmishes with the REM and students who accused him of harsh behavior, Stieve explained that his nervousness resulted from the stress of having to ensure the students' secrecy,[28] despite the fact that the connection between the anatomical department and Plötzensee was more or less public knowledge.

Not only in Berlin but all over Germany, medical students were confronted with the bodies of NS victims. Hedwig Wallis, a pediatric psychiatrist, had started her medical studies in Hamburg in 1941. She opposed the regime, but she and her colleagues understood exactly how far they could go in their careful show of resistance. While they exchanged whispered political jokes during anatomical dissection exercises, they did not share news overheard on "enemy radio" in this venue. And when Wallis noticed an entrance wound with traces of gunpowder in the neck region of a fresh young body laid out on a dissection table, she only dared communicate through eye movements with her colleagues. An anatomy assistant appeared immediately and started to distract her attention from the victim. The students assigned to the table quietly moved away, and the body vanished by the next day.[29] Wallis concluded: "Resistance was a lonely, deathly serious and pointless venture. [...] One cannot imagine the loneliness and desperation which accompanies such a situation [the death sentence of a dissident friend]. In the midst of a terrible war, threatened by daily bomb raids and fearing for friends and relatives at the front lines, hoping for a victory of the so-called enemies, while only capable of imagining a horrible end."[30]

Despite the dangers, some succeeded in working against the regime. Hiltgunt Zassenhaus also studied medicine in Hamburg during those years. As an interpreter for Scandinavian languages, she was able to help prisoners by smuggling food and messages during her visits to the prisons. In the dissection room she was very much aware of the provenance of the bodies. One of her professors pointed out that the bodies lacked fat, indicating that they had been executed or came directly from prisons; another body's loose cervical spine was the result of a vertebral fracture from hanging.[31] In her memoirs Zassenhaus described her impressions from an early morning study visit to the quiet dissection room: "I was the only living soul in this hall. The yellow emaciated bodies, bodies of deceased prisoners and executed persons lay there. [...] No pain was in his features. I stared at the eyes. They were empty, asked no questions, gave no answers. One body looked like another. Still, they all had once been living human beings, like Frederick, like Björn, like—me. They had waited, despaired, and still with hope in their hearts, this spark of hope that stays with us until the last breath."[32]

Another student who realized that she could easily become one of those lying on the dissection tables was physician Ingeborg Lötterle.[33] She started her medical studies in September 1940 and later became a student assistant in Stieve's histology course. Seventy years after the event she recalled vividly Stieve's introduction to the anatomical dissection course in the labs. He cautioned them: "Dear young colleagues, you find before you the beautiful bodies of young, decapitated human beings, who had to lose their lives for various reasons, and who now make their bodies available to you so that you can learn from them for the welfare of sick people. Please be aware that this can happen to any of you. There is no laughing or joking here, only serious work in gratitude and respect."[34] Lötterle and her colleagues were thoroughly shaken by Stieve's words, as they all knew that these bodies of men as young as themselves came from Plötzensee. For her the experience was doubly troubling, as she felt vulnerable because she loathed the National Socialists and had never joined the Hitler youth. Thus Lötterle felt drawn to Stieve's "wisdom and humanity"—as she put it—and because Stieve was not a party member either. She admired him because he demanded respect toward the dead, to be expressed by the students' industrious learning and orderliness and enforced through his active and helpful presence in the dissection rooms.

Lötterle was deeply disturbed when she experienced the exact opposite situation during a semester of studies in Tübingen. Robert Wetzel, director of the dissection course, was too busy with his NS political work to pay much attention to the dissection rooms, where an atmosphere of disrespect—with flying body parts, coarse language, and bad behavior—prevailed. Lötterle soon returned to Berlin where she received her clinical training amidst the grief for her partner, who was killed at the front, and the devastation of the bombing raids. She lost her notes on the experiments for her first doctoral thesis when her parents' house was destroyed. Much worse was the loss of neighbors and patients. Medical students were recruited for all health services in the continuous emergency situation of the war, and Lötterle worked in a hospital. After a bombing raid, she heard a fourteen-year-old boy cry out to her from a line of waiting patients: "Help, *Doktorle!*[35] Just help me! I can't breathe!"[36] Moments later, he died in her arms, blood pouring from his mouth. This and similar fates never left her. At the end of the war, Lötterle, still a medical student, was solely responsible for an entire ward at a neurological hospital in Berlin-Buch. Having lost her home, she slept at the hospital and was on call around the clock. She finished her second thesis at night. By the end of the war in May 1945 she had not yet taken her last exams and spent the rest of the summer trying to find a university at which she could complete her studies. A year later she finally received her MD from the University of Tübingen. Looking back on the NS period, she concluded: "This was a cruel time that nobody who has not actually experienced it can truly understand."[37]

Death was everywhere—colleagues, neighbors, patients, friends, lovers, and relatives could vanish from one day to the next. Physician Margarethe von Zahn, a niece of *Rote Kapelle* members Arvid and Mildred von Harnack, was also one of Stieve's students.[38] While von Zahn herself felt safe, she lost nine members of her family. Shortly after the war, in 1945 or 1946, Stieve called von Zahn into his office. He asked her to take a seat and declared, "Here I have an urn with the ashes of your aunt, Mildred Harnack-Fish. I want to return the ashes to the family. I have saved her from being dissected." Von Zahn accepted the vessel and put it in her satchel, carrying it around during her afternoon activities and feeling rather strange in doing so. She handed the urn over to her aunt's family, who buried Mildred's ashes in a family grave some years later.[39] Von Zahn believed that Stieve must have acted heroically by saving her aunt's body. However, Mildred von Harnack's name was on Stieve's list of research subjects, so he probably used tissues from her body.[40]

Rarely did students not realize where the bodies came from. One of the unsuspecting ones was Horst-Eberhard Richter, a leading psychoanalyst. He reported in his 1986 autobiography that it simply did not occur to him to ask about the bodies' provenance while he was busy trying to learn and integrate all information necessary to become a physician. He suppressed these questions as much as his discomfort and nausea in the dissection laboratories. Richter had just returned from war service at the Eastern Front during the winter of 1941–42. He had to leave his company, which was on the way to Stalingrad, because he had come down with a postdiphtheric polyneuritis that had rendered him temporarily paralyzed. While recuperating in a hospital, he heard of the death of most of his comrades in Stalingrad, and a bed neighbor informed him about the deportation of the Jews. In this situation Richter, who at age nineteen was no longer deemed fit for front service, decided to study medicine.[41] It is quite possible that Richter's lack of curiosity concerning the bodies had to do with the physical and emotional trauma that he had just barely survived. Knowing the fate of the persons whose bodies he was dissecting might have been too much for him.

Many other medical students had also just returned from the front. Among them was Harald Bräutigam, professor of gynecology. Late in 1942 he was flown out of Stalingrad and returned home to medical studies and bombing raids in Berlin. In his 1998 autobiography he recalled distinctly his and his fellow students' silence concerning any potential criticism of authorities, whether they were NS officials or university professors.[42] Bräutigam later contemplated whether this silence was due to cowardice or the early impressions of the brutality of the regime and the general climate of fear for one's own life in war.[43] He remembered a dissection course interrupted by air-raid sirens, and with bodies of young women that showed no signs of disease but strangulation marks on their necks. Bräutigam claimed that Stieve himself had told the students that these women were Polish concentration camp inmates and that he had determined their execution date based on their menstruation cycle on days when no ovulation was to be expected. Further, Bräutigam remembered Stieve saying that the data from these women had confirmed his findings on paracyclical ovulation.[44] While Bräutigam's description of these activities is quite specific and bears similarities to many other postwar rumors surrounding Stieve, the accusations have not yet been borne out by any documents. While there were Polish women among the executed from Plötzensee, they had been decapitated, not hanged. Also, there is no evidence that bodies from concentration camps were delivered to the Berlin anatomical department. Margarete von Zahn was asked about this passage from Bräutigam's memoir but had never heard of Polish women or Stieve's influence on execution dates.[45] If in 1943 Bräutigam did see the bodies of Polish women who had been hanged, they must have hailed from sources other than Plötzensee, conceivably from forced laborer camps. As the body registries from the anatomical departments have vanished, this assumption cannot currently be verified.

Hoimar von Ditfurth, a scientific journalist, began his medical studies in Berlin in 1940. He recalled that there were always enough bodies for dissection. The majority were those of healthy young men without heads, which had been cleanly removed just above the shoulders. Von Ditfurth knew exactly where these bodies came from, as he had noticed the bright red posters distributed all over Berlin announcing executions and reasons for verdicts. And everyone else knew, too: Stieve, von Ditfurth's fellow students, and the general public. Fifty years later von Ditfurth asked himself whether his silence, his lack of resistance against the criminal regime, made him guilty. He concluded that, while it could be explained by weakness arising from fear, the silence led to a moral failure that left him guilty. He also noted that he had never considered his own guilt at the time, as the war had changed all standards.[46]

Theologian and social ethicist Stephan Pfürtner started medical studies in Breslau in 1940 after having served as a soldier at the Eastern Front. On his first encounter with bodies in the dissection course, he experienced several "shock-seconds" but suppressed any signs of emotion, remembering his encounters with bloody bodies in surgery during his war service.[47] However, in his memoirs he recalled having many questions while dissecting:

> Why did we assume to have the right to snip away at the mortal remains of human beings? Did not this man whom I was supposed to "dissect" also have a name before? Who was he? What was his name? Of course, these were only

the "mortal remains", not the human being he was before. Did this "speci-men" represent nothing other than a piece of chemically treated flesh, an ob-ject or thing, with which one may do as one pleases? Hadn't humans from time immemorial—ultimately into the present—demanded something like piety to-wards the dead and honored them in cultural rites like for example a burial?

I could not answer these questions for myself, and certainly not at that moment. I simply connected with the attitude I had learned during my medical service during the Poland campaign. It had occurred to me that one could not work as a physician if one could not abstract oneself from one's own emotions during certain situations of suffering and emergency, in order to be able to at-tend with a level head to that which was factually necessary.[48]

Pfürtner's questions were very much the same as the ones asked by modern medical students when they reflect on the dissection course. In National Socialist Germany and under the conditions of war these issues remained untouched and suppressed for Pfürtner and his fellow students. Like so many of their other experiences in the Third Reich, they became part of the big silence that lasted through their professional lives after the war. However, it seems inconceivable that these experiences did not pro-foundly affect them. The group of former students quoted here is a small one, consisting of those who finally broke the silence, a mixed assembly of general practitioners, professors, and journalists. They all had vivid memo-ries of the dissection course and of the secrecy surrounding it. The silence about their wartime experiences is all the more notable, as narratives of the fates of medical students in the resistance, especially the members of the *Weisse Rose*, were omnipresent in postwar Germany. But then, tales of self-sacrificing righteousness are easier on the soul than critical self-reflection. As the postwar German public celebrated the dissident medical students, these shining examples of courage probably contributed to the feelings of guilt in their surviving fellow students. One of the typical reac-tions to guilt is silence; the alternative, confession, was not very popular at the time.

As a group these medical students have never been studied. Neither is there much information on the many other distinct sets of medical stu-dents who received training under the conditions of war: Jewish students in the Warsaw ghetto underground medical school,[49] Norwegian students in the Buchenwald concentration camp,[50] and SS medical students trained at the Dachau concentration camp.[51] A comparative study of medical edu-cation under war conditions is necessary to understand the postwar devel-opment of the medical professions in the countries affected by this history.

If little is known about medical students who received their training during the war, there is even less information on those students who were taught during the 1950s and 1960s by anatomists who had worked under the NS regime. Currently only one statement is available, which comes from Karl-Heinz Roth, one of the physician activists who pioneered the first inquiries into the history of medicine in the Third Reich. During a panel discussion at the *Gesundheitstag* Berlin 1980, the first official meeting on the subject, Roth reported about his motivation for this research: "I started my medical studies with an anatomist who said during a lecture that the brain specimens for our studies were unfortunately not very good; that sadly, executions did not exist any longer and thus there were no fresh brains. This was a traumatic start for me, during one of my first pre-clinical semesters, one that I have never forgotten."[52] Roth studied with Neubert in Würzburg from 1963 to 1965. He remembered Neubert as jovial but strict and thus feared by the students. Roth, who was deeply disturbed by Neubert's attitude toward the executed, did not feel ready to openly confront his professor at the time. However, Roth's experience was one of the reasons why he later left Würzburg,[53] and it ultimately became one of the motivating forces for research on the history of medicine in National Socialism.

International Criticism of Stieve, 1945 and 1946

While many anatomical institutes came under the investigation of occupying forces, several others had to face direct questions about their use of bodies of NS victims. Some of these inquiries became widely publicized in the early postwar years. Hirt had escaped any consequences of his deeds by committing suicide in the summer of 1945, several months after the discovery of his victims in Strasbourg had been published by the British newspaper *Daily Mail*.[54] Spanner's activities in Danzig were exposed in Poland, and Stieve's work became the topic of an international discussion in 1945 and 1946. Stieve provided "material" from executed persons to the Swedish doctoral student Sten Floderus, who had visited Stieve's laboratory in 1938 for training in histological techniques. In his investigation on the morphology of pituitary glands he used "material" from seven men executed in Berlin. In 1944 Floderus defended his thesis publicly, as is the Swedish custom. Apparently, the Swedish press then became aware of Floderus' special "material," as an article questioning the ethics of the use of tissues from NS victims appeared in the daily newspaper *Aftontidningen* on 18 May 1945.[55] Floderus explained in an interview that he had only used tissues from executed men but had refused "material" from executed women. He had known—so he claimed—that the tissues came "from women, who had been executed in a specific, prearranged manner. These were women, who had been murdered on certain days of their menstruation, e.g. on the 1., 3. or the 5. day".[56] This article in *Aftontidingen*[57] represents the first currently known public and international criticism of the use of bodies of executed NS victims for anatomical purposes, published only ten days after the end of the war in Europe. A response from Stieve is not documented. Floderus' claim as to Stieve's determination of the execution date according to menstruation details is highly unlikely for various reasons, among them the rigidity of the NS legislative apparatus and the fact that many women were not menstruating at all or were executed on the same day.[58]

The Swedish article was followed by another criticism from Switzerland in 1946. Physician Hans Jacob Gerster was a fervent supporter of Hermann Knaus and his method of natural birth control based on the timing of a woman's ovulation.[59] As such Gerster was also one of Stieve's most outspoken critics, as Stieve purported to have found evidence of so-called "paracyclical ovulations" and disputed Knaus' work vehemently.[60] In 1946 Gerster published a paper in which he discussed criticisms of Knaus' method, among them Stieve's.[61] While Gerster's analysis of Stieve's argument was lucid—Stieve's theory of "paracyclical ovulations" was indeed wrong—Gerster also mentioned "articles published in the daily press" reporting that Stieve had received organs removed from concentration camp prisoners by vivisection and questioned whether data retrieved from such sources could be scientific.[62]

Stieve answered with a rebuttal in 1947, in which he accused Gerster of using slander in aid of a scientific argument.[63] Stieve declared that he had never set foot in a concentration camp or received bodies from such a source, and that he used "material" from surgical cases and bodies of women who had been executed following "orderly court sentences" for "heinous crimes, murder, looting and professional abortion."[64] Though Stieve had to defend his activities to the university authorities in Berlin,[65] the local press portrayed him and his colleagues as "serious men of science, whose hearts had not hardened."[66] The new government of the German Democratic Republic decided to avoid discussions of Stieve's contentious past and supported him in his role as leader of the Berlin anatomical department. Stieve continued to use the "material" from NS victims even after all these debates. He told his colleague Albert Hasselwander on 11 November 1949, "One of the main reasons why I stayed in Berlin are the specimens which I collected over the last 40 years; specimens that don't exist anywhere else in the world any more. I hope to be able to continue working with these at least for a few more years."[67] He published several more investigations based on the tissues of NS victims. The last one appeared posthumously in 1953.[68] Stieve's use of bodies of NS victims was not discussed again until Aly's publication of Voss's diary in 1987.[69]

Herrlinger: A Secret Discussion on the Ethics of Anatomy, 1957 to 1959

One postwar discussion on the use of bodies of NS victims in anatomy was only recently discovered because it was kept secret at the University of Würzburg. Götz Aly's 1987 publication on Voss mentioned a controversy surrounding the recruitment of Robert Herrlinger as anatomist and stated that Herrlinger was said to have later regretted his research on NS victims.[70] Newly discovered files at the university archives in Würzburg revealed the full documentation of a debate that began in 1957 and ended with Herrlinger's promotion to *ausserordentlicher* professor of the history of medicine in 1960.[71] Herrlinger never applied for employment in anatomy but exclusively in the history of medicine, a discipline he represented in Würzburg as acting director of the Georg Sticker Institute of the History of Medicine from 1953 to 1962. However, his promotion was significantly delayed by a *Sondervotum* (dissenting opinion) voiced by three academics who questioned Herrlinger's ethical attitude based on his work during the war. A unique and highly controversial discussion ensued on the ethics of using bodies of the executed for anatomical purposes during the Third Reich. It involved participants on a political spectrum from active or silent supporter to dissident and victim of the NS regime. As this controversy touched on many of the essential ethical questions still actively discussed in modern anatomy, it is documented here in some detail.

Herrlinger was born in 1914 and studied medicine and art history, and received his MD in 1938 for anatomical studies on rat spleens under Hoepke's supervision in Heidelberg. Clinical training and art historical studies in Mannheim and London in 1939 were cut short by the beginning of the war. Herrlinger returned to Germany and accepted a position as scientific assistant in physiology in Jena in October 1939, but transferred to the institute of anatomy under Rüdiger von Volkmann in 1940. At that time he also received a PhD in art history.[72] According to Herrlinger his relocation from Jena to Posen followed a personal recommendation by Jena anatomy professor Fritz Körner for a more senior position with Voss at the newly opened German university of Posen.[73] Herrlinger worked in this capacity from August 1941 until the end of the war in 1945, at the same time serving as military physician on the Eastern Front from May 1941 to October 1943. He received his *Venia legendi* in anatomy on 17 November 1944, based on histological studies of fresh human spleens procured from prisoners executed at the Posen/Poznan execution site. Together with Voss, Herrlinger wrote an anatomy compendium adapted for the accelerated medical education during wartime. The book was finally published in

1946 and became, due to its compact style, a popular resource for medical students for the following forty years.[74]

At the end of the war Herrlinger fled Poland and settled in Bavaria, where he became subject to the mandatory denazification procedures and reported memberships in the NSDAP, the SA, and several other minor NS organizations.[75] A first hearing of his case on 20 January 1947 ended in a verdict of "fellow traveler," but this was revised on 14 December 1948 to "exonerated." Copies of Herrlinger's NSDAP membership cards do not support his self-reported cessation of membership on 3 April 1941.[76] From 1945 to 1955 Herrlinger worked as a family physician in Münchsteinach, while he served as a part-time lecturer in the history of medicine at the University of Regensburg from 1949 to 1952 and published the results of his wartime research.[77] Herrlinger's career at the University of Würzburg began in 1951 with work as a part-time lecturer in the history of medicine. His *Venia legendi* was transferred to Würzburg and changed from anatomy to history of medicine. He led the institute for history of medicine as acting director, first as a part-time lecturer, from 1955 on as a *Privatdozent* and as *ausserplanmässiger* professor in 1957. On 25 April 1960 he was promoted to *ausserordentlicher* professor, and was recruited as chairman of the newly founded institute for the history of medicine at the University of Kiel in the summer of 1962, where he died on 8 February 1968.

A decisive period in Herrlinger's career were the years 1957 to 1959, when the controversy ensued around his suitability as a teacher of history and ethics in medicine. Among the main participants of the debate was Kurt Neubert, chair of anatomy of the University of Würzburg since 1952,[78] who served as expert witness on the uses of bodies of the executed in anatomy. Neubert had published several studies based on this "material."[79] He had been a NSDAP party member since 1 May 1933, held several leading positions in the NS lecturers' union and the SA, and was member of other political groups.[80] The men opposing Herrlinger's promotion included psychiatrist Heinrich Scheller, internist Ernst Wollheim, and pediatrician Josef Ströder, all chairmen of their disciplines. Scheller[81] was a psychiatrist who had experienced the reality of the NS sterilization laws during his work at the psychiatric hospital of the Charité, which was chaired by Germany's most prominent psychiatrist Karl Bonhoeffer and later by the leading National Socialist Max de Crinis. While Bonhoeffer did not advocate NS eugenic laws, he and his staff delivered 1,991 diagnostic expert reports on patients, which in 862 cases confirmed diagnoses that could lead to forced sterilizations under NS law.[82] Wollheim worked in internal medicine at the Charité under the guidance of the leading internist Friedrich Kraus.[83] His productive academic career in cardiology was disrupted in 1933, when Wollheim was dismissed from his position by the new regime for so-called "racial" reasons.[84] He immigrated to Sweden and continued his scientific work at the University of Lund from 1936 to 1948, when he was recruited by the University of Würzburg. Ströder[85] was a devout Catholic and was drafted as physician to the air force. He was on the verge of a military trial because of oppositional remarks to colleagues when he was saved by a superior and transferred to Cracow in occupied Poland. As acting director of the children's hospital he was responsible for German and Polish children and resisted orders by the military to deny his wards medication and food. This together with his cordial relationships with Polish colleagues led to the threat of another military trial in 1944, which he was able to evade again. After the war he resumed his position in Düsseldorf and was recruited by the American occupying forces to take care of displaced children at the former concentration camp Buchenwald. He accepted the chair in Würzburg in 1948.

In response to the *Sondervotum,* the university senate decided on 31 July 1957 to form a commission to investigate the claims,[86] naming as its members Ulrich Stock, professor of law and president of the university; Heinz Fleckenstein, vice president of the university and professor of theology; Wilhelm Arnold, professor of psychology and dean of the philosophical faculty; and Heinrich Ott, professor of physics and dean of natural sciences; Georges Schaltenbrand, dean of the medical school and professor of neurology, was chosen as observing member of the commission.[87] Fleckenstein was a catholic theologian and the only commission member described as opposed to the NS regime.[88] Schaltenbrand was internationally well respected, but later became notorious for his work on the potential viral origin of multiple sclerosis, using patients from a psychiatric institution as experimental subjects in 1940. The "Schaltenbrand experiment" has become an important teaching example for studies of an unethical scientific basis within a continuum of a worldwide practice of unethical experiments.[89]

The controversy started in July 1957 when the dean of the medical faculty recommended Herrlinger as the main candidate for the position of *ausserordentlicher* professor at the history of medicine department.[90] Wollheim, Scheller, and Ströder opposed this nomination. They argued that Herrlinger was an unacceptable representative because of his previous work in anatomy and art history, which they considered unethical.[91] Under the premise that no external observers would be allowed in any proceedings, the senate decided to form a committee made up of men from diverse disciplines within the University of Würzburg.[92] After Herrlinger's first hearing in December 1957, Neubert was asked to give an expert witness report on Herrlinger's experiments and his possible offense against medical ethics, specifically anatomical ethics, which Neubert delivered in February

1958.[93] A second senate meeting was called for June 1958 and in January 1959 the final statement of the committee followed Herrlinger's explanations on all counts, and he was appointed as *ausserplanmässiger* professor for the history of medicine on 24 April 1960.[94]

The authors of the *Sondervotum* renewed their protest, claiming that Herrlinger's personality was unsuitable for teaching history of medicine and the ethical foundations of the medical profession. They based their opinion on his anatomical work and art history essays,[95] specifically criticizing him for never doubting the legality and fairness of NS court verdicts during his time in Posen. They believed the NS justice system to have been neither just nor legal, and they had good reason for this belief. Ströder had firsthand experience of life in occupied Poland during the war, and knew intimately how quickly Polish and German dissident individuals were persecuted and killed by NS authorities. Wollheim had been an early victim of the NS discriminatory legislation through the disruption of his career, and Scheller had been at least a witness to if not a collaborator in the preparation of diagnostic reports on patients who were to be sterilized according to the NS laws. The authors correctly assumed that there were political prisoners among the executed,[96] even though Herrlinger did not believe this. Polish citizens were not always executed by hanging, as Herrlinger assumed, but also by decapitation.

Still, not all men of the *Sondervotum* criticized the use of bodies of executed persons as research subjects, as long as those had been criminal offenders. However, they did point out that even minor crimes had been punished by execution in the Third Reich. While they accepted the content of Neubert's expert report, which described the traditional use of bodies of the executed in anatomy, they found these findings irrelevant for the questions at hand, which they considered to reside not on a level of anatomical procedure but on one of political perceptions and ethical behavior. Scheller held the most radical, and in many ways most modern, position, in that he thought the use of any executed person for anatomical purposes to be objectionable.

The men of the *Sondervotum* also asked whether Herrlinger was a convinced National Socialist because of his posting to the *Reichsuniversität* Posen, but he insisted that his recruitment was not politically motivated, and was instead based on a personal recommendation. They also doubted that it had been necessary to use fresh "material" for Herrlinger's studies, and Wollheim questioned the permissibility of dissecting the freshly dead on a more philosophical level by presenting the example of pathologists who had to adhere to the legal limits of waiting to perform an autopsy until six hours after death. He explained that this rule had been introduced to create a distance between the person's time of death and the body's becoming an object of scientific inquiry. This question had certainly never been publicly discussed in anatomy.

The authors questioned Herrlinger's ethical attitude—which they saw in the wording of his publications, both anatomical and art historical—as well as his style of procedure. Wollheim found Herrlinger's detailed description of the procedure within the execution chambers particularly repulsive. This included references to sampling blood from "pulsating carotids" which was in danger of being contaminated by "food pulp" exuding immediately from the esophagus, and simultaneous laparotomy for procuring the spleen and blood from the "stump of the splenic vein."[97] The authors perceived this phrasing as a sign of callousness in Herrlinger's character. They believed their opinion was supported by the discussion of the sequence of events in the execution chamber, in which he talked about "grabbing" and "throwing" the prisoner's body. They also read the language in the art history essay as expressing Herrlinger's close association with NS ideology. They did not accept his explanation that he had to employ the National Socialist idiom if he wanted to be published at all, and that he had done so to express ideas that ran contrary to NS thinking. Rather, Wollheim saw Herrlinger's argument concerning the influence of "race" on culture as consistent with NS ideology and criticized Herrlinger for not being able to perceive the impermissibility of mixing science and "racial" theories. The medical historian Eduard Seidler believes that especially younger scientists of all academic disciplines sometimes used NS terminology in order to avoid jeopardizing their budding careers.[98]

An important point of criticism by the men of the *Sondervotum* was why Herrlinger had never asked about the reasons leading to the execution of the defendants. After Herrlinger repeatedly stated his belief that these had been criminal defendants, legally convicted by regular courts, Scheller pointed out that his professed lack of curiosity concerning the origin of the bodies and absence of doubt toward the justice system rested on an eminently political background. Thus Herrlinger's professional responsibility and political attitude were in question. The authors perceived evidence of Herrlinger's putative National Socialist convictions in his publications, his attitude toward the executed, his activities in Posen, and his arguments in the controversy. Even after extensive discussions, the men of the *Sondervotum* did not change their evaluation of Herrlinger's character, but their vote was ultimately overruled by a senate majority.

One remaining question is, why they had tolerated Herrlinger in his position as representative of the discipline of history of medicine until 1957? They all had been present at the university when he was first recruited as lecturer for the history of medicine in 1951 and had not publicly complained about his teaching or subsequent promotions until 1957. It is not known whether they had only then become aware of Herrlinger's history and publications, or whether this final promotion was one that they felt called upon to prevent.

Among Herrlinger's most vocal defenders were Neubert and Schaltenbrand. Neubert's expert witness statement from 1958 represents a rare explanation of anatomists' rationale for the use of bodies of the executed. It is also remarkable for the questions it does not address, namely those of a more political and philosophical nature.[99] Neubert based his argument on the central paradigm of anatomy that declares dissection of human bodies to be the best way for acquiring anatomical knowledge. From this he deduced the need for "fresh human material" in histology, which could only be procured from the freshly executed. He stated that the practice had been an accepted procedure in all "civilized nations" since the early nineteenth century. Neubert took offense to the critics' notion that Herrlinger's use of a traditional source could be interpreted as a lack of moral fiber. He did not comment on the anatomist's potential political responsibility and addressed any political issues only obliquely, if at all. Neubert included historical examples of anatomists working with "material" from the executed to underline the fact that this was not specifically a National Socialist method, and independent of prevailing politics.

He did not mention, however, that this practice changed during National Socialism, possibly because he had not realized the enormity of these changes. He must have been at least partly aware of these as he introduced Stieve's research on executed women into the discussion of the second senate meeting. Should he truly not have realized that this work marked a change not just in the quantity of studies and bodies available, but also in the ethical quality of the research? Neubert constructed a historical continuity that did not exist by ignoring the distinct transformations of body supply during the NS period, and stated that over "the last 50 years, [...] hardly any anatomist had not used opportunities to procure fresh 'material' from the bodies of the executed."[100] Was Neubert's apparent "blindness" for the change in method, a "blindness" shared by most anatomists of his time, reinforced by his own political convictions as a former active National Socialist? Those convictions would have made it impossible for him to declare the NS judicial system as anything other than just, and all NS executions would have been legal to him, independent of the crime committed. It could have made no ethical difference to him whether the executed person had been a criminal or a political prisoner. This latter point turned out to become a central one in the criticism from the men of the *Sondervotum*. While Wollheim accepted the anatomical method of procuring "material" from the bodies of executed criminals, he objected to the exploitation of bodies of political prisoners. In 1958 Neubert professed to be of the same opinion, but it is doubtful that he would have acted on this insight during the Third Reich.

The fact that the men of the *Sondervotum*, with the exception of Scheller, did not contest the use of bodies of the executed criminals for anatomical purposes may be the reason why Herrlinger's anatomical work was never compared to that of his contemporaries in this discussion. Indeed, Herrlinger, as one of the younger colleagues in his discipline, had produced only a few publications based on "material" of the executed compared to many of his peers, including those with thriving postwar careers in anatomy like Bargmann and von Hayek. Apparently, having worked with the bodies of persons executed by the NS regime did not affect anatomical careers after the war. While Herrlinger consistently denied having dissected the bodies of executed political prisoners, his critics reproached him for not even having asked about the verdicts. Herrlinger answered in his own defense that "it was a general human desire that, when one already had to deal with such tasks, one tried to internally shield oneself against them."[101] He added that, while he had respect for the bodies on which he was working, he had no relationship with the people the bodies had belonged to in life.

Herrlinger described here the classic definition of a distancing process physicians move through in their training that results in "clinical detachment," the ability to neutrally observe patients in order to help them. It has been hypothesized by historians that anatomists, like psychiatrists, have a greater need for "clinical detachment" than other medical disciplines because of the character of their work.[102] However, the case can be made that during National Socialism the omission of asking about the origin of bodies was not just overreaching "clinical detachment" but an expression of what the German anatomist Christoph Viebahn called a "political" and ultimately "moral detachment."[103] The authors of the *Sondervotum* also developed a unique point of criticism that addressed the ethics not only of Herrlinger but of all anatomists active in Germany during the Third Reich, and indeed all anatomists globally at any time period. Wollheim argued that "any collaboration in an act of execution—and the dissection of a body was a collaboration—would have to be considered as an expression of the fact that one identified with the judicial methods on which the verdict rested."[104] This was a truly modern argument, which Herrlinger strongly opposed. He countered that "if one took the point of view that during the war no verdict had been just or legal, then no anatomist should have been allowed to touch a dead body."[105] Scheller's answer "that he certainly would not have done so" illuminated the deep division of political as well as ethical attitudes between the men of the *Sondervo-*

tum on the one side and the anatomist and his defenders on the other. Ultimately this disagreement remained irreconcilable, and still exists in current anatomy.[106]

Herrlinger's colleague Eduard Seidler, who knew him as a kind and supportive mentor, asked him once about his activities in Posen, and Herrlinger answered briefly, "What could I have done, it was not pleasant." Seidler never considered Herrlinger to have been a convinced National Socialist and thought it quite possible that he might have regretted his Posen work, but never heard him say so.[107] Another indication of Herrlinger's state of mind comes from a 1968 article he wrote, which was interpreted by Daniela Bohde as a reflection on his Posen experience. In this art history essay Herrlinger remarked, "With the subject of anatomy one always stands with one leg on or beyond the edge of a state of little culture".[108] At this point in his life Herrlinger seemed to have distanced himself completely from the discipline of anatomy. After the war Herrlinger never sought a career in this field but one in the history of medicine. The men of the *Sondervotum* protested not because Herrlinger was an anatomist, but because he was a medical historian. As such they held him to a high ethical professional standard, which they thought he did not represent. Their criticism aimed at the political responsibility of anatomists in the Third Reich, which they thought Herrlinger lacked. They also missed in his personality the sensitivity for medical ethics they deemed necessary in a teacher. It is possible that they might not have objected to his promotion as an anatomist, as no one had ever protested against the promotion of those anatomists who had been active during the Third Reich. The university senate did not follow the argument of the *Sondervotum* and promoted Herrlinger in 1960. The sole mild criticism expressed in the senate report referred to Herrlinger's unquestioning trust in the NS judicial system.

The Herrlinger controversy is exceptional as it addressed the ethical prerequisites for a medical historian and not those for an anatomist. It speaks to the perception of anatomy at the time, that the men of the *Sondervotum* had lesser expectations of the ethical behavior of anatomists. The involvement of medical historians with the NS regime was not well known in the late 1950s.[109] This postwar debate on the work of anatomists during National Socialism remains unique, and the university was successful in keeping the situation secret and contained within its walls. The discussion was significant in that it addressed central questions of ethics in anatomy, specifically those of the provenance of bodies for anatomical dissection and the political responsibility of the scientist, anatomist, and physician. How much farther could these issues have been advanced if controversies like this one had been openly debated and not kept from the public?

The Pernkopf Controversy

Further discussions on the role of ethics in anatomy came much later when Pernkopf's atlas became a focal point of international debate. The popular *Topographische Anatomie des Menschen* was first published in 1937, with the first American edition published in 1963.[110] Surgeons valued its intricacy, as it contained detailed and extensively annotated color illustrations enhanced in quality by the beauty of the coloration, which was the product of an innovative printing technique.[111] The origins of the bodies in these illustrations were not publicly discussed until 1980, when physician Gerald Weissmann inquired into the political changes at the Vienna Medical School in 1938 and its new dean Pernkopf.[112] David J. Williams, a professor of medical illustration, published the first detailed investigation into the background of the creation of the atlas in 1988. During a sabbatical in Vienna, Williams studied the more than 800 original paintings for the atlas and conducted interviews with Franz Batke, at that time the last living contributing artist to the atlas, from whom he had hoped to learn his painting technique. Williams learned from Batke that not only Pernkopf, but also illustrators Erich Lepier, Ludwig Schrott, Karl Endtresser, and Batke were either active members of the NSDAP or participants in the war. Evidence of the illustrators' NS sympathies was visible in the first edition of the atlas: Lepier had added a swastika to his signature in some of the plates created between 1938 and 1945, while Endtresser signed the double "ss" in his name in the shape of the typical SS symbols and Batke shaped the 44 in the date 1944 like the SS runes. Williams also found evidence that the Viennese Anatomy Institute regularly received the bodies of victims of executions.

In 1995 Edzard Ernst, former faculty member at the Vienna Medical School, reported that Pernkopf, as dean of the medical faculty, had been personally responsible for the removal of all Jewish faculty members, spouses of Jews, and political dissidents, 153 of the total of 197 members of the medical faculty.[113] Ernst claimed that Pernkopf used material from children killed in a Viennese hospital in his atlas and bodies of executed persons for teaching purposes. In direct response to Ernst's publication, physicians Panusch and Briggs[114] asked their medical center to remove the Pernkopf atlas from circulation and entered into a discussion about the ethics of a continued publication of the atlas with the distributors of the atlas. Edward E. Hutton Jr., as spokesman for the publisher Waverly Inc. for the German subsidiary Urban and Schwarzenberg, stated that, in spite of their own inquiries into the matter, they continued publishing the Pernkopf atlas "because of its scientific merit and the fact that, to date, no concrete evidence exists to substantiate Pernkopf's use of cadavers originating from Nazi concentration camp victims," and that they tried to "separate Pernkopf, the man, from the work."[115] Hutton stated that the publisher supported the request for an inquiry conducted by the University of Vienna, put to the Austrian authorities and the publishers by the Israel Holocaust and Martyrs Remembrance Authority, Yad Vashem. Authors Howard Israel and William Seidelman reported these events,[116] and supported Yad Vashem's demand for a commemoration of potential victims of NS terror and an acknowledgment documenting the history of Pernkopf in future editions of the atlas. This opinion was endorsed by Daniel Cutler, a medical illustrator at the University of Michigan.[117]

At this point in 1997, the president of the University of Vienna Alfred Ebenbauer[118] admitted for the first time publicly that the university and specifically the department of anatomy had systematically suppressed and even denied its NS past, and that relevant investigations had not been performed. Ebenbauer, together with a number of new university faculty members from a younger generation without NS ties, explained that the attitude of the university had changed due to "increasing pressure from abroad" and a new political atmosphere in Austria after former chancellor Vranitzky's public recognition of Austria's responsibility for the events of 1938–45. They gave a preliminary report of the history as far as it was known, and announced a research project by the senate of the university named "The Anatomical Sciences 1938–1945." This was followed by a lively discussion in the general media.[119]

The senatorial project of the University of Vienna investigated two sets of circumstances: first, those dealing with the origin and destiny of the bodies used by Pernkopf; second, those concerning Pernkopf's political activity. The design of this project was based in part on the Tübingen project, which dealt with the history of the anatomical institute during the NS period at that university. The Vienna project revealed that throughout his tenure Pernkopf was actively involved in the acquisition of bodies for his institute. During the war the influx of bodies increased to such an extent that the anatomy institute's storage rooms sometimes became overfilled, and executions were postponed because of this. Pernkopf applied for an increase of the institute's budget for 1943 in order to handle the rising number of bodies.[120] The study further disclosed the origin of the bodies delivered to the anatomical institute from 1938 to 1945.[121] Among them were 3,964 unclaimed or, rarely, donated bodies from hospitals and geriatric and charitable institutions; about 7,000 bodies of fetuses and children, including miscarriages and premature and stillborn babies; and there were at least 1,377 bodies of executed persons, including eight Jews, who had been decapitated at the Vienna assize court or shot by the GeStapo at a rifle range. Due to incomplete documentation, it was impossible to obtain the exact number of all executed persons. There was no evidence that bodies from the concentration camp Mauthausen or the affiliated camp Gusen were brought to Vienna, but such bodies seem to have been transported to the anatomical institute at Graz. More than half of the executions had been carried out for political reasons, including 526 verdicts of "high treason." Of the bodies of the eight Jews, one was handed over to his family, while the other seven were delivered to the anatomical institute. The investigations of the anatomical collections at different institutes of the University of Vienna revealed the existence of specimens from NS victims that were then removed and interred in a grave of honor provided by the city of Vienna in 2012.[122] Previous findings encountered in Pernkopf's and Lepier's biographies were confirmed.[123]

After the war Lepier continued his highly praised work as an anatomical illustrator, contributing to other popular atlases such as the Sobotta/Becher atlas,[124] and the one by Carmine Clemente. Clemente initially used the Pernkopf plates, including those drawn by Lepier, for his own atlas.[125] Lepier received the title of professor in 1959 in recognition of his contribution to science.[126] About half of the original 791 illustrations in the Pernkopf atlas were created during the Nazi years; the other half either predated 1937 or were produced after 1945. Forty-one plates were definitely signed with dates from the Nazi period, and it is likely that at least some of the persons depicted were executed NS victims. For the remaining 350 plates the date of creation as well as the provenance of the bodies used as models is unclear.[127]

After the results of the Vienna Senatorial Project were disclosed, Howard Spiro, director of the Yale Program for Humanities in Medicine, felt that the "silence of words" had finally been broken.[128] Early reports on the NS activities in Austrian medical schools had not found a wide audience, with the exception of the controversy surrounding Heinrich Gross, the physician implicated in NS "euthanasia."[129] The critiques by Seidelman, Israel, and Weissmann concerning the lack of historical analysis of the origin of the Pernkopf atlas and its authors[130] created a "push from the other side of the Atlantic and from Yad Vashem"[131] that initiated a "belated [...] research into this shameful era."[132] In addition to this impulse several other factors contributed to a new openness for the discussion of the NS past and ethics in anatomy. Many scientists active during the NS period had died and the general political climate in Austria had changed, initiated by the international controversy in the 1980s surrounding former president Kurt Waldheim's NS affiliation. The country did not represent itself any longer only as a victim of the Nazi regime but also as a collaborator in NS crimes.[133] In addition, the ethical debate concerning body acquisition and demonstration had become very active in Germany among anatomists, philosophers, artists, lawyers, physicians, theologians, sociologists, and journalists following the controversial "Body World" exhibitions by Gunther von Hagens in the 1990s.[134]

The Pernkopf controversy also questioned whether it was ethical to continue to use the atlas, spurring many arguments.[135] On one side, detractors wanted the books removed from all libraries.[136] Arguments for complete banishment included, among other things, that fundamental evil contributed to the creation of the atlas; that nobody should profit from the exploitation of human life, especially of victims of the NS regime; that the active use of results from research by NS scientists could justify the atrocities committed; that a work cannot be separated from its creator (thus if the creator is evil, the work is too); that the use of NS data might initiate society's slide down a "slippery slope" toward amorality; and that the atlas is easily replaceable by other anatomical atlases or more modern means of medical imaging. On the other side, supporters of the atlas argued for its continued use as a historical document, preferably in its original form (including the NS symbols), and with the addition of a historical note commenting on the origin of the work. These arguments included the opinion that good may derive from evil in providing new doctors with the means to perform better operations; that victims of the NS regime and their sacrifice are best honored by a continued use of the atlas; that publishing the atlas in its original form, including NS symbols and information about the historical context, can be used not only for the anatomical but also ethical and historical education of future physicians; that eliminating or suppressing books is a symptom of totalitarian systems; and that the atlas is a work of great aesthetic value.

On balance it seems justifiable to continue using Pernkopf's book under the condition that information on its historic background is made available at the time of use. To see the atlas as a masterwork of greatest aesthetic value or as the evil manifestation of NS science[137] seems to ascribe this book too much power. The atlas is neither of these things, but the product of an obsessive perfectionist who would have pursued his work under any political circumstances. Indeed, the first and the last parts of the atlas were not created during the time of the NS regime in Austria, but before and after it and under very different political and material circumstances. The atlas is still one of the best in terms of accuracy, showing levels of detail that are of direct relevance for the actual dissection process.

Pernkopf's story remains an object lesson for modern anatomy in that the inquiry into the sources of human bodies cannot be careful enough and that rigorous standards have to be formulated and followed. Meanwhile, the publisher has stopped printing the atlas, citing the possible use of NS victims in its creation as the reason for this decision.[138]

The Legacy of History: Eponyms and Awards

The long reach of the legacy of medicine during National Socialism becomes vividly apparent in discussions on the continued use of eponyms and names of awards that are linked to physicians and scientists with NS pasts. Eponyms are a convenient method used in medicine to quickly and specifically describe structures, diseases, and technologies by the name of the person most associated with them. Generally, bestowing a name to a medical term is considered to be an honor. However, using eponyms of those with NS pasts has become a topic of active discussions. Two issues are at the center of this controversy: the general usefulness of eponyms in medicine, and whether the use of certain eponyms should be discontinued depending on the moral worthiness of the person whose name is attached. A full account of the debate would go beyond the scope of this book; however, a few significant examples will be discussed.

The surgeons Fargen and Hoh have recently reviewed all arguments for and against the use of eponyms.[139] On the pro side they list the pervasive use of eponyms throughout history and physicians' emotional attachment to them, the ease of communication, the chance to teach medical history and celebrate discoveries or discoverers, and the indispensability of popular eponyms. Claims against the use of eponyms include that they are nondescript and inaccurate, the fact that they are sometimes undeserved or celebrate only one of many discoverers, and that they can create confusion. Anatomist Tom Gest argues for the abandonment of all eponyms, as he sees them as a means of "obfuscation of the nomenclature" that leads to confusion in the minds of medical students.[140] In an evidence-based study of the use of 453 anatomical eponyms, Winkelmann found their occurrence highly variable.[141] He concluded pragmatically that medical students need not be familiar with all of these eponyms, but that some "must be actively retained to understand clinicians and efficiently research medical literature".[142] Olry comes to a similar conclusion after reviewing all arguments for and against the retention of eponyms.[143]

Changing or abandoning eponyms derived from the names of scientists who are morally unworthy of such an honor is a hotly debated issue. Several authors have reviewed eponyms that refer to physicians and scientists who actively supported and collaborated with the NS regime, or who became its victims or protested it.[144] They all agree that eponyms associated with victims and political dissidents "should be remembered and even strengthened, as opposed to those of the perpetrators, which should be obliterated".[145] Most eponyms in anatomy stem from an older period of scientific discovery and have not come under discussion. An exception is the Clara cell, a secretory cell in bronchioles first described by Clara in 1937[146] and associated with his name since the 1950s.[147] Winkelmann and Noack recommend changing this internationally used eponym to the descriptive term "club cell", which refers to the cell's morphology.[148] Pernkopf's atlas is the only other example in this discussion, and the argument has already been made that it seems advisable to continue using it an educational tool in the history and ethics of anatomy. The same is true for the use of eponyms: if eponyms derived from NS members were purged from the medical textbooks and conversation, medical education would lose an important opportunity for teaching about medicine during the Holocaust. The elimination of the names and books that remind us of human failure and crimes would very much resemble National Socialists burning books of dissenting authors in the spring of 1933. It would be a form of censorship. History should not be "purified" but remembered and actively discussed.

Maybe it is time to view the meaning of eponyms in a different light altogether: no longer should it be "one of the finest ways" to recognize achievement with "one's name as an eponym".[149] This interpretation smacks of hero worship, which is no longer an adequate way to express the ethical values of modern medicine and life in a society based on equality. Rather, eponyms should be recognized as simple historic markers, inherent to their time and intricately connected to their historical framework. As such, they should live on in the medical conversation and help ground modern medicine in its history. And in talking about perpetrators, their victims could be remembered, too, as the medical ethicists Fangerau and Krischel point out. They support this contextualizing approach to eponyms and conclude: "Even if this is a laborious and sometimes cumbrous approach, it would [...] create the opportunity to remember the victims and point out the wrongdoings of the perpetrators."[150]

Discussions similar to those in the eponym debate concern honorary awards named after persons active in National Socialism. Among those who have taken a step away from honoring members of the NS regime are political scientists, who eliminated the Theodor Eschenburg Award in 2013,[151] and musicians, who renamed the Emil Berlanda Award.[152] One of the latest controversies revolved around the Hubertus Strughold Award in space medicine. An extensive public discussion,[153] as well as a subsequent investigation, led to the retirement of the award. The Space Medicine Association explained its decision in the following manner: "While there was no evidence to demonstrate that Dr. Strughold was directly involved in any atrocities during World War II or that he was a member of the Nazi party, the award created controversy which was distracting to the main work and goals of the Space Medicine Association".[154] In anatomy the Wolfgang Bargmann Award, which had been given to students for excellent doctoral theses by the *Anatomische Gesellschaft* until 2011, was renamed the Young

Talent Award in 2012 in order to separate the recipients of the award from Bargmann's early political biography and use of bodies of NS victims for his research before 1945. The ethical argument for renaming awards is somewhat different from the one for eponyms. Eponyms should be used so that they can be connected with the full biographies of those who no longer exclude the dark NS chapter of their lives, and thereby offer young scientists the opportunity to learn about this history and to question their own political and professional decisions. Awards, on the other hand, are generally meant to be associated with the names of prominent role models within professions, not with persons under critical discussion. The name of the award will be linked to recipients who are meant to be honored and not to be connected with a questionable role model.[155] Thus, following Heiner Fangerau's argument, the benefit from using the name of an award as a learning opportunity for history and ethics of medicine is superseded by the potential harm for the recipient of that award, whose name will be linked with a controversial professional. Wolfgang Bargmann's early political biography and use of bodies of NS victims for his research before 1945 became a focus of discussions within the *Anatomische Gesellschaft*, thus the society decided to rename the prize to Young Talent Award in 2012.[156]

This overview of the postwar developments in German anatomy can only be the beginning of a more detailed investigation of the influence of the NS period on German medicine since 1945. As has been pointed out, there are very few systematic studies on the topic. Much of the information gathered here was found in newly accessed archival sources or autobiographies of physicians and scientists, and many questions remain. What were the long-term effects for German medicine of having lost 20 percent of its academic faculty through the discriminatory practices of the NS regime? What was the effect of the remaining 80 percent who stayed in their positions, watching their colleagues' exclusion and continuing in their careers? What did the daily exposure to work with the bodies of victims of the NS regime mean for anatomists and medical students in the long run? How could it not have traumatized and possibly corrupted their souls? Robert Herrlinger appears as a tragic figure, one of the few who had to break their silence through the forceful inquiry of the men of the *Sondervotum*. He seemed completely at sea when asked to comment on his lack of doubt of the NS justice system. He did his work to earn a living for his growing family, work that happened to be at Posen, where Polish resistance fighters were executed. He was a sensitive man, as his writings on art history reveal. How can we explain his apparent blindness for the tragedy around him? Can the daily grind of brutality make any person oblivious to the reality of blatant injustice and violence? These questions need to be investigated; not to judge men like Herrlinger, but to learn about the circumstances that allowed all of this to happen. The answers to these questions might help modern students and health professionals realize how they themselves are wrapped up in a political reality that might also need to be questioned.

Notes

1. Rilke 1963, 68.
2. Biddiscombe 2006; Taylor 2011.
3. Vollnhals 1991.
4. Hildebrandt 2013b.
5. Arias 2004, 347.
6. Winkelmann and Noack 2010; Schütz et al. 2014.
7. Buddecke 2011.
8. Noack 2008, 23.
9. Letter Bayerisches Staatsministerium für Unterricht und Kultus to Reichsminister für Wissenschaft, Erziehung und Volksbildung, 22 September 1944; BayHSTA MK 43712.
10. UAG Med Fak I 107 2, f. 84–120.
11. BayHSTA MK 43712.
12. E.g., Letter Dr. Med. J. Hett to Dekanat Medizinische Fakultät Erlangen, UAE C3/5 Nr. 21 Johannes Hett Anatomie 1846–1959, f. 492.
13. Ortmann 1986, 22.
14. Perschke 2007, 105–6.
15. Schiebler 1982, 1000; Bogner et al. 2007.
16. Oehler-Klein et al. 2012; Noack 2012; Czech 2015.
17. Noack 2008, 20–21.
18. Letter from Benninghoff to Starck, 11 November 1947; estate Benninghoff.
19. Letter from Benninghoff to Dr. A. Heyn, Kreiskrankenhaus Bad Hersfeld, 21 February 1951; estate Benninghoff.
20. Letter from Benninghoff to Zenker, 1 December 1952; estate Benninghoff.
21. Professor Reinhard Putz, retired chairman of Anatomical Institute I, Ludwig-Maximilians-Universität Munich; personal communication via electronic mail, 30 April 2014.
22. Winkelmann 2012, 248.
23. Ferner 1967.
24. Hayek 1957.
25. Kudlien 1985, 226.
26. Ibid., 237; Baader and Schultz 1987, 225–26; Kidder 2012.
27. Noack 2008, 17–18.
28. Zimmermann 2007, 33.
29. Wallis 1989, 402–3.
30. Ibid., 402–4.
31. Zassenhaus 1974, 116, 172.
32. Zassenhaus 1974, 171.
33. The author is deeply grateful to Dr. Ingborg Lötterle for sharing chapters from her unpublished memoirs and personal memories of Hermann Stieve.
34. Lötterle, personal communication via electronic mail, 11 April 2014.
35. "Little doctor."
36. Lötterle, unpublished memoirs.
37. Lötterle, personal communication via electronic mail, 11 April 2014.
38. Wulfert 2010; Wonschik 2005; Waldinger 2011.
39. Wonschik 2005.

40. Hildebrandt 2013a.
41. Richter 1986, 30–40.
42. Bräutigam 1998.
43. Ibid., 8.
44. Ibid., 8–9.
45. Wonschik 2005.
46. Ditfurth 1993, 166–169.
47. Pfürtner 2001, 283.
48. Ibid., 284.
49. Roland 1989.
50. Hirte and Stein 2003, 383.
51. Testimony of Dr. Franz Blaha, prisoner physician at Dachau, see Czech 2015, 175.
52. Baader and Schultz 1987, 13.
53. Karl-Heinz Roth, personal communication, electronic mail 25 September 2013.
54. Lang 2007, 187.
55. Hansson and Hildebrandt 2014.
56. Anonymous 1945.
57. Högberg 2013, 248.
58. Winkelmann and Schagen 2009.
59. Gerster 1955.
60. Marx 2003; Winkelmann and Schagen 2009.
61. Gerster 1946.
62. Ibid., 372.
63. Stieve 1947.
64. Ibid., 783.
65. Schagen 2005.
66. Brammer 1945; the author would like to thank Dr. Andreas Winkelmann for sharing a copy of this article.
67. BayHSta MK 43752, personnel file Albert Hasselwander.
68. Stieve 1953.
69. Aly 1987.
70. Ibid., 84; Aly 1994, 152.
71. Hildebrandt 2013c.
72. Self-reported curricula vitae from 1950, 1954, and 1955 in UWü ZV PA Herrlinger, f. 166–68, 169, 184–85; Röhrich 1968; Seidler 1968.
73. UWü ZV PA Herrlinger, letter from Herrlinger to *Rektor* Stock, 3 July 1958.
74. Voss and Herrlinger 1946.
75. NSDAP 1 May 1937 to 3 April 1941, membership number 5262603; the SA (mandatory student-SA) May 1933 to 1935 (excluded from SA on 4 October 1935); and several other minor NS organizations; see Hildebrandt 2013c.
76. BA-ehem. BDC-NSDAP-Zentralkartei, BA-ehem. BDC-NSDAP-Gaukartei: Robert Herrlinger, Nr. 5262603.
77. Herrlinger 1947; Herrlinger 1949.
78. Wüstenfeld 1972; Buddrus 2007.
79. Neubert 1922; Neubert 1928; Neubert 1950.
80. Buddrus 2007, 296–97.
81. Zutt 1973; Ley 1999.
82. Beddies 2005.
83. Schneider 1965; Fischer et al. 1994.
84. Schottlaender 1988.
85. Ströder 1985; Künzer 1972.
86. UWü ZV PA Herrlinger H 31, Auszug aus dem Senatsprotokoll, 31 July 1957.

87. UWü ZV PA Herrlinger H 31, Auszug aus dem Senatsprotokoll, 6 November 1957.

88. Bausewein 2010; Laumer 2000, 42–43.

89. Hopf 1980; Peiffer 1998; Shevell and Evans 1994; Collmann 2008; for background on global unethical medical experimentation: Griesecke et al. 2009; Lafleur et al. 2007.

90. UWü ZV PA Herrlinger H 31, letter Saar, 25 July 1957.

91. UWü ZV PA Herrlinger H 31, letter, 26 July 1957.

92. UWü ZV PA Herrlinger H 31, Auszug aus dem Senatsprotokoll, 6 November 1957.

93. UWü ZV PA Herrlinger H 31, Gutachten 3. Feb. 1958; letter from Neubert to Rektor 4. Feb. 1958; note by Rektor, 6. Feb. 1958, letter from Dean Schaltenbrand to Rektor, 13 June 1958.

94. UWü ZV PA Herrlinger H 31, letter from Wollheim to Bayerisches Staatsministerium für Unterricht und Kultur, 17 January 1959; letter from Neubert to Bayerisches Staatsministerium für Unterricht und Kultur, 28 January 1959; UWü ZV PA Herrlinger, folium 25: Empfangsbestätigung signed by Herrlinger, 6 May 1960.

95. Anatomical studies: Herrlinger 1947 and 1949; essay on art history: Herrlinger 1943, 1944.

96. Aly 1987, 49; Aly 1994, 136.

97. Herrlinger 1947, 228.

98. Professor Eduard Seidler, personal communication via electronic mail, 1 and 7 February 2012.

99. Hildebrandt 2013c.

100. Page 4, expert witness statement in Würzburg, 3 February 1958, UWü ZV PA Herrlinger.

101. Page 6, proceedings of the senate commission, 30 June 1958, UWü ZV PA Herrlinger.

102. Schönhagen 1992; Peiffer 1997.

103. Quoted in Hildebrandt 2011.

104. Page 9, UWü ZV PA Herrlinger, proceedings of the senate commission, 30 June 1958.

105. See note 105.

106. Hildebrandt 2008.

107. Seidler, see note 99.

108. Bohde 2005, 332.

109. Bruns 2009.

110. Second volume 1942, third volume 1952, fourth volume 1956–57, 1961; first American edition: Pernkopf 1963.

111. Williams 1988.

112. Weissmann 1985.

113. Mühlberger 1998b.

114. Panush and Briggs 1995; Panush 1996; Panush 1997.

115. Hutton 1996.

116. Israel and Seidelman 1996; Israel and Seidelman 1997.

117. Cutler 1997.

118. Ebenbauer and Schütz 1997.

119. Examples: *Michigan Daily Online* 1997; McManus 1996; Williams 1999.

120. Mühlberger 1998a.

121. Malina and Spann 1999; Angetter 2000.

122. Malina and Spann 1999; Angetter 2000; Seidelman, personal comunication.

123. Malina 1997; Malina and Spann 1999; Angetter 2000.

124. Ferner and Staubesand 1973; Atlas 2001.

125. Clemente 1975.

126. Urban and Schwarzenberg 1977.

127. Angetter 2000.
128. Spiro 1998.
129. Hubenstorf 2000; Neugebauer 1998.
130. Israel and Seidelman 1996 and 1997; Cutler 1997; Seidelman 1996; Seidelman 1999.
131. Holubar 2000.
132. Ebenbauer and Schütz 1997; Schütz et al. 1998; Malina and Spann 1999.
133. Ebenbauer and Schütz 1997.
134. Röbel and Wassermann 2004; Peuker and Schulz 2004; Wetz and Tag 2001.
135. See Atlas 2001; Williams 1999; Field 1999; Marcuse 2002; Spiro 1998.
136. Panush and Briggs 1995.
137. Paterniti 2003.
138. Hubbard 2001; personal communication via electronic mail from the editorial director of Elsevier GmbH, Urban & Fischer Verlag, 9 August 2005.
139. Fargen and Hoh 2014.
140. Gest 2014.
141. Winkelmann 2012.
142. Winkelmann 2012, 241.
143. Olry 2014 a and b.
144. Strous and Edelman 2007; Kondziella 2009; Cohen 2010.
145. Strous and Edelman 2007, 207.
146. Clara 1937.
147. Winkelmann and Noack 2010.
148. Ibid., 726.
149. Strous and Edelman 2007, 213.
150. Fangerau and Krischel 2011, 24.
151. DVPW 2014.
152. Mittelstaedt 2014.
153. Lagnado 2012.
154. Space Medicine Association 2013.
155. Fangerau, personal communication.
156. Anatomische Gesellschaft 2012.

Archival Sources

Bundesarchiv
– BA-ehem. BDC-NSDAP-Zentralkartei, BA-ehem. BDC-NSDAP-Gaukartei: Robert Herrlinger, Nr. 5262603
Bayerisches Hauptstaatsarchiv
– BayHSTA MK 43712, personnel file Albert Hasselwander
Universitätsarchiv Erlangen
– UAE C3/5 Nr. 21 Johannes Hett Anatomie 1846–1959
Universitätsarchiv Greifswald
– UAG Med Fak I 107 2
Universitätsarchiv Marburg
– Estate (*Nachlass*) Benninghoff
Universitätsarchiv Würzburg
– UWü ZV PA Herrlinger
– UWü ZV PA Herrlinger H 31

Bibliography

Aly, Götz. 1987. "Das Posener Tagebuch des Anatomen Hermann Voss." In: *Biedermann und Schreibtischtäter: Materialien zur deutschen Täter-Biographie*, ed. Götz Aly, Peter Chroust, and Christian Pross, 15–66. Berlin: Rotbuch Verlag.

———. 1994. "The Posen Diaries of the Anatomist Hermann Voss." In: *Cleansing the Fatherland: Nazi Medicine and Racial Hygiene*, ed. Götz Aly, Peter Chroust, and Christian Pross, 99–155. Baltimore: Johns Hopkins University Press.

Anatomische Gesellschaft. 2012. "Awards." Accessed 7 May 2014. http://anatomische-ge sellschaft.de/preise-ag3/stazung-nachwuchspreis.html.

Angetter, Daniela C. 2000. "Anatomical Science at University of Vienna 1938–45." *Lancet* 355: 1445–57.

Anonymous. 1945. "Kvinnor dödades för äggstocksexperiment: Svensk läkare fick hypofyser från halshuggna tyska fångar." *Aftontidningen*, 18 May.

Arias, Ingrid. 2004. "Entnazifizierung an der Wiener Medizinischen Fakultät: Bruch oder Kontinuität? Das Beispiel des Anatomischen Institutes." *Zeitgeschichte* 6(31): 339–69.

Atlas, Michel C. 2001. "Ethics and Access to Teaching Materials in the Medical Library: The Case of the Pernkopf Atlas." *Bulletin of the Medical Library Association* 89(1): 51–58.

Baader, Gerhard, and Ulrich Schultz. 1987. *Medizin und Nationalsozialismus. Tabuisierte Vergangenheit-Ungebrochene Tradition? Dokumentation des Gesundheitstages Berlin 1980*. 3rd ed. Frankfurt: Dr. Med. Mabuse.

Bausewein, Ulrich. 2010. "Diözesanpriester des 20. Jahrhunderts (4): Professor Heinz Fleckenstein (1907–1995). Ein milder Mahner. Nachrichtenüberblick Bistum Würzburg 20th July 2010." Accessed 22 May 2012. http://www.nachrichten.bistum-wuerz burg.de/bwo/dcms/sites/bistum/service/nachrichten/index.html.

Beddies, Thomas. 2005. "Universitätspsychiatrie im Dritten Reich. Die Nervenklinik der Charité unter Karl Bonhoeffer und Maximinian de Crinis." In: *Die Charité im Dritten Reich. Zur Dienstbarkeit medizinischer Wissenschaft im Nationalsozialismus*, ed. Sabine Schleiermacher and Udo Schagen, 55–72. Paderborn: Ferdinand Schöningh.

Biddiscombe, Perry. 2006. *The Denazification of Germany: A History 1945–1950*. Chalford: Tempus Publishing Limited.

Bogner, Agnès, Pierre-Henri Jouneau, Gilbert Thollet, D. Basset, and Catherine Gauthier. 2007. "A History of Scanning Electron Microscopy Developments: Towards 'wet-STEM' Imaging." *Micron* 38: 390–401.

Bohde, Daniela. 2005. "Pellis memoriae: Die Moralisierung der Haut in Frontispizen und Anatomietheatern der Niederlande im 17. Jahrhundert—ein blinder Fleck in der Medizingeschichte nach 1945." In: *Zergliederungen· Anatomie und Wahrnehmung in der frühen Neuzeit*, ed. Albert Schirrmeister, 327–58. Frankfurt: Vittorio Klostermann.

Brammer, K. 1945. "Im Schatten des Scharfrichters. Besuch in der Anatomie-Schreckensrekorde der Henker." *Neue Zeit* 75 (October 17): 3.

Bräutigam, Hans H. 1998. *Beruf: Frauenarzt. Erfahrungen und Erkenntnisse eines Gynäkologen*. Hamburg: Hoffmann und Campe.

Bruns, Florian. 2009. *Medizinethik im Nationalsozialismus*. Stuttgart: Franz Steiner Verlag.

Buddecke, Julia. 2011. *Endstation Anatomie: Die Opfer nationalsozialistischer Vernichtungsjustiz in Schleswig-Holstein*. Hildesheim: Georg-Olms Verlag.

Buddrus, Michael, and Sigrid Fritzlar. 2007. *Die Professoren der Universität Rostock im Dritten Reich. Ein biographisches Lexikon*. München: KG Saur.

Clara, Max. 1937. "Zur Histologie des Bronchalepithels." *Zeitschrift für miskroskopisch-anatomische Forschung* 41: 321–47.

Clemente, Carmine D. 1975. *Anatomy: A Regional Atlas of the Human Body*. München, Berlin, Wien: Urban und Schwarzenberg.

Cohen, Michael M., Jr. 2010. "Overview of German, Nazi and Holocaust Medicine." *American Journal of Medical Genetics Part A* 152A: 687–707.

Collmann, Hartmut. 2008. "Georges Schaltenbrand (26.11.1897–24.10.1979)." *Würzburger medizinhistorische Mitteilungen* 27: 63–92.

Cutler, Daniel S. 1997. "Origins of the Pernkopf Atlas." *Journal of the American Medical Association* 277(14): 1122.

Czech, Herwig. 2015. "Von der Richtstätte auf den Seziertisch: Zur anatomischen Verwertung von NS-Opfern in Wien, Innsbruck und Graz." *Jahrbuch des Dokumentationsarchiv des österreichischen Widerstandes*, 2015: 141–90.

Ditfurth, Hoimar von. 1993. *Innenansichten eines Artgenossen. Meine Bilanz.* München: Deutscher Taschenbuch Verlag.

DVPW. 2014. "Eschenburg Debatte. Deutsche Vereinigung für politische Wissenschaft." Accessed 16 March 2015. http://www.dvpw.de/eschenburg-debatte.html.

Ebenbauer, Alfred, and Wolfgang Schütz. 1997. "Origins of the Pernkopf Atlas: In Reply." *Journal of the American Medical Association* 277(14): 1123–24.

Fangerau, Heiner, and Mathis Krischel. 2011. "Der Wert des Lebens und das Schweigen der Opfer: Zum Umgang mit den Opfern nationalsozialistischer Verfolgung in der Medizinhistoriographie." In: *NS-"Euthanasie" und Erinnerung. Vergangenheitsaufarbeitung-Gedenkformen-Betroffenenperspektiven,* ed. Stephanie Westermann, Richard Kühl, and Tim Ohnhäuser, 19–28. Berlin: Lit Verlag.

Fargen, Kyle M., and Brian L. Hoh. 2014. "The Debate Over Eponyms." *Clinical Anatomy* 27: 1137–40.

Ferner, Helmut. 1967. "Max Clara." *Anatomischer Anzeiger* 121: 220–30.

Ferner, Helmut, and Jochen Staubesand. 1973. *Sobotta/Becher: Atlas der Anatomie des Menschen.* 16th ed. München, Berlin, Wien: Urban und Schwarzenberg.

Field, Richard. 1999. "A Practical Guide to Ethical Theory." Accessed 16 March 2015. http://catpages.nwmissouri.edu/m/rfield/274guide/title.htm.

Fischer, Wolfram, Klaus Hierholzer, and Michael Hubenstorf. 1994. *Exodus von Wissenschaften aus Berlin.* Berlin: De Gruyter.

Gerster, Hans Jacob. 1946. "Die 'Versagerfrage' in der Lehre Knaus." *Schweizer Medizinische Wochenschrift* 76: 371–75.

———. 1955. *Kinderzahl nach Wunsch und Willen.* Neunte Auflage. Rüschlikon bei Zürich: Albert Müller Verlag.

Gest, Thomas R. 2014. "Anatomical Nomenclature and the Use of Eponyms." *Clinical Anatomy* 27: 1141.

Griesecke, Birgit, Marcus Krause, Nicolas Pethes, and Katja Sabisch. 2009. *Kulturgeschichte des Menschenversuchs im 20. Jahrhundert.* Frankfurt am Main: Suhrkamp Verlag.

Hansson, Nils, and Sabine Hildebrandt. 2014. "Swedish-German Contacts in Anatomy 1930–1950: The Example of Gösta Häggqvist and Hermann Stieve." *Annals of Anatomy* 196: 259–67.

Hayek, Heinrich. 1957. "Prof. Dr. Med. Georg Politzer." *Wiener Klinische Wochenschrift* 69(5): 86–87.

Herrlinger, Robert. 1943. "Die ostische Rassenseele und der weiche Stil. Teil 1." *Volk und Rasse* 18: 90–97.

———. 1944. "Die ostische Rassenseele und der weiche Stil. Teil 2." *Volk und Rasse* 19: 6–13.

———. 1947. "Das Blut in der Milzvene des Menschen." *Anatomischer Anzeiger* 96: 226–35.

———. 1949. "Neue funktionell-histologische Untersuchungen an der menschlichen Milz." *Zeitschrift für Mikroskopisch-Anatomische Forschung* 114(4): 341–65.

Hildebrandt, Sabine. 2008. "Capital Punishment and Anatomy: History and Ethics of an Ongoing Association." *Clinical Anatomy* 21: 5–14.

————. 2011. "First Symposium on 'Anatomie im Nationalsozialismus' ('Anatomy in National Socialism'), Würzburg, Germany, September 29, 2010." *Clinical Anatomy* 24: 97–100.

————. 2013a. "The Women on Stieve's List: Victims of National Socislism Whose Bodies Were Used for Anatomical Research." *Clinical Anatomy* 26: 3–21.

————. 2013b. "Anatomische Gesellschaft from 1933 to 1950: A Professional Society under Political Strain—The Benninghoff Papers." *Annals of Anatomy* 195: 381–92.

————. 2013c. "The Case of Robert Herrlinger: A Unique Postwar Controversy on the Ethics of the Anatomical Use of Bodies of the Executed During National Socialism." *Annals of Anatomy* 195: 11–24.

Hirte, Ronald, and Harry Stein. 2003. "Die Beziehungen der Universität Jena zum Konzentrationslager Buchenwald." In: *"Kämpferische Wissenschaft":Studien zur Universität Jena im Nationalsozialismus*, ed. Uwe Hossfeld, Jürgen John, Oliver Lemuth, Rüdiger Stutz, 361–400. Köln: Böhlau Verlag.

Högberg, Ulf. 2013. *Vita rockar och bruna skjortor: nazimedicin och läkare på flykt*. Malmö: Universus.

Holubar, Karl. 2000. "The Pernkopf Story: The Austrian Perspective of 1998, 60 Years after It All Began." *Perspectives in Biology and Medicine* 43(3): 382–88.

Hopf, Hanns Christian. 1980. "Georges Schaltenbrand (1897–1979)." *Journal of Neurology* 223: 153–58.

Hubbard, Chris. 2001. "Historical Note: Eduard Pernkopf's Atlas of Topographical and Applied Human Anatomy: The Continuing Ethical Controversy." *Anatomical Record* 265(5): 207–11.

Hubenstorf, Michael. 2000. "Anatomical Science in Vienna, 1938–1945." *Lancet* 355: 1385–86.

Hutton, Edward B., Jr. 1996. "Nazi Origins of an Anatomy Text: The Pernkopf Atlas: In Reply." *Journal of the American Medical Association* 276: 1634.

Israel, Howard A., and William E. Seidelman. 1996. "Nazi Origins of an Anatomy Text: The Pernkopf Atlas." *Journal of the American Medical Association* 276(20): 1633.

————. 1997. "Origins of the Pernkopf Atlas: In Reply." *Journal of the American Medical Association* 277(14): 1123.

Kidder, Annemarie S. 2012. *Ultimate Price: Testimonies of Christians Who Resisted the Third Reich*. New York: Orbis Books.

Kondziella, Daniel. 2009. "Thirty Neurological Eponyms Associated with the Nazi Era." *European Neurology* 62: 56–64.

Kudlien, Fridolf. 1985. *Ärzte im Nationalsozialismus*. Köln: Kiepenheuer und Witsch.

Künzer, Wilhelm. 1972. "J. Ströder zum 60. Geburtstag." *Klinische Pädiatrie* 184: 157–58.

Lafleur, William L., Gernot Böhme, Susumu Shimazono. 2007. *Dark Medicine: Rationalizing Unethical Research*. Bloomington: Indiana University Press.

Lagnado, Lucette. 2012. "A Scientist's Nazi-Era Past Haunts Prestigious Space Prize." *Wall Street Journal* 260(129), Weekend, pp. 1, 12. Accessed 16 March 2015. http://www.wsj.com/articles/SB10001424052970204349404578101393870218834.

Lang, Hans-Joachim. 2007. *Die Namen der Nummern: Wie es gelang, die 86 Opfer eines NS-Verbrechens zu identifizieren. Überarbeitete Ausgabe*. Frankfurt am Main: S. Fischer Verlag.

Laumer, August. 2005. *Heinz Fleckenstein (1907–1995) Pastoral- und Moraltheologe in Regensburg und Würzburg. Leben und Werk*. Würzburg: Echter-Verlag.

Ley, Astrid. 1999. "Teil 2. Medizinische Fakultät." In: *Die Professoren und Dozenten der Friedrich-Alexander-Universität Erlangen, 1743–1960*, ed. Renate Wittern-Sterzel. Erlangen: Verlagsdruckerei Schmidt.

Malina, Peter. 1997. "Eduard Pernkopf's Anatomie oder: die Fiktion einer "reinen" Wissenschaft." *Wiener Klinische Wochenschrift* 109(24): 935–43.

Malina, Peter, and Gustav Spann. 1999. "Das Senatsprojekt der Universität Wien 'Untersuchungen zur Anatomischen Wissenschaft in Wien 1938–1945.'" *Wiener Klinische Wochenschrift* 111(18): 743–53.

Marcuse, Harold. 2002. "Pernkopf's Atlas: Ethics of Choice." Accessed 16 March 2015. http://www.history.ucsb.edu/faculty/marcuse/classes/33d/prevyears/33d02/33d02Lectures/33d02l13.htm.

Marx, Jörg. 2003. "'Der Wille zum Kind' und der Streit um die physiologische Unfruchtbarkeit der Frau. Die Geburt der modernen Reproduktionsmedizin im Kriegsjahr 1942." In: *Biopolitik und Rassismus*, ed. Martin Stingelin, 112–59. Frankfurt am Main: Suhrkamp Verlag.

McManus Rich. 1996. "A Tainted Classic: Anatomy Text Draws Criticism." *NIH Record* 48(0). 24 September. Accessed 2 April 2014. http://nihrecord.od.nih.gov/newsletters/09_24_96/story01.htm.

Michigan Daily Online. 1997. "Vienna University Apologizes for Nazi Involvement, Plans Investigation, 13 February 1997." Accessed 1 June 2005. http://www.pub.umich.edu/daily/1997/feb/02-13-97/news/news20.html.

Mittelstaedt, Katharina. 2014. "NS-Aufarbeitung: Musikpreis wird umbenannt." *Der Standard*, 20 January 2014. Accessed 7 May 2014. http://derstandard.at/1389857658924/NS-Aufarbeitung-Musikpreis-wird-umbenannt.

Mühlberger, Kurt. 1998a. "II. Die Belieferung des anatomischen Instituts der Universität Wien mit Studienleichen in der Zeit von 1938–1946." In: *Senatsprojekt der Universität Wien: Untersuchungen zur anatomischen Wissenschaft in Wien 1938–1945*, ed. Akademischer Senat der Universität Wien, 29–66. Unpublished manuscript.

———. 1998b. "Enthebungen an der medizinischen Fakultät 1938–1945. Professoren und Dozenten." *Wiener Klininische Wochenschrift* 110(4–5): 115–20.

Neubert, Kurt. 1922. "Der Übergang der arteriellen in die venöse Blutbahn bei der Milz." *Zeitschrift für Anatomie und Entwicklungsgeschichte* 66: 424–50.

———. 1928. "Zur Morphologie der Talgdrüsen." *Verhandlungen der Anatomischen Gesellschaft* 37: 124–31.

———. 1950. "Die Basilarmembran des Menschen und ihr Verankerungssystem." *Zeitschrift für Anatomie und Entwicklungsgeschichte* 114(5): 539–88.

Neugebauer, Wolfgang. 1998. "Zum Umgang mit sterblichen Resten von NS-Opfern nach 1945." In: *Senatsprojekt der Universität Wien: Untersuchungen zur anatomischenWissenschaft in Wien 1938–1945*, ed. Akademischer Senat der Universität Wien, 459–65. Unpublished manuscript.

Noack, Thorsten. 2008. "Begehrte Leichen. Der Berliner Anatom Hermann Stieve (1886–1952) und die medizinische Verwertung Hingerichteter im Natinoalsozialismus." *Medizin, Gesellschaft und Geschichte. Jahrbuch des Instituts für Geschichte der Medizin der Robert Bosch Stiftung* 26: 9–35.

———. 2012. "Anatomical Departments in Bavaria and the Corpses of Executed Victims of National Socialism." *Annals of Anatomy* 194: 286–92.

Oehler-Klein, Sigrid, Dirk Preuss, and Volker Roelcke. 2012. "The Use of Executed Nazi Victims in Anatomy: Findings from the Institute of Anatomy at Giessen University, Pre- and Post-1945." *Annals of Anatomy* 194: 293–97.

Olry, Regis. 2014a. "Anatomical Eponyms, Part 1: To Look on the Bright Side." *Clinical Anatomy* 27: 1145–48.

———. 2014b. "Anatomical Eponyms, Part 2: The Other Side of the Coin." *Clinical Anatomy* 27: 1142–44.

Ortmann. Rolf. 1986. *Die jüngere Geschichte des anatomischen Instituts der Universität Köln 1919–1984*. Köln: Böhlau Verlag..

Panush, Richard S. 1996. "Nazi Origins of an Anatomy Text: The Pernkopf Atlas." *Journal of the American Medical Association* 276(20): 1633–34.

———. 1997. "Origins of the Pernkopf Atlas." *Journal of the American Medical Association* 277(14): 1123.

Panush, Richard S., and Robert M. Briggs. 1995. "The Exodus of a Medical School." *Annals of Internal Medicine* 123(12): 963.

Paterniti, Michael. 2003. "The Most Dangerous Beauty." In: *The Best American Magazine Writing in 2003*, ed. American Society of Magazine Editors, 2–31. New York: HarperCollins.

Peiffer, Jürgen. 1997. *Hirnforschung im Zwielicht: Beispiele verführbarer Wissenschaft aus der Zeit des Nationalsozialismus.* Husum: Mathiessen Verlag.

———. 1998. "Zur Neurologie im 'Dritten Reich' und ihren Nachwirkungen." *Nervenarzt* 69: 728–33.

Pernkopf, Eduard. 1963. *Atlas of Topographical and Applied Human Anatomy. Volume 1: Head and Neck.* Philadelphia and London: W. B. Saunders.

———. 1964. *Atlas of Topographical and Applied Human Anatomy. Volume 2: Thorax, Abdomen and Extremities.* Philadelphia and London: W. B. Saunders.

Perschke, Birgit. 2007. ""Studentenausweis statt Arbeitspass!" Jobs neben dem Studium, 1943–1948." In: *Zwischen "Endsieg" und Examen. Studieren an der Universität Köln 1943–1948. Brüche und Kontinuitäten*, ed. Margit Szöllösi-Janzen, 98–111. Nümbrecht: KIRSCH-Verlag.

Peuker, Torsten, and Christian Schulz. 2004. *Der über Leichen geht.* Berlin: Ch. Links Verlag.

Pfürtner, Stephan H. 2001. *Nicht ohne Hoffnung: erlebte Geschichte 1922–1945.* Stuttgart: W. Kohlhammer GmbH.

Richter, Horst-Eberhard. 1986. *Die Chance des Gewissens. Erinnerungen und Assoziationen.* Hamburg: Hoffmann und Campe.

Rilke, Rainer Maria. 1963. *Die Aufzeichnungen des Malte Laurids Brigge.* Frankfurt am Main: Insel Verlag. First published in 1910.

Röbel, Sven, and Andreas Wassermann. 2004. "Händler des Todes." *Der Spiegel* 4 (19 January): 36–50.

Röhrich, H. 1968. "In Memoriam Robert Herrlinger." *Anatomischer Anzeiger* 123(5): 573–75.

Roland, Charles. 1989. "An Underground Medical School in the Warsaw Ghetto, 1941–1942." *Medical History* 33: 399–419.

Schiebler, Theodor Heinrich. 1982. "Anatomie in Würzburg (von 1593 bis zur Gegenwart)." In: *Vierhundert Jahre Universität Würzburg: Eine Festschrift*, ed. Peter Baumgart, 985–1004. Neustadt an der Aisch: Degener und Co.

Schneider, Klaus W. 1965. "Ernst Wollheim zum 65. Geburtstag." *Archiv für Kreislaufforschung* 46: 1–6.

Schönhagen, Benigna. 1992. "Das Gräberfeld X auf dem Tübinger Stadtfriedhof. Die verdrängte 'Normalität' nationalsozialistischer Vernichtungspolitik." In: *Menschenverachtung und Opportunismus, Tübingen: Zur Medizin im Dritten Reich*, ed. Jürgen Peiffer, 69–92. Tübingen: Attempto.

Schottlaender, Rudolf. 1988. *Verfolgte Berliner Wissenschaft. Ein Gedenkwerk.* Berlin: Edition Hentrich.

Schütz, Wolfgang, Karl Holubar, and Wilfred Druml. 1998. "On the 60th Anniversary of the Dismissal of Jewish Faculty Members from the Vienna Medical School." *Wiener Klinische Wochenschrift* 110(4–5): 113–14.

Schütz, Mathias, Maximilian Schochow, Jens Waschke, Georg Marckmann, and Florian Steger. 2015. "Anatomische Vitamin C-Forschung im Nationalsozialismus und in der Nachkriegszeit: Max Claras Humanexperimente an der Anatomischen Anstalt München." *Medizinhistorisches Journal*, 50: 330–355.

Seidelman, William E. 1996. "Nuremberg Doctors' Trial: Nuremberg Lamentation: For the Forgotten Victims of Medical Science." *British Medical Journal* 313: 1463–67.

―――. 1999. "Medicine and Murder in the Third Reich." In: *Jewish Virtual Library*. Accessed 11 September 2014. http://www.jewishvirtuallibrary.org/jsource/Holocaust/med murder.html.

Seidler, Eduard. 1968. "Robert Herrlinger, 1914–1968." *Deutsche Medizinische Wochenschrift* 93(21): 1078–79.

Shevell, Michael I., and Bradley K. Evans. 1994. "The 'Schaltenbrand Experiment,' Würzburg, 1940: Scientific, Historical, and Ethical Perspectives." *Neurology* 44: 350–56.

Space Medicine Association. 2013. "Strughold Award." Accessed 7 May 2014. http://www .spacemedicineassociation.org/strughold.htm.

Spiro, Howard M. 1998. "The Silence of Words: Some Thoughts on the Pernkopf Atlas." *Wiener Klinische Wochenschrift* 110(4–5): 183–84.

Stieve, Hermann. 1947. "Die 'Versagerfrage' in der Lehre Knaus- Eine Richtigstellung zum Aufsatz von H. J. Gerster." *Schweizer Medizinische Wochenschrift* 77: 782–83.

―――. 1953. "Cyclus, Physiologie und Pathologie (Anatomie)." *Archiv für Gynäkologie* 183: 178–203.

Ströder, Josef. 1985. *Angeklagt wegen Polenfreundschaft. Als Kinderarzt im besetzten Krakau.* Freiburg: Herderbücherei.

Strous, Rael D., and Morris C. Edelmann. 2007. "Eponyms and the Nazi Era: Time to Remember and Time for Change." *Israel Medical Association Journal* 9: 207–14.

Taylor, Frederick. 2011. *Exorcising Hitler: The Occupation and Denazification of Germany.* London: Bloomsbury Publishing.

Urban & Schwarzenberg. 1977. *The Urban & Schwarzenberg Collection of Medical Illustrations since 1896.* Baltimore, Munich: Urban & Schwarzenberg.

Vollnhals, Clemens. 1991. *Entnazifizierung: Politische Säuberung und Rehabilitierung in den vier Besatzungszonen 1945–1949.* München: Deutscher Taschenbuch Verlag.

Voss, Hermann, and Robert Herrlinger. 1946. *Taschenbuch der Anatomie.* 3 Bände. 1st ed. Jena: Gustav-Fischer-Verlag.

Waldinger, Joel. 2011. "The Mildred Fish-Harnack Story." *Wisconsin Public Radio.* Accessed 4 May 2014. http://www.wpr.org/shows/mildred-fish-harnack-story-joel-waldinger.

Wallis, Hedwig. 1989. "Medizinstudentin im Nationalsozialismus." In: *100 Jahre Universitätskrankenhaus Eppendorf 1889–1989*, ed. Ursula Weisser, 399–404. Tübingen: Attempto.

Weissmann, Gerald. 1985. "Springtime for Pernkopf." Reprinted 1987 in: *They All Laughed at Christopher Columbus,* ed. Gerald Weissmann, 48–69. New York: Times Books.

Wetz, Franz J., Brigitte Tag, and Klaus Tiedemann. 2001. *Schöne Neue Körperwelten.* Stuttgart: Klett-Cotta.

Williams, David J. 1988. "The History of Eduard Pernkopf's *Topographische Anatomie des Menschen.*" *Journal of Biocommunication* 15(2): 2–12.

Williams, Robyn. 1999. "Nazi Science, Transcript of a Radio Program: *Ockham's Razor,* 29 August. Accessed 2 April 2014. http://www.abc.net.au/radionational/programs/ock hamsrazor/nazi-science/3558712 .

Winkelmann, Andreas. 2012a. "The Anatomische Gesellschaft and National Socialism: A Preliminary Analysis Based on Society Proceedings." *Annals of Anatomy* 194: 243–50.

―――. 2012b. "Should We Teach Abernethy and Zuckerkandl?" *Clinical Anatomy* 25: 241–245.

Winkelmann, Andreas, and Thorsten Noack. 2010. "The Clara Cell: A 'Third Reich Eponym'?" *European Respiratory Journal* 36: 722–27.

Winkelmann, Andreas, and Udo Schagen. 2009. "Hermann Stieve's Clinical-Anatomical Research on Executed Women during the 'Third Reich.'" *Clinical Anatomy* 22(2): 163–71.

Wonschik, Helmut. 2005. *Mildreds Asche.* Radio Program Südwestdeutscher Rundfunk.

Wüstenfeld, Ewald. 1972. "Nachruf Kurt Neubert." *Anatomischer Anzeiger* 132: 435–39.

Wulfert, Tatjana. 2010. "Margarete von Zahn (geb. 1924)." *Der Tagesspiegel* 18 November 2010. Accessed 26 February 2014. http://www.tagesspiegel.de/berlin/nachrufe/marga rete-von-zahn-geb-1924/2892332.html.

Zassenhaus, Hiltgunt. 1974. *Ein Baum blüht im November. Bericht aus den Jahren des zweiten Weltkrieges.* Hamburg: Hoffmann und Campe. English version: Zassenhaus, Hiltgunt. 1974. *Walls. Resisting the Third Reich—One Woman's Story.* Boston: Beacon Press.

Zimmermann, Susanne. 2007. "… er lebt weiter in seinen Arbeiten, die als unverrück-bare Steine in das Gebäude der Wissenschaft eingefügt sind": Zum Umgang mit den Arbeiten des Anatomen Hermann Stieve (1886–1952) in der Nachkriegszeit. In: *Täterschaft-Strafverfolgung-Schuldentlastung. Ärztebiographien zwischen nationaler Ge-waltherrschaft und deutscher Nachkriegsgeschichte,* ed. Boris Böhm, and Norbert Haase, 29–40. Leipzig: Leipziger Universitätsverlag.

Zutt, Jürg. 1973. "In Memoriam Heinrich Scheller 1901–1972." *Nervenarzt* 44: 386–87.

Chapter 10

DEVELOPMENTS IN PROFESSIONAL ETHICS IN ANATOMY

The same suspension of empathy that was so necessary a part of a physician's task was also, in other contexts, the root of all monstrosity.

Pat Barker[1]

The public discourse on ethics in anatomy is a relatively recent phenomenon, as "anatomy and ethics have traditionally been viewed as inhabiting different conceptual worlds on the assumption that the practice of anatomy is ethically neutral."[2] However, a science that so intimately involves dealing with the bodies of the dead has always had ethical implications for its practitioners as well as for the society it is performed in. Discussions about ethics in anatomy have become part of the wider field of bioethics, which has evolved since its inception in the 1970s.[3]

Several tenets lie at the core of professional ethics in anatomy. The central one, the basic paradigm of anatomy, claims that anatomical knowledge has to be gained by the dissection of dead human bodies. This assertion leads directly to subsequent issues, such as the sources of body procurement, proper respect for the dead, and the balance of clinical detachment and empathy to save the dissector's humanity. Anatomy in NS Germany brought with it specific ethical challenges that led not only to the collaboration of anatomists with the NS regime through the use of bodies of NS victims, but also to some anatomists' ethical transgressions in traditional anatomy, which is limited to work with the dead, in which they experimented on living human beings, or the "future dead." A new perception of anatomy sees it as a discipline that not just provides knowledge of the structure and function of the human body, but also an introduction to medical professionalism and ethics in medicine. This development is accompanied by an increased interest in the person of the body donor and the socialization of the physician within the dissection course. A generally heightened sensitivity to questions of ethics and responsibility may have contributed to German anatomists' readiness in the new millennium to discuss the history of anatomy during National Socialism.

The Anatomical Method: The Initial Transgression of a Taboo

Anatomy as a medical discipline is based on the premise that knowledge of the structure of the human body is essential for effective medical prac- tice. Traditionally, anatomists postulate that anatomical knowledge must be gained by the dissection of the dead human body. In most societies, "cutting up" bodies is considered an unbearable violation under any other than these scientific circumstances, because respect for the dead human body and its integrity are part of all human culture throughout history. The newly dead are often seen as existing in an intermediate, ambiguous, and undetermined space where they are no longer alive but not yet dead, and in need of special treatment.[4] Most societies have religious rituals that often include concern for the physically inviolate state of the dead. Cere- monies acknowledge the special status of the deceased, and have at heart the care for the dead in their transitional state as well as the need to care for the survivors.[5] Disrespect, including physical abuse, of the dead body is taboo, as "a corpse and its survivors are entitled to maintain absolute cadaveric integrity."[6]

Thus, anatomists need to be granted permission by society to transgress this taboo under the special circumstances of the advancement of medical knowledge.[7] The legal status of the human body remains ambiguous and prone to conflicting interpretations of so-called "posthumous interests" in many societies.[8] As a result, not all members of society have found the ana- tomical method acceptable over time, especially in situations when body procurement was perceived as coercive[9] or the scientific reductionist and mechanistic view of human life was not shared.[10]

The work of the anatomist thus begins by breaking a taboo, of which beginners and sensitive dissectors are still aware each time they put scalpel to skin.[11] The anatomist learns to overcome this initial reluctance with the development of clinical detachment, which is one of the core elements in the socialization of a future physician, and without which objective observation and effective medical practice are impossible.[12] Indeed, the phenomenon is so distinct that William Hunter, the eighteenth-century anatomist, called this change of thinking and emotional state "a certain inhumanity."[13] A few years later John Warren, first professor of anatomy at Harvard Medical School, echoed this sentiment in his *Lectures upon Anatomy*: "By frequent dissecting dead bodies, the surgeon is armed with an useful, a necessary inhumanity. At the first view the stomach is apt to turn, but custom wears off such impression."[14] Historian Ruth Richardson argues that clinical detachment was responsible for the "transformation of the human corpse" from "the object of veneration and supernatural power which the corpse was in popular culture, into an object of scientific study."[15]

Clinical detachment was certainly not the only driving force behind the objectification of the human body, whether dead or alive. Rather, it was part of the more complex integration of the scientific method into medicine. However, the strong term "necessary inhumanity" implies the potential for further transgressions unless the societal boundaries of the special permission given to anatomists are clearly defined. Herrlinger felt this danger clearly when he teetered "on or beyond the edge of a state of little culture" with anatomy,[16] a sensation reflective of the potential for the transgression of ethical boundaries that Winkelmann sees as inherent to the anatomical occupation.[17] Medical ethicist Michael Grodin describes the possible dangers within the development of clinical detachment, which can lead to an overreaching compartmentalization in the clinical practitioner and the dehumanization of the patient, and states, "Medicine as a profession contains the rudiments of evil, and some of the most humane acts of medicine are only small steps away from real evil."[18]

The potential "inhumanity" of the anatomist stems from two sources: first, from the destructive nature of the anatomical method that "disassembles" a human body against all taboos; second, from the scientific reductionist view of human life. This "inhumanity" or clinical detachment is sometimes interpreted as a distinct loss of empathy. However, in the ideal case, the anatomist and physician will learn to counteract a potentially overbearing clinical detachment by self-reflection and a balance with natural empathy in order to become a fully effective and humane practitioner.[19] In the worst-case scenario, societal permission under a criminal regime like National Socialism can amplify this inhumanity and lead to medical atrocities.

Stages of Ethical Transgressions in NS Anatomy: New "Opportunities"

The new "opportunities" provided by the NS regime with the ever-increasing supply of bodies of its victims became the basis for ethical transgressions unprecedented in the documented history of anatomy. Anatomists did not only passively receive these bodies, but also actively petitioned the authorities for them. In the first years of the Third Reich the general situation of body procurement for anatomy had not changed much compared to earlier years; a lack of bodies persisted and anatomists complained and asked for more support.[20] They lobbied for adequate body procurement and competed against each other, especially with respect to the bodies of the executed.[21] Some thought that the lack of bodies might be due to relatives refusing to give their consent to dissection in case the body should reveal signs of an inherited disease that might alert the hereditary health courts and endanger the family.[22] At the beginning of the war several anatomists inquired into the possibility of having some of the increasing number of bodies from various camps shipped to their departments. Pernkopf's petition to use the bodies of Polish prisoners was denied, but Hirt successfully requested those of Russian prisoners of war for Strasbourg.[23] Spanner welcomed the building of execution chambers in Danzig that facilitated a more "convenient" body supply for his department there. The anatomical department of the University of Hamburg applied successfully for the delivery of bodies from the nearby concentration camp Neuengamme.[24] In 1944 Görttler in Heidelberg and Nauck in Freiburg asked the authorities to provide the bodies of prisoners of war for their departments.[25] However, during the war the number of executions had so greatly increased in other places that anatomical departments ran out of storage space for the bodies and began to refuse them,[26] such as in 1943 in Jena and Innsbruck.[27] The supply of bodies to the University of Königsberg was so plentiful that at some point only the bodies of women and young persons were accepted,[28] and the situation at the University of Vienna was similar.[29] A medical illustrator remembered up to thirty-three executions per day, and described the storage spaces as filled with the bodies of executed persons "stacked like cords of wood" and "severed heads" swimming in "vats filled with formalin."[30]

Before 1933 and early on in the NS regime relatives of executed persons did not always release these bodies for dissection. This led in some cities to the practice of anatomists' removing fresh tissues secretly from the bodies inside the execution facility, before they were collected by the family. Johannes Hett, Spanner, and Clara used these sources of "material."[31] When Hett was not allowed to do the same in his new position in Erlangen and had to wait for the bodies to be delivered from the execution site in Nuremberg, he complained to the authorities about the lack of "freshness."[32] His was not the only criticism of the "quality of material." A Frankfurt anatomist protested in a letter to the authorities in 1940 that his subject had obviously been decapitated with a blunt blade, which caused contusions in the neck area and made this region unsuitable for dissection. He asked that "in the interest of science, which alone we all should serve," the apparatus should be repaired or replaced.[33]

Human bodies were generally seen simply as another type of "material" within the range of tissues of interest to anatomists, especially in matters of comparative anatomy. Whereas this objectifying approach to the human body had been established before 1933, the casual and routine inclusion of details of executions was specific to anatomy during the Third Reich. A typical example of this perception was a publication by Kick from the anatomical department in Breslau, who reported in the "material and methods" section of his paper: "Under investigation were the livers of 16 mice, 4 guinea pigs and 6 healthy human beings. The latter died suddenly on different days at 6:30 pm".[34] Furthermore, this careless acceptance of the use of NS victims was shared by all anatomists, independent of their political convictions. Alternative behavior was possible, but at a high cost, as the example of Charlotte Pommer shows.

Apart from the increase of studies with the now freely available and very sensitive "fresh material" from NS victims, small verbal additions in the publications from this time seem to indicate a gradual transgression into new and ethically dubious territory. As early as 1935, Stöhr wrote that until then the known difficulties of conserving a greater number of intact adrenal glands had caused a lack of data concerning ganglion cells in this organ, but he was now able to present results on twelve such organs from the bodies of executed persons.[35] Clearly, he seemed to imply, times had changed, and for the better. Von Hayek reported in his 1940 treatise on lung structure, "Of course, best suitable are lungs from younger executed persons, of which I had several at my disposal."[36] And whereas in 1931 Bargmann had to collect tissue samples from six different departments all over Germany to produce results on fourteen executed persons, he was able to work with ten pituitary glands from the executed in 1942 and eighteen samples of intestinal tissues in 1944, all procured locally from the execution site in Königsberg.[37]

Stöhr, Bargmann, von Hayek, and others were following the traditional anatomical procedure with the highest standard of work in mind. However, a new quality had crept into their work and was expressed in their wording, one that might be called "increased clinical detachment." An expert of this quality and the reductionist-mechanistic view of human life was Hermann Stieve. He realized that the environment in which he conducted his animal experiments on the influences of chronic and acute stress on reproductive organs was perfectly reflected in the setting of prisoners on death row, thereby performing, as it were, a human experiment, if only in thought. He saw it as his professional duty to use the newly available "material," and justified his use of bodies of the executed after the war: "Never before has an anatomist been reproached for this practice. The anatomist has no dealings with court proceedings or court sentences. He only tries to gain insights [...] that otherwise cannot be obtained at all. [...] Our knowledge of the human body and its functions is built to a large extent on such investigations of the executed. The facts discovered herein benefit all physicians and thus ultimately all of humanity."[38]

A dangerous convergence of executive and anatomical practice in the later years of the Third Reich occurred when authorities tried to move executions to the anatomical institute. While many anatomists were familiar with the reality of capital punishment at specific execution sites in prisons, there was a strict distinction between the activities at the place of execution and their workplace at the anatomical institutes. The two spaces were clearly demarcated and separate in terms of their location and function. Anatomists only entered the execution sites for the removal of "fresh" tissues. While they had accepted this situation as part of their work long before the advent of the NS regime, they were perfectly unprepared for executions inside an anatomical institute. The anatomists reacted with horror and revulsion to this perceived assault on their own space, as two well-documented episodes illustrate. Both cases involve prisoners who were to be executed by the GeStapo.

The first of these incidents occurred at the University of Giessen in spring 1942,[39] and was described by Ernst von Herrath in his denazification trial.[40] Twenty-one-year-old Polish forced laborer Wladislaw Kaczmarek had been accused of fathering the child of a German girl who worked at the same farm, and he was sentenced to death for "racial defilement." On 9 April 1942 Kaszmarek was hanged by the GeStapo on a mobile gallows in a forest near the city of Wetzlar. Fellow Polish citizens were forced to act as executioners and witness the spectacle. The body was picked up by technicians from the anatomical department, who realized that the victim was still alive when they unloaded him at the institute. After the men notified the police, GeStapo officials came to the institute and instructed one of the technicians to kill Kaszmarek, but the man repeatedly refused to do this. When von Herrath appeared at the scene the GeStapo men ordered him to kill Kaszmarek with an injection. Instead, the anatomist demanded a transfer of the injured prisoner to the hospital. In response, one of the GeStapo officials drew his pistol intending to shoot Kaszmarek immediately, but von Herrath intervened by insisting on his superior authority in the institute building. The GeStapo men then took Kaszmarek away and killed him in a forest near Giessen. His body was transported to Frankfurt and cremated. The GeStapo officials returned later to the institute and ordered the staff and faculty to keep this incident absolutely secret.

The second episode of an intended execution at an anatomical institute happened in Erlangen in 1944. Various reports of the exact dates and events differ, but the most comprehensive documentations of this event were assembled in the senior prosecutor's files at the regional court of Nuremberg-Fürth and by historian Alfred Wendehorst.[41] In early July a transport of executed Polish citizens was delivered to the anatomical institute. The vehicle with the coffins was accompanied by a GeStapo official, a Polish translator, two Polish executioners, a police physician, and a convicted Polish prisoner named Dybel, whose first name was not recorded. On arrival at the institute the official explained to the technician Heider that they intended to kill Dybel right there in the basement of the institute so that Hett's request for "fresh material" could be fulfilled. Heider immediately informed Hasselwander, the director of the institute, about the situation, who strongly objected to this plan and declared that such an execution would be an "eternal disgrace" for his institute. Apparently Hett was also appalled by this idea. The executioners thus led Dybel from the institute, murdered him at a sporting ground in Erlangen, and returned his body to the anatomical institute.

Both incidents had postwar repercussions for the anatomists. For von Herrath, these were quite positive, as his intervention on behalf of the injured prisoner was interpreted as an act of resistance against the NS regime and eased his denazification process.[42] In Hasselwander's case the senior prosecutor at the regional court of Nuremberg-Fürth started a murder investigation in early 1949. The anatomist was accused of having kept a Polish citizen as prisoner in the basement of the anatomical institute during the last period of the war, for purposes of studying the topography of the internal organs in relation to the body's position. This accusation was based on a rumor circling the university in 1945 and 1946, which probably originated in misunderstandings of the incident surrounding Dybel's murder, but was otherwise unsubstantiated. The investigation was discontinued in September 1949.[43]

While neither von Herrath nor Hasselwander or Hett were averse to using bodies of executed prisoners for their work, and although Hett was familiar with execution sites from his work in Halle,[44] they all had an immediate, visceral, and hostile reaction against executions at their anatomical institutes. This was one boundary they were not willing to cross. It is possible that they saw executions as necessary but demeaning events that were completely separate and alien to the higher calling of their work. Their professional ethics were bound to the perceived purity of their science and where it was housed—in the institutional building—and they were committed to keeping both the science and its physical home immaculate. However, this attitude did not deter the anatomists in Erlangen from accepting Dybel's body after the man had been murdered outside the institute. It is very likely that Hett saw it as his professional duty to use Dybel's body, once the man was dead.

The two incidents raise several questions. What if this was just the "tip of the iceberg"? Did similar incidents at other anatomical institutes go unreported? And what might have happened if the GeStapo executioners had met with anatomists and anatomical staff who were not averse to lending their institutes for the murder of NS victims? What if this had happened, for example, in Strasbourg, were Hirt was in charge? These questions remain unanswered for now.

Stages of Ethical Transgressions in NS Anatomy: Research on the "Future Dead"

The traditional paradigm of anatomy is that anatomists work with dead human bodies, with the exception of studies of surface and functional anatomy of living humans. Otherwise, experimentation is strictly postmortem. Historically anatomists used unaltered tissues, injected organs before removal in some cases, and, at most, posed the joints or the whole body when studying disciplines such as body mechanics.[45] They were rarely interested in the history of the persons whose bodies they investigated. Should they require any premortem information, they received it from relatives and medical records or from prison personnel. German anatomists of the early twentieth century did not contact or approach living persons whose body they would later dissect, and did not make their death part of their research design.

Beginning in 1942 there is evidence of three anatomists performing human experiments. They crossed the boundaries of the traditional anatomical paradigm of work with the dead to work with the "future dead." They experimented on living human beings, prisoners, and concentration camp inmates whose death became part of their research design. At this point it is unclear what Clara's motivation was for this or how he gained access to the prisoners in Dresden. It is possible that he used the same method that he later proposed in his plans for Munich, where he intended to continue his research by having vitamin C secretly added to the prisoners' food by the prison staff.[46] Kremer's and Hirt's experiments on concentration camp inmates had their roots in the year 1939, when Himmler, who had an interest in experimental medicine, first granted Sigmund Rascher access to prisoners in Dachau for human experiments, an "opportunity" more widely used by several SS physicians beginning in 1941.[47] While Clara's experiments were relatively harmless, they were still coercive, as there is no evidence that the prisoners had been asked for their consent to the administration of the vitamin.[48] Kremer's and Hirt's experiments represent an escalation in the ethical transgressions of anatomists, as Kremer contributed directly to the deaths of prisoners selecting "interesting" subjects, and Hirt participated directly in the murder of his victims.

While Hirt's human experiments with poison gas were enough to qualify him as a war criminal, his ideas for the future of anatomy and his first experimental steps in this direction mark him as a truly innovative and dangerous thinker, indeed the mastermind of work with the "future dead." Initial proposals for a "racial" skeleton or skull collection went back to November 1941,[49] and in early 1942 he formulated his idea of creating a skeleton collection by selecting still living but "future dead" persons. These plans, which became known as "*Auftrag Beger*" (commission Beger), included a "commissioner [who] will be charged with safeguarding the material" and the instructions that "when the death of these Jews has been effected—the head must not be injured—he severs the heads from the bodies and sends them on to their destination."[50] In 1942, while Hirt was still waiting for the implementation of this plan, he had a greater vision for all of German anatomy in sight, an idea apparently discussed at the meeting of German anatomists in Tübingen in November 1942 (see chapter 4). In a report about the conference, Hirt wrote to the *Ahnenerbe*: "There a proposal was made that anatomists should collect and process materials as already specified in 'Auftrag Beger.' Others are gradually becoming aware that something could happen here." He added in handwriting: "I have been commissioned to compose guidelines for the collection of materials for all German anatomists."[51] Little is known about this meeting, but Winkelmann recently discovered more information in the estate papers of von Eggeling and Elze. These reveal that speakers at the meeting were Freerksen, Goerttler, Hirt, Nagel, Neubert, Petersen, and Wetzel, and that Pernkopf, Elze, Niessing, Bargmann, and university officials were among the participants. The meeting was financed by the NSDAP entirely. It is not known if Hirt truly presented his new ideas there, and Winkelmann believes that Hirt might have simply bragged in his letter to Wolfram Sievers, the manager of the Ahnenerbe.[52] However, judging from the general escalation of human experimentation during the later war years, and specifically in view of Hirt's murder of the eighty-six victims from Auschwitz in the summer of 1943, Hirt's firm intent for the implementation of his plans for anatomy is evident. It also is highly likely that his letter to Sievers contained at least a kernel of truth, and that a certain subsection of German anatomists did indeed discuss Hirt's new ideas of body procurement from the "future dead" in an open conversation and even commissioned him to create guidelines for such research. The "future dead" became a research source that any anatomist willing to commit this final transgression could exploit. Hans-Joachim Lang was the first to point out this decisive paradigm shift in German anatomy.[53] Other anatomists remained within its traditional boundaries. It is probably no coincidence that Clara, Kremer, and Hirt committed this last ethical transgression in the middle of the war, when German society had become imbued with inhumanity and brutality ranging from NS governmental violence to the carnage of a multifront conflict and the devastating bombardments of cities. The turbulence of the later stages of war may have prevented the further use of "extraordinary opportunities for cadaver delivery"[54] after 1943.

Stages of Ethical Transgressions in NS Anatomy: Conclusion

Acknowledging the initial transgression in anatomy—the societally accepted violation of the dead human body for scientific reasons—an escalation of ethical transgressions specific to anatomists in National Socialist Germany can be identified:

- Victims of the NS regime became part of the traditional sources of body procurement in anatomy and were used for purposes of anatomical education and research by all German anatomical departments.
- Many anatomists used and even sought out the "opportunities" given by increasing numbers of executed victims for research. They perceived and presented this new development as positive.
- Stieve cold-bloodedly interpreted the situation of women on death row in terms of his strictly planned research, thereby engaging in a human experiment, but in thought only. There is no documentation that he sought access to prisoners before death or caused any victim's death.
- Clara used still-living prisoners, whose execution dates were already determined and who were thus "future dead" persons, for planned experiments, but did not cause these persons' deaths.
- Kremer selected living prisoners after arriving in the camp without a premeditated research plan. He created "future dead" persons by his selection. He contributed indirectly to their death because, even in Auschwitz, they might have survived otherwise.
- Hirt had living prisoners selected according to a strict research plan, thus creating "future dead" persons by causing their death. He also participated directly in their murder.

The anatomical research presented here, even in its various stages of transgression, was still based on the scientific method and cannot be dismissed as scientifically irrelevant. The histological publications by Bargmann, von Hayek, and Clara were typical for their field; Stieve was integrating innovative ideas of functional morphology in his work. Even Hirt's plan for a "racial" skull collection still fit within the realm of physical anthropology in the 1930s. However, much of this anatomical research had become an "experimental medicine in epistemologically conclusive form [... which] had lost its moral boundaries."[55] Whereas the transition from animal to human experiment in Stieve's anatomical research was scientifically justified, he never considered possible ethical shortcomings in terms of his responsibility for the prisoners in his research design. In that respect, he was very much a scientist of his time. Medical historian Volker Roelcke called this phenomenon a "methodological carefulness" that was without consideration for any ethical implications and neglected any "moral care,"[56] thereby emphasizing these scientists' utmost concern for correct scientific methodology in combination with a complete lack of care for the human "objects" of their studies. The same "care-less" attitude toward their victims can be observed in Kremer's and Hirt's transition from animal experiments to coercive human experiments. "Methodological care" trumped "moral care."[57]

This phenomenon is not specific to either the NS period or anatomists alone. However, the murderous nature of the NS regime and the specific susceptibility of anatomists complemented each other. Anatomists, in need of societal permission for their work through legislative authority, were officially and legally given the opportunity to not only work with bodies procured through traditional sources, but also with bodies of all those persons who were eliminated from society according to the biologistic and racist ideology of the regime. Furthermore, they were given official permission by authorities such as Himmler to work with the "future dead," and some anatomists did so without any empathy for the victims. In these situations, the anatomists became morally complicit with the NS regime in their use of "material" gained from wrongfully executed victims, because an "indissoluble ethical link" exists between the "origin of the material" and "what later can be done with it," and who does it.[58] At the same time, anatomists were familiar with the process of clinical detachment through their work, so that, for many of them, a "suspension of empathy"[59] toward the persons whose bodies they dissected may not have been hard to attain.

Anatomy's development during the Third Reich can be interpreted as an extreme manifestation of the "experimentalization of biology and medicine in the 19th and 20th century" when a "categorical separation and disconnected perception of science, ethics and politics in the modern biosciences developed."[60] The specific conditions of the NS regime unleashed the "aggressive and destructive potential [of a] medical science that is primarily focused on the gain of scientific knowledge and blanks out the humanity [...] of the research subject."[61] In 1947 the psychiatrist Alice von Platen-Hallermund, an observer at the Nuremberg Doctors' Trial, noticed this phenomenon and remarked that atrocities are possible when scientists lose their focus on the well-being of individual patients and instead allow the scientific question alone to dominate their thinking.[62]

In historian Robert Proctor's analysis, NS physicians' "research integrity" lacked the respective "research ethics" due to a "failure of physicians to challenge the rotten, substantive core of Nazi values."[63] Most anatomists had indeed not questioned these values. However, they were not lacking research ethics as Proctor claims, but had instead embraced the new ethics of National Socialism and its repercussions for medicine.[64] In 1940 Wetzel explained these new ideals:

> Empathy and human kindness shall still be the prerequisites for the true qualification of a physician—but they are not the only supporting idea; that is the great change. It is exactly through examples of medical thought and practice that one can show that the general 'absolutes' of love, compassion, and kindness, become a curse and cause disaster if they alone determine the actions of a community bound together by blood relationships. [...] The individual is a subordinate entity compared to [...] the German people and the race, and thus care for the individual can never alone be the sole determining idea of the medical profession.[65]

The new ethics was an exclusionary one that barred whole sections of society from the preservation, indeed health care, of the "whole of the German people."[66] The change of focus in medical ethics from the individual to the "whole" as a patient became one of the central causes of evil in National Socialism, and remained not only unchallenged but supported by many within the medical profession, including anatomists. Anatomists became the agents of evil through the convergence of their own reductionist view of human life, the NS exclusionary medical ethics, and the new "opportunities" provided by the regime.

Professional Ethics in Anatomy after National Socialism

In the first years after the war German anatomists were fully occupied with rebuilding their lives and not intent on reflecting on the recent past. They were not particularly enthusiastic about inquiries into the identity of the bodies they had received during NS times or were still using for teaching purposes. While they readily prepared the lists of known NS victims demanded by the various military authorities, some acted defensively, especially when confronted with family members who were trying to locate their loved ones. In Kiel, Bargmann quoted a lack of documentation as the reason for responding late or not at all to inquiries,[67] and Anna Hanika quoted Franz Bauer in Vienna as complaining: "This eternal search for

bodies! [...] This has to stop! 70% of the executed were criminals any-
way; they do not have to be pitied. And nobody can work this hard on
account of the other 30%. Anatomy simply needs bodies!"[68] In September
1945 Bauer was approached about the identification of the bodies of forty
executed political dissidents. He asked for a list of names so that he could
"separate the bodies of criminals from those of political prisoners."[69] This
attitude toward differentiating between so-called "criminals" and political
dissidents was shared by Bauer's colleague Krause, who referred to this
matter in a letter responding to a press publication in January 1962 that
had accused the anatomical institute of withholding bodies of NS victims
from burial until 1957.[70] Krause had not wanted to use these bodies for
teaching purposes, as he thought they might remind students of "those
times", and that such memories were apt to "poison the souls of our young
students".[71] Von Hayek, his chairman, publicly declared that he considered
the use of bodies of NS victims as "impious."[72] He must have changed
his mind, unless he was differentiating between educational and research
purposes, as he himself had used these bodies avidly even after the war. In
most of these instances concern for the identity of the bodies was forced on
the anatomists from outside and did not originate with them.

Only a few direct statements by anatomists concerning the NS period
and their attitude toward the bodies they dissected are known. Common
to all of them is the fact that the scientists did not inquire into the background of the persons whose bodies they used for their work, and that they
felt justified in using bodies provided by the NS regime (see chapter 9). A
comparable view is revealed in more recent interviews with Viennese contemporaries of Pernkopf Krause and Werner Platzer.[73] When asked whether
the use of bodies of the executed had bothered him, Krause answered,
"Nobody cared, and why should we care?" His colleague Platzer responded
"Nobody cared." In a similar fashion, a German emeritus professor of anatomy wrote in 2012: "It is stressful to know this [the possible tragic fate of
persons whose bodies were used in anatomy], and thus understandable
if one did not seek complete insight into the background of bodies for
anatomy, given that this background was irrelevant for the fulfillment of
educational duties."[74] The statements reveal a strong clinical detachment,
even when these anatomists were confronted directly with families looking
for their relatives, an attitude still prevalent among anatomists for many
years after the war.

Since the late twentieth century, this mindset has started to change,
a shift closely linked to the success of body-donation programs. The de-
creasing availability of unclaimed bodies, the change in public opinion
concerning modern medicine, and economic considerations have driven
the establishment of such programs. By now some countries receive most

of their bodies in anatomy from donations, while others are still working on the introduction of such programs.[75] In the United States the combination of an improvement in general health as well as better burial benefits led to a decrease of unclaimed bodies by the middle of the twentieth century.[76] The German situation was particularly dire in the first years after the war. Throughout the following decades the shortage of bodies for dissection continued at many German medical schools until body donation programs became successful. While sporadic body donations occurred in Europe and the United States in the eighteenth and nineteenth centuries, these were usually individual donations from anatomists, doctors, and prominent individuals. Among those who donated their bodies were English philosopher Jeremy Bentham in 1832, English religious prophetess Joanna Southcott,[77] and German anatomists Philipp Friedrich Theodor Meckel in 1803 and Wilhelm von Waldeyer-Hartz in 1921.[78] With the rise of better public education as well as the increasing success of modern medicine, especially of transplantations, the public opinion toward scientific medicine and anatomy changed favorably throughout the twentieth century, so that the concept of body donation started to find wider acceptance. University medical schools in Europe, the United States, Australia, New Zealand, and many other countries started to introduce body donation programs in the 1950s, which supplied most of the bodies for anatomical dissection by the end of the twentieth century. Nonacademic body donation programs introduced by "cadaver merchants" who work on a for-profit basis have also been introduced.[79]

Discussions around the ethics of body procurement for anatomical dissection and respective legislation are still going on around the world. Many anatomists will only work with donated bodies on principle,[80] while others argue for an inclusion of unclaimed bodies.[81] In addition, the practice of body procurement in general and the understanding of "body donation" specifically are under debate globally.[82] However, a full documentation of the various issues, which include the legal ambiguity and dignity of the dead human body and posthumous interests, goes beyond the scope of this book,[83] especially as systematic research on the history and ethics of body donation is still missing. In Germany, medical schools allow only donated bodies for anatomical purposes according to legislation in the individual German states.[84] Each anatomical institute has its own willed donor program, with such good response that several medical schools have had to temporarily suspend acceptance of new prospective donors.[85] Apart from educating the public and the rise of altruistic donors, the increase in body donations in Germany may be explained by the decrease in traditional burial benefits, making body donation an economically attractive alternative to burial for some.[86]

Much has changed in anatomy since the end of the war, not only the body supply through donation programs, but also the general perception of the role that anatomy plays in medical education. Even the traditional standard of anonymization of the body donor is being questioned. While anonymization is still the norm at most medical schools, alternative practices are developing in some cultures and anatomical programs, where body donors are identified by name.[87] Many contemporary anatomists and their students believe that the history of the person whose body is being dissected is indeed an important part of the "fulfillment of educational duties in anatomy." This "educational duty" has expanded from teaching the structure and function of the human body to "humanist values in medical school."[88] Students want to learn more about the lives and motivations of body donors, a need that is met at some medical schools where pupils view recorded interviews with prospective donors or meet the families of donors.[89] Furthermore, the professional development and psychological well-being of students during the dissection course have become prominent topics of research in medical education in Germany and elsewhere.[90] One of the most important outward signs of the change in attitude are the memorial services to remember and honor the donors and their families, which have become a regular feature of dissection courses around the world.[91]

The last decades have also brought a greater public awareness and sensibility concerning the handling of human remains.[92] An international debate surrounding the restitution of human remains hailing from indigenous populations started in the 1970s and 80s. Major anthropological collections, such as those in Germany and Austria, received restitution requests from former colonies, which were met with extensive research efforts and the return of human remains to their native countries. This work is still ongoing.[93] All of these developments contributed to new guidelines regarding the use of human bodies, organs, and tissues not only in museums and other collections but also in anatomical body acquisition.[94]

Controversial discussions still occur about the public display of plastinated bodies, as do scandals involving inappropriate handling of bodies.[95] They highlight the importance of a continued effort to define ethical guidelines in contemporary anatomy.[96] The German anatomical community has participated in developing recommendations for "good practices" in body donation for Europe,[97] and an international effort has resulted in global guidelines in anatomy by the International Federation of the Association of Anatomists.[98] Among the proposed standards are "lectures in ethics relating to the bequest of human remains" and the renunciation of commercial gain from donated human remains.[99] The *Anatomische Gesellschaft* issued a statement on the practices of trade with anatomical specimens on the international market.[100] However, all of these guidelines and recommendations are naught if they are not enforced by an agency responsible for their application, as the mishandling of bones and other specimens discovered on the grounds of the former KWI Anthropology in Berlin exemplifies. In the summer of 2014 these bones were discovered during construction work, routinely handed to the police, who was unaware of the historic connection between the KWI and potential victims of Mengele, and further miscommunications between the parties involved led to the anonymous cremation and burial of these human remains.[101] An oversight agency responsible for the proper identification and memorialization of potential remains from NS victims could prevent such unfortunate mistakes.

One of the most significant postwar developments in German anatomy was the decision of the *Anatomische Gesellschaft* to finally take a stand concerning its history in National Socialism. In 2010 a first symposium on the topic was organized by Christoph Redies, chairman of anatomy at the University of Jena, and held in Würzburg. The proceedings of this international meeting were published in a special edition of the *Annals of Anatomy*.[102] In 2011 the 125-year anniversary of the *Anatomische Gesellschaft* was celebrated with an anniversary edition of the *Annals* including a first German-language overview of the history of anatomy in the Third Reich. A table honoring the victims named the scholars of anatomy whose careers were disrupted by NS policies and commemorated the NS victims whose bodies were used for anatomical purposes.[103] In 2014 the society published a statement on its website which officially acknowledges its role during National Socialism:

> The Anatomische Gesellschaft admits the historical fact that numerous members of the Anatomische Gesellschaft willingly used the bodies of victims of National Socialism for educational and research purposes. In this manner they cooperated de facto with the iniquitous National Socialist regime. Two members, August Hirt and Johann Paul Kremer, thereby became murderers.
>
> We are appalled by these actions and remember with deepest sorrow the many, frequently nameless victims of National Socialism whose bodies were delivered to anatomical institutes and treated in an inhumane manner. Our sympathy is with all their relatives, especially those who, to this day, have no certain knowledge about the fate of their loved ones' bodies.
>
> In memory of this past history the Anatomische Gesellschaft and its members strongly support the exclusive use of voluntarily donated bodies for anatomical purposes. The Anatomische Gesellschaft and its members promote the ethically appropriate handling of bodies through willed body donation programmes on the national and the international level.
>
> We also commemorate the members of the Anatomische Gesellschaft whose lives and careers were disrupted or destroyed by the National Socialist regime for "racial" or political reasons.

The Anatomische Gesellschaft commits to the support of systematic research on its history during the time of National Socialism and to the publication of the results of these investigations.[104]

Notes

1. Barker 1995, 165.
2. Jones 1998, 100.
3. Jones and Whitaker 2009.
4. Prüll 2000, 65–66; Winkelmann 2007a
5. Prüll 2000, 65–66.
6. Cantor 2010, 209.
7. Winkelmann 2007a, 62.
8. Bundesärztekammer 2005; Sperling 2008.
9. Prüll 2000; Sappol 2002; Hildebrandt 2008.
10. Bergmann 2008.
11. Sukol 1995.
12. Warner and Rizzolo 2006; Böckers et al. 2010.
13. Richardson 2000, 31.
14. Warren 1783–1785.
15. Richardson 2000, 51.
16. Bohde 2005, 332.
17. Winkelmann 2007a, 62.
18. Grodin 2010, 58.
19. Swick 2006,; Warner 2009; Böckers et al. 2010; Montross 2007; Hildebrandt 2010.
20. Noack and Heyll 2006; Bussche 1989, 156; Forsbach 2006; Alvermann 2015.
21. Noack 2012.
22. Bussche 1989, 156.
23. Mühlberger 1998; Dr. Raphael Toledano, personal communication, July 2013.
24. Bussche 1989, 156.
25. Seidler and Leven 2007, 512–13.
26. Noack 2008, 15.
27. Bussche 1989, 156.
28. Maier 1973, 178.
29. Mühlberger 1998.
30. Lehner 1990, 110.
31. Noack and Heyll 2006, 12–13.
32. Noack 2008, 12.
33. Schmid 1965, 104–5.
34. Kick 1944, 311.
35. Stöhr 1935.
36. Hayek 1940, 405.
37. Bargmann 1931; Bargmann 1942; Bargmann and Scheffler 1943.
38. Stieve 1947, 783.
39. Porezag 2002, 294–300.
40. UAG, Med Fak II-52, Bl. 119.
41. *Der Oberstaatsanwalt beim Landgericht Nürnberg-Fürth*, 1 September 1949, BayHSTA MK 43712; Wendehorst 1993, 23.
42. UAG, Med Fak II-52, Bl. 119.
43. *Der Oberstaatsanwalt beim Landgericht Nürnberg-Fürth*, 1 September1949, BayHSTA MK 43712.

44. Noack und Heyll 2006, 138, 139.
45. E.g., Hayek 1935; Hewel 1941.
46. Schütz et al. 2015.
47. Kater 2006, 99.
48. Winkelmann and Noack 2010.
49. Kasten 1991, 183.
50. Mitscherlich and Mielke 2007, 81–82.
51. Quoted after Lang 2013, 373–74.
52. Winkelmann 2015.
53. Lang 2013.
54. Ibid., 373.
55. Roelcke 2009, 46.
56. Ibid., 43, 47; Roelcke 2014, 62–63.
57. Roelcke 2009, 43.
58. Jones 2007, 339; on the subject of moral complicity: Miller 2012.
59. Barker 1995.
60. Roelcke 2009, 16, 47.
61. Roelcke 2012, 101.
62. Quoted in Schleiermacher and Schagen 2012, 174.
63. Proctor 2000, 343, 344.
64. Compare Rütten 1997; Schmidt 2009.
65. Wetzel 1940, 13, 14; see also, Peiffer 1991, 129.
66. Fritz Bauer Institut 2009; Gross 2010.
67. Buddecke 2010.
68. Neugebauer 1998, 460.
69. Ibid.
70. Kurier1962.
71. Letter from Walter Krause, copy received by von Hayek, 19 January 1962, Akademischer Senat der Universität Wien 1998, 475–76.
72. Kurier 1962.
73. Aharinejad and Carmichael 2012.
74. Arnold 2012, 49.
75. Compare Jones and Whitaker 2009; Gangata et al. 2010.
76. Garment et al. 2007; Warner 2009.
77. Marshall 1995.
78. Schultka and Göbbel 2005; Winkelmann 2007b.
79. Champney 2014.
80. Jones 2014.
81. Wilkinson 2014a; Wilkinson 2014b.
82. McHanwell et al. 2008, Riederer et al. 2012, Goodwin 2013.
83. E.g., Bleyl 1999; Sperling 2008; Herrmann 2011; Madoff 2013.
84. Bleyl 1999; Bundesärztekammer 2005; McHanwell et al. 2008; Riederer et al. 2012.
85. Schäfer and Gross 2009; Spiegel 2008; Seipel 2010.
86. Schäfer and Gross 2009, 43.
87. Winkelmann and Güldner 2004; Lin et al. 2009; Talarico and Prather 2007; Talarico 2013.
88. Dyer and Thorndike 2000, 369; Sukol 1995; Goddard 2003; Rizzolo 2002.
89. Trotman 2009; Bohl et al. 2011; Bohl et al. 2013; Crow et al. 2012.
90. E.g., Hafferty 1991; Nnodim 1996; Abu-Hijelh et al. 1997; Blumenthal-Barby and Stefenelli 1998; Neuhuber 1998; Böckers et al. 2010; Swick 2006; Pawlina 2006; Lachman and Pawlina 2006; Grochowski et al. 2014.

91. Pabst and Pabst 2006; Lin et al. 2009; Hildebrandt 2010; Jones et al. 2014; Zhang et al. 2014.
92. E.g., Skloot 2010; Fine-Dare 2002.
93. For a review see, Stoecker et al. 2013.
94. Bundesärztekammer 2003; Leiden Declaration 2012; Deutscher Museumsbund e.V. 2013.
95. E.g., Jones 2014; Wilkinson 2014b; Himmelrath 2012; Spiegel 2014; Roach 2003; MacDonald 2006, 186–89; Cheney 2006; Schmitt et al. 2014; Hildebrandt 2014; Terry 2014.
96. E.g., Champney 2011; Schmitt et al. 2014.
97. McHanwell et al. 2008; Riederer et al. 2012.
98. IFAA 2012.
99. McHanwell et al. 2008, 24.
100. Anatomische Gesellschaft 2010.
101. Aly 2015; Kühne 2015.
102. Hildebrandt and Redies 2012.
103. Hildebrandt and Aumüller 2012.
104. Anatomische Gesellschaft 2014.

Archival Sources

Bayerisches Hauptstaatsarchiv
– Signatur MK 43712, personnel file, Albert Hasselwander
Universitätsarchiv Greifswald
– UAG, Med Fak II-52, Bl. 119

Bibliography

Abu-Hijelh, Marwan F., Nazih A. Hamdi, Satei T. Moqattash, Philipp F. Harris, Gilbert F. D. Heseltine. 1997. "Attitudes and Reactions of Arab Medical Students to the Dissecting Room." *Clinical Anatomy* 10: 272–78.
Aharinejad, Seyed Hossein, and Stephen W. Carmichael. 2012. "First Hand Accounts of Events in the Laboratory of Prof. Eduard Pernkopf." *Clinical Anatomy* 26: 297–303.
Akademischer Senat der Universität Wien. 1998. *Senatsprojekt der Universität Wien: Untersuchungen zur anatomischen Wissenschaft in Wien 1938–1945.* Unpublished manuscript.
Alvermann, Dirk. 2015. "'Praktisch begraben': NS-Opfer in der Greifswalder Anatomie 1935–1947." In: *"… die letzten Schranken fallen lassen": Studien zur Universität Greifswald im Nationalsozialismus,* ed. Dirk Alvermann, 311–51. Köln Weimar Wien: Böhlau Verlag.
Aly, Götz. 2015. "Geistlos und roh." *Berliner Zeitung,* 2 February. Accessed 2 February 2015. http://www.berliner-zeitung.de/meinung/kolumne-zur-freien-universitaet-geistlos-und-roh-an-der-fu-berlin,10808020,29728282.html.
Anatomische Gesellschaft. 2010. "Statement with Regard to the Practice of Selling Anatomical Specimens on the International Market." Accessed 23 May 2014. http://anatomische-gesellschaft.de/informationen-ag3/stellungnahme-2010-englisch-body-donation-ag3hen-modellstudiengaenge.html.
———. 2014. "Statement by the Anatomische Gesellschaft on the History of Anatomy in National Socialism." Accessed 22 May 2014. http://anatomische-gesellschaft.de/ethik-ag3/statement-on-the-history-of-anatomy-in-national-ag3.html.

Arnold, Michael. 2012. "Anatomie im Zwielicht." In: *Anatomische Gesellschaft: Jubiläum-sheft 125 Jahre Anatomische Gesellschaft*, ed. Wolfgang Kühnel, 41–48. Lübeck: Kaiser & Mietzner.

Bargmann, Wolfgang. 1931. "Über Struktur und Speicherungsvermögen des Nierenglomer-ulus." *Zeitschrift für Zellforschung* 14: 73–137.

———. 1942. "Über Kernsekretion in der Neurohypophyse des Menschen." *Zeitschrift für Zellforschung* 28: 99–102.

Bargmann, Wolfgang, and A. Scheffler. 1943. "Über den Saum des menschlichen Darme-pithels." *Zeitschrift für Zellforschung* 33(1–2): 5–13.

Barker, Pat. 1995. *The Eye in the Door.* London: Plume–Penguin. First published in 1993.

Bergmann, Anna. 2008. "Taboo Transgressions in Transplantation Medicine." *Journal of the American Physicians and Surgeons* 13(2): 52–55.

Bleyl, Uwe. 1999. "Der Mensch muss immer Zweck an sich selbst bleiben: Der rechtsethi-sche Vorbehalt gegen die sog. Anatomiekunst." *Annals of Anatomy* 181: 309–16.

Blumenthal-Barby, Kay, and Norbert Stefenelli. 1998. "Auswirkungen der Beschäftigung mit Leichen oder Leichenteilen auf Teilnehmer allgemeiner Sezierübungen." In: *Kör-per ohne Leben: Begegnung und Umgang mt Toten*, ed. Nobert Stefenelli, 607–18. Wien: Böhlau Verlag.

Böckers, Anja, Lucia Jerg-Bretzke, Christoph Lamp, Anke Brinkmann, Harals C. Traue, and Tobias M. Böckers. 2010. "The Gross Anatomy Course: An Analysis of Its Impor-tance." *Anatomical Sciences Education* 3: 3–11.

Bohde, Daniela. 2005. "Pellis memoriae: Die Moralisierung der Haut in Frontispizen und Anatomietheatern der Niederlande im 17. Jahrhundert—ein blinder Fleck in der Medi-zingeschichte nach 1945." In: *Zergliederungen- Anatomie und Wahrnehmung in der frühen Neuzeit*, ed. Albert Schirrmeister, 327–58. Frankfurt: Vittorio Klostermann.

Bohl, Michael, Peter Bosch, Sabine Hildebrandt. 2011. "Medical Students' Perceptions of the Body Donor as a 'First Patient' or 'Teacher': A Pilot Study." *Anatomical Sciences Education* 4: 208–13.

Bohl, Michael, A. Holman, Dean A.Mueller, Larry D. Gruppen, Sabine Hildebrandt. 2013. "The Willed Donor Interview Project: Medical Student and Donor Expectations." *An-atomical Sciences Education* 6(2): 90–100.

Buddecke, Julia. 2011. *Endstation Anatomie: Die Opfer nationalsozialistischer Vernichtungsjus-tiz in Schleswig-Holstein.* Hildesheim: Georg-Olms Verlag.

Bundesärztekammer. 2003. "Arbeitskreis 'Menschliche Präparatesammlungen': Empfeh-lungen zum Umgang mit Präparaten aus menschlichem Gewebe in Sammlungen, Mu-seen und öffentlichen Räumen." *Deutsches Ärzteblatt* 8: 378–83, English translation. Accessed 5 February 2015. http://www.aemhsm.net/ressources/actus/TranslationGuide lines_final.doc.

———. 2005. "Rechtsfragen beim Umgang mit der Leiche." Accessed 21 May 2014. http:// www.bundesaerztekammer.de/page.asp?his=0.7.47.3179.3180.3186.

Bussche, Hendrik van den. 1989. *Im Dienste der "Volksgemeinschaft": Studienreform im Na-tionalsozialismus am Beispiel der ärztlichen Ausbildung.* Berlin/Hamburg: Dietrich Reimer Verlag.

Cantor, Norman L. 2010. *After We Die: The Life and Times of the Human Cadaver.* Washing-ton, DC: Georgetown University Press.

Champney, Thomas H. 2011. "A Proposal for a Policy on the Ethical Care and Use of Ca-davers and their Tissues." *Anatomical Sciences Education* 4(1): 49–52.

———. 2014. "The Business of Bodies: Ethical Issues with For-Profit Cadaver Merchants." *Annals of Anatomy* 196S: 261.

Cheney, Annie. 2006. *Body Brokers: Inside America's Underground Trade in Human Remains.* New York: Broadway Books.

Crow, Sheila M., Dan O'Donoghue, Jerry B. Vannatta, and Britta M. Thompson. 2012. "Meeting the Family: Promoting Humanism in Gross Anatomy." *Teaching and Learning Medicine* 24(1): 49–54.

Deutscher Museumsbund e.V., ed. 2013. "Recommendations for the Care of Human Remains in Museums and Collections." Accessed 28 June 2014. http://www.muse umsbund.de/fileadmin/geschaefts/dokumente/Leitfaeden_und_anderes/2013__Recom mendations_for_the_Care_of_Human_Remains.pdf.

Dyer, George S. M., and Mary E. L. Thorndike. 2000. "Quidne Mortui Vivos Docent? The Evolving Purpose of Human Dissection in Medical Education." *Academic Medicine* 75: 969–79

Fine-Dare, Kathleen S. 2002. *Grave Injustice: The American Indian Repatriation Movement and NAGPRA.* Lincoln: University of Nebraska Press.

Forsbach, Ralf. 2006. *Die Medizinische Fakultät der Universität Bonn im "Dritten Reich."* München: R. Oldenbourg Verlag.

Fritz Bauer Institut, ed. 2009. *Moralität des Bösen. Ethik und nationalsozialistische Verbrechen. Jahrbuch 2009 zur Geschichte und Wirkung des Holocaust.* Frankfurt/New York: Campus Verlag.

Gangata, Hope, Phateka Ntaba, Princess Akol, and Graham Low. 2010. "The Reliance on Unclaimed Cadavers for Anatomical Teaching by Medical School in Africa." *Anatomical Sciences Education* 3: 174–83.

Garment, Ann, Susan Lederer, Naomi Rogers, and Lisa Boult, L. 2007. "Let the Dead Teach the Living: The Rise of Body Bequeathal in 20th-Century America." *Academic Medicine* 82: 1000–1005.

Goddard, Shannon. 2003. "A History of Gross Anatomy: Lessons for the Future." *University of Toronto Medical Journal* 80: 145–47.

Goodwin, Michele, ed. 2013. *The Global Body Market: Altruism's Limits.* Cambridge: Cambridge University Press.

Grochowski, Colleen O'Connor, Jerry Cartmill, Jerry Reiter, Jean Spaulding, James Haviland, Fidel Valea, Patricia L. Thibodeau, Stacey McCorison, and Edward C. Halperin. 2014. "Anxiety in First Year Medical Students Taking Gross Anatomy." *Clinical Anatomy* 27: 835–38.

Grodin, Michael A. 2010. "Mad, Bad, or Evil: How Physician Healers Turn to Torture and Murder." In: *Medicine after the Holocaust: From the Master Race to the Human Genome and Beyond,* ed. Sheldon Rubenfeld, 49–65. New York: Palgrave MacMillan.

Gross, Raphael. 2010. *Anständig geblieben: Nationalsozialistische Moral.* Frankfurt: Fischer Verlag.

Hafferty, Francis W. 1991. *Into the Valley: Death and Socialization of Medical Students.* New Haven: Yale University Press.

Hayek, Heinrich. 1935. "Das Verhalten der Arterien bei Beugung der Gelenke." *Zeitschrift für Anatomie und Entwicklungsgeschichte* 105(1): 25–36.

———. 1940. "Die Läppchen und Septa interlobaria der menschlichen Lunge." *Zeitschrift für Anatomie und Entwicklungsgeschichte* 110(3): 405–11.

Herrmann, Beate. 2011. *Der menschliche Körper zwischen Vermarktung und Unverfügbarkeit: Grundlinien einer Ethik der Zelbstverfügung.* Freiburg: Verlag Karl Alber.

Hewel, Julius. 1941. "Über die Beweglichkeit der menschlichen Leber." *Anatomischer Anzeiger* 90(22/24): 273–96.

Hildebrandt, Sabine. 2008. "Capital Punishment and Anatomy: History and Ethics of an Ongoing Association." *Clinical Anatomy* 21: 5–14.

———. 2010. "Developing Empathy and Clinical Detachment during the Dissections Course in Gross Anatomy." *Anatomical Sciences Education* 3: 216.

———. 2014. "What Is Happening in Our Dissection Rooms?" *Clinical Anatomy* 27: 833–34.

Hildebrandt, Sabine, and Gerhard Aumüller. 2012. "Anatomie im Dritten Reich." In: *Anatomische Gesellschaft: Jubiläumsheft 125 Jahre Anatomische Gesellschaft*, ed. Wolfgang Kühnel, 41–48. Lübeck: Kaiser & Mietzner.

Hildebrandt, Sabine, and Christoph Redies. 2012. "Anatomy in the Third Reich." *Annals of Anatomy* 194: 225–27.

Himmelrath, Armin. 2012. "Anatomie-Skandal in Köln: Chaos im Leichenkeller." *Spiegel online* March 8 2012. Accessed 22 May 2014. http://www.spiegel.de/unispiegel/studium/anatomie-skandal-in-koeln-chaos-im-leichenkeller-a-820001.html.

IFAA. 2012. "Guidelines for Body Donation." Accessed 17 March 2015. http://www.ifaa.net/index.php/ficem.

Jones, David G. 1998. "Anatomy and Ethics: An Exploration of Some Ethical Dimensions of Contemporary Anatomy." *Clinical Anatomy* 11: 100–105.

———. 2007. "Anatomical Investigations and Their Ethical Dilemmas." *Clinical Anatomy* 20: 338–43.

———. 2014. "Using and Respecting the Dead Human Body: An Anatomist's Perspective." *Clinical Anatomy* 27: 839–43.

Jones, David G., and Maya I. Whitaker. 2009. *Speaking for the Dead: The Human Body in Biology and Medicine*. Farnham, UK: Ashgate.

Kasten, Frederick H. 1991. "Unethical Nazi Medicine in Annexed Alsace-Lorraine: The Strange Case of Nazi Anatomist Professor Dr. August Hirt." In: *Essays in Modern German History and Archival Policy*, ed. George O. Kent, 173–208. Fairfax, VA: George Mason University Press.

Kater, Michael H. 2006. *Das "Ahnenerbe" der SS 1935–1945. Ein Beitrag zur Kulturpolitik des Dritten Reiches*. 4th ed. Munich: R. Oldenbourg Verlag.

Kick, U. 1944. "Beitrag zur Frage des bestmöglichen geweblichen Glykogennachweis." *Anatomischer Anzeiger* 95: 310–26.

Kühne, Anja. 2015. "Umgang mit den Skelettfunden in Dahlem. Einfach eingeäschert." *Tagespiegel*, 26 January. Accessed 2 February 2015. http://www.tagesspiegel.de/wissen/umgang-mit-den-skelettfunden-in-dahlem-einfach-eingeaeschert/11278454.html.

Kurier. 1962. "Geköpfte lagen 15 Jahre im anatomischen Institut." *Kurier*, 17 January 1962.

Lachman, Nirusha, and Wojciech Pawlina. 2006. "Integrating Professionalism in Early Medical Education: The Theory and Application of Reflective Practice in the Anatomy Curriculum." *Clinical Anatomy* 19: 456–60.

Lang, Hans-Joachim. 2013. "August Hirt and 'Extraordinary Opportunities for Cadaver Delivery' to Anatomical Institutes in National Socialism: A Murderous Change in Paradigm." *Annals of Anatomy* 195: 373–80.

Lehner, Martina. 1990. "Die Medizinische Fakultät der Universität Wien 1938–1945." Diplomarbeit zur Erlangung des Magistergrades der Philosophie, Geisteswissenschaftliche Fakultät der Universität Wien.

Leiden Declaration. 2012. "The Leiden Declaration on Human Anatomy/Anatomical Collections: Concerning the Conservation and Preservation of Anatomical and Pathological Collections." Accessed 21 August 2014. http://media.leidenuniv.nl/legacy/leiden-declaration.pdf.

Lin, Steven C., Julia Hsu, Victoria Y. Fan. 2009. "'Silent Virtuous Teachers': Anatomical Dissection in Taiwan." *British Medical Journal* 339: b5001.

MacDonald, Helen. 2006. *Human Remains: Dissection and Its Histories*. New Haven and London: Yale University Press.

McHanwell, S., E. Brenner, A. R. M. Chirculescu, Jan Drukker, H. van Mameren, G. Mazzotti, Diego Pais, Friedrich Paulsen, Odile Plaisant, M. Maître, E. Caillaud, E. Laforêt, Beat M. Riederer, J. R. Sañudo, F. Bueno-López, P. Doñate-Oliver, P. Sprumont, G. Teofilovski-Parapid, Bernard J. Moxham. 2008. "The Legal and Ethical Framework Governing Body Donation in Europe: A Review of Current Practice and Recommendations for Good Practice." *European Journal of Anatomy* 1: 1–24.

Madoff, Ray D. 2013. "The Perverse History of Dead Bodies under American Law." In: *The Global Body Market: Altruism's Limits*, ed. Michele Goodwin, 127–47. Cambridge: Cambridge University Press.

Maier, A. 1973. "Erlebnisse eines Gefängnisseelsorgers. Königsberg 1939–1945." In: *Unser Ermlandbuch 1973*, ed. Bischof-Maximilian-Kaller-Stiftung, 169–87. Osnabrück: A. Fromm KG.

Marcum, James A. 2008. *Humanizing Modern Medicine: An Introductory Philosophy of Medicine*. Dordrecht: Springer Science + Business Media B.V.

Marshall, Tim. 1995. *Murdering to Dissect: Grave-Robbing, Frankenstein and the Anatomy Literature*. 1st ed. Manchester, UK: Manchester Press.

Miller, Franklin G. 2012. "Research and Complicity: The Case of Julius Hallervorden." *Journal of Medical Ethics* 38: 53–56.

Mitscherlich, Alexander and Fred Mielke. 2007. *Doctors of Infamy. The Story of the Nazi Medical Crimes*. Whitefish, MT, USA: Kessinger Publishing. First published in 1949.

Montross, Christine. 2007. *Body of Work: Meditations on Mortality from the Human Anatomy Lab*. New York: Penguin Press.

Mühlberger, Kurt. 1998. "II. Die Belieferung des anatomischen Instituts der Universität Wien mit Studienleichen in der Zeit von 1938–1946." In: *Senatsprojekt der Universität Wien: Untersuchungen zur anatomischen Wissenschaft in Wien 1938–1945*, ed. Akademischer Senat der Universität Wien, 29–66. Unpublished manuscript.

Neugebauer, Wolfgang. 1998. "Zum Umgang mit sterblichen Resten von NS-Opfern nach 1945." In: *Senatsprojekt der Universität Wien: Untersuchungen zur anatomischen Wissenschaft in Wien 1938–1945*, ed. Akademischer Senat der Universität Wien, 459–65. Unpublished manuscript.

Neuhuber, Winfried L. 1998. "Die Bedeutung der Beschäftigung mit dem toten Körper für die Entwicklung des Medizinstudenten zum Arzt." In: *Körper ohne Leben Begegnung und Umgang mit Toten*, ed. Norbert Stefenelli, 619–22. Wien: Böhlau Verlag.

Nnodim, Joseph O. 1996. "Preclinical Student Reactions to Dissection, Death, and Dying." *Clinical Anatomy* 9:175–82.

Noack, Thorsten. 2008. "Begehrte Leichen. Der Berliner Anatom Hermann Stieve (1886–1952) und die medizinische Verwertung Hingerichteter im Natinoalsozialismus." *Medizin, Gesellschaft und Geschichte. Jahrbuch des Instituts für Geschichte der Medizin der Robert Bosch Stiftung* 26: 9–35.

———. 2012. "Anatomical Departments in Bavaria and the Corpses of Executed Victims of National Socialism." *Annals of Anatomy* 194: 286–92.

Noack, Thorsten, and Uwe Heyll. 2006. "Der Streit der Fakultäten. Die medizinische Verwertung der Leichen Hingerichteter im Nationalsozialismus." In: *Geschichte der Medizin-Geschichte in der Medizin*, ed. Jörg Vögele, Heiner Fangerau, and Thorsten Noack, 133–42. Hamburg: Literatur Verlag.

Pabst, Vera Christina, and Reinhard Pabst. 2006. "Danken und Gedenken am Ende des Präparierkurses." *Deutsches Ärzteblatt* 103(45): A3008–3010.

Pawlina, Wojciech. 2006. "Professionalism and Anatomy: How Do These Two Terms Define Our Role?" *Clinical Anatomy* 19: 391–92.

Peiffer, Jürgen. 1991. "Neuropathology in the Third Reich: Memorial to Those Victims of National-Socialist Atrocities in Germany Who Were Used by Medical Science." *Brain Pathology* 1: 125–31.

Porezag, Karsten. 2002. *Zwangsarbeit in Wetzlar: Der "Ausländer-Einsatz" 1939–1945. Die Ausländerlager 1945–1949*. Wetzlar: Porezag.

Proctor, Robert N. 2000. "Nazi Science and Nazi Medical Ethics: Some Myths and Misconceptions." *Perspectives in Biology and Medicine* 43(3): 335–46.

Prüll, Cay-Rüdiger. 2000. "Der Umgang mit der menschlichen Leiche in der Medizin: die historische Perspektive." In: *Zum Umgang mit der menschlichen Leiche in der Medizin*, ed. Hans-Konrat Wellmer, and Gisela Bockenheimer-Lucius, 59–69. Lübeck: Schmidt-Römhild.

Richardson, Ruth. 2000. *Death, Dissection and the Destitute.* 2nd ed., with a new afterword. Chicago and London: University of Chicago Press. First published in 1987

Riederer, Beat M., S. Bolt, E. Brenner, J. L. Bueno-López, A. M. Chirculescu, D.C. Davis, R. De Caro, P. O. Gerrits, S. M. Hanwell, Diego Pais, Friedrich Paulsen, Odile Plaisant, Erdogan Sendemir, I. Stabile, Bernard J. Moxham. 2012. "The Legal and Ethical Framework Governing Body Donation in Europe: 1st Update on Current Practice." *European Journal of Anatomy* 16: 13–33.

Rizzolo, Lawrence J. 2002. "Human Dissection: An Approach to Interweaving the Traditional and Humanistic Goals of Medical Education." *Anatomical Record* 269(6): 242–48.

Roach, Mary. 2003. *Stiff: The Curious Lives of Human Cadavers.* New York: W. W. Norton & Company.

Roelcke, Volker. 2009. "Tiermodell und Menschenbild. Konfigurationen der epistemiologischen und ethischen Mensch-Tier-Grenzziehung in der Humanmedizin zwischen 1880 und 1945." In: *Kulturgeschichte des Menschenversuchs im 20. Jahrhundert*, ed. Griesecke, Birgit, Marcus Krause, Nicolas Pethes, and Katja Sabisch, 16–47. Frankfurt: Suhrkamp.

———. 2012. "Fortschritt ohne Rücksicht. Menschen als Versuchskaninchen bei den Sulfonamid-Experimenten im Konzentrationslager Ravensbrück." In: *Geschlecht und "Rasse" in der NS-Medizin*, ed. Insa Eschebach and Astrid Ley, 101–14. Berlin: Metropol Verlag.

———. 2014. "Sulfonamide Experiments on Prisoners in Nazi Concentration Camps: Coherent Scientific Rationality Combined with Complete Disregard of Humanity." In: *Human Subjects Research after the Holocaust*, ed. Sheldon Rubenfeld and Susan Benedict. Cham: Springer, 51–66.

Rütten, Thomas. 1997. "Hitler with—or without—Hippocrates? The Hippocratic Oath during the Third Reich." *Korot* (12): 91–106.

Sappol, Michael. 2002. *A Traffic of Dead Bodies: Anatomy and Embodied Social Identity in Nineteenth-Century America.* 1st ed. Princeton, NJ: Princeton University Press.

Schäfer, Gereon, and Dominik Gross. 2009. "Körperspende oder Tauschgeschäft? Der geldwerte Vorteil 'gespendeter' Leichname und seine Bedeutung für die Einordnung des toten Körpers als Ressource." In: *Die dienstbare Leiche. Der tote Körper als medizinische, soziokulturelle und ökonomische Ressource*, ed. Dominik Gross, 42–45. Kassel: Kassel University Press.

Schleiermacher, Sabine, and Udo Schagen. 2012. "Semantik als Strategie: Die Klassifizierung der Medizinverbrechen in Konzentrationslagern als Pseudowissenschaft. In: *Geschlecht und "Rasse" in der NS-Medizin*, ed. Insa Eschebach and Astrid Ley, 157–77. Berlin: Metropol Verlag.

Schmid, Armin. 1965. *Frankfurt im Feuersturm. Die Geschichte der Stadt im Zweiten Weltkrieg.* Frankfurt am Main: Verlag Frankfurter Bücher.

Schmidt, Ulf. 2009. "Medical Ethics and Nazism." In: *The Cambridge World History of Medical Ethics*, ed. Robert Baker, and Laurence B. McCullough, 595–608. Cambridge: Cambridge University Press.

Schmitt, Brandi, Charlotte Wacker, Lise Ikemoto, and Frederick Meyers. 2014. "A Transparent Oversight Policy for Human Anatomical Specimen Management: The University of California, Davis Experience." *Journal of Anatomy* 89: 410–14.

Schütz, Mathias, Maximilian Schochow, Jens Waschke, Georg Marckmann, and Florian Steger.2015. "Anatomische Vitamin C-Forschung im Nationalsozialismus und in der Nachkriegszeit: Max Claras Humanexperimente an der Anatomischen Anstalt München." *Medizinhistorisches Journal,* 50: 330–355.

Schultka, Rüdiger, and Luminita Göbbel. 2005. *Die Hallesche Anatomie und ihre Sammlungen. Ein Instituts- und Sammlungsführer.* 2nd ed. Reinbek: LAU-Verlag.

Seidler, Eduard, and Karl-Heinz Leven. 2007. *Die Medizinische Fakultät der Alberts-Ludwigs-Universität Freiburg im Breisgau. Grundlagen und Entwicklungen.* Freiburg: Verlag Karl Alber.

Seipel, Regine. 2010. "Geschenkte Körper." *Frankfurter Rundschau,* 29 July 2010. Accessed 21 May 2014. http://www.fr-online.de/rhein-main/anatomie-geschenkte-koer per,1472796,4516346.html.

Skloot, Rebecca. 2010. *The Immortal Life of Henrietta Lacks.* New York: Random House.

Sperling, Daniel. 2008. *Posthumous Interests: Legal and Ethical Perspectives.* Cambridge: Cambridge University Press.

Spiegel. 2008. "Zu viele Leichen für die Unis." *Spiegel* online, 4 July 2008. Accessed 21 May 2014. http://www.spiegel.de/unispiegel/studium/koerperspenden-zu-viele-leichen-fuer-die-unis-a-563908.html.

———. 2014. "Skandalfund in Madrid. Universität soll Leichen verkauft und vermietet haben." *Spiegel* online, 20 May 2014. Accessed 22 May 2014. http://www.spiegel.de/ unispiegel/studium/leichenfund-an-madrider-universitaet-complutense-koerper-ver kauft-a-970428.html.

Stieve, Hermann. 1947. "Die 'Versagerfrage' in der Lehre Knaus: Eine Richtigstellung zum Aufsatz von H.J. Gerster." *Schweizer Medizinische Wochenschrift* 77, 782–83.

Stoecker, Holger, Thomas Schnalke, and Andreas Winkelmann, Andreas, ed. 2013. *Sammeln, erforschen, zurückgeben? Menschliche Gebeine aus der Kolonialzeit in akademischen und musealen Sammlungen.* Berlin: Ch. Links Verlag.

Stöhr, Philipp. 1935. "Zur Innervation der menschlichen Nebenniere." *Zeitschrift für Anatomie und Entwicklungsgeschichte* 104(5): 475–90.

Sukol, Roxanne B. 1995. "Building on a Tradition of Ethical Consideration of the Dead." *Human Pathology* 26: 700–705.

Swick, Herbert M. 2006. "Medical Professionalism and the Clinical Anatomist." *Clinical Anatomy* 19: 393–402.

Talarico, Ernest F. 2013. "A Change in Paradigm: Giving Identity to Donors in the Anatomy Laboratory." *Clinical Anatomy* 26: 161–72

Talarico, Ernest F., and Andrew Prather. 2007. "A Piece of My Mind: Connecting the Dots to Make a Difference." *Journal of the American Medical Association* 298: 381–82.

Terry, Michael. 2014. "Dear Joseph." *Pulse: Voices from the Heart of Medicine.* 21 February 2014. Accessed 21 August 2014. http://pulsevoices.org/archive/stories/352-dear-joseph.

Trotman, Paul. 2009. *Donated to Science* (DVD). Dunedin, New Zealand: Paul Trotman Films.

Warner, John Harley. 2009. "Witnessing Dissection: Photography, Medicine and American Culture." In: *Dissection: Photographs of a Rite of Passage in American Medicine 1880–1930,* ed. John Harley Warner, and James M. Edmonson, 7–29. New York: Blast Books.

Warner, John Harley, and Lawrence J. Rizzol. 2006. "Anatomical Instruction and Training for Professionalism from the 19th to the 21st Centuries." *Clinical Anatomy* 19: 403–14.

Warren, John. 1783–1785. *Lectures upon Anatomy: Manuscript 1783–1785.* Boston: Countway Library of Medicine Collections.

Wendehorst, Alfred.1993. *Geschichte der Friedrich-Alexander-Universität Erlangen-Nürnberg 1743–1993.* München: Verlag C. H. Beck.

Wetzel, Robert. 1940. "Theorie und Wissenschaft im ärztlichen Beruf." *Deutschlands Erneuerung* 24(1): 12–17.

Wilkinson, Timothy M. 2014a. "Respect for the Dead and the Ethics of Anatomy." *Clinical Anatomy* 27: 286–90.

———. 2014b. "Getting Consent into Perspective." *Clinical Anatomy* 27: 844–46.

Winkelmann, Andreas. 2007a. "Die menschliche Leiche in der heutigen Anatomie." In: *Grenzen des Lebens. Beiträge aus dem Institut Mensch, Ethik und Wissenschaft.* Band 5, ed. Sigrid Graumann, and Katrin Grüber, Katrin, 62–74. Berlin: LIT Verlag Dr. W. Hopf.

———. 2007b. "Wilhelm von Waldeyer-Hartz (1836–1921): An Anatomist Who Left His Mark." *Clinical Anatomy* 20: 231–34.

———. 2015 "The Anatomische Gesellschaft and National Socialism: An Analysis Based on Newly Available Archival Material." *Annals of Anatomy* 201: 17–30.

Winkelmann, Andreas, and Fritz H. Güldner. 2004. "Cadavers as Teachers: The Dissecting Room Experience in Thailand." *British Medical Journal* 329: 1455–57.

Winkelmann, Andreas, and Thorsten Noack. 2010. "The Clara Cell: A 'Third Reich Eponym'?" *European Respiratory Journal* 36: 722–27.

Zhang, Luqing, Ming Xiao, Mufeng Gu, Yongyie Zhang, Yongjie, Jianliang Jin, and Jiong Ding. 2014. "An Overview of the Roles and Responsibilities of Chinese Medical Colleges in Body Donation Programs." *Anatomical Sciences Education* 7: 312–20.

Chapter 11

ANATOMY—ON THE EDGE OF CULTURE

Is evil insanity? Or a really unpleasant version of sanity?

Laura Lippman[1]

Nay, be what thou wilt; but I will bury him.

Antigone[2]

Gareth Jones, a pioneer of ethics in anatomy, stated that "dissection and autopsies exist on the edge of the cultures that allow them, walking a fine line between acceptance and repulsion" and there "are limits on what a culture will accept."[3] Unfortunately, there were hardly any limits on what the National Socialist regime allowed scientists to do with the living and dead bodies of those not considered part of the "body of the German people." Looking at the history of anatomy in National Socialism, it becomes evident that evil is "a really unpleasant version of sanity."[4] Medicine and anatomy did not "go mad"[5] and German society did not fall into temporary insanity. No such easy excuse exists. Rather, as recent historical research and this investigation of anatomy reveal, mechanisms within medical science itself and its practitioners colluded with a criminal political system in creating the all-encompassing destruction wreaked by National Socialism.

While many German anatomists collaborated with the regime—and most biographies still need to be investigated further—others suffered disruptions of their careers for "racial" or political reasons. Some of the persecuted scholars of anatomy were just starting out in their professional lives; others were at the height of their skills; still others were looking back upon productive careers in anatomy. All of them suffered the humiliation and frustration that accompany the thwarting of one's life plans. And worse, some were imprisoned and murdered by the NS regime. In order to augment the biographies that have already been written and compose those that have not, further work, especially exploration of potential archival sources, is necessary. A detailed recounting of each individual life is called for "as though they were of your own family, or even you yourself."[6]

Anatomy and the Destruction of Memory

What happened in German execution chambers during the Third Reich was horrific, and the exploitation of the newly dead by anatomists was shameful and without excuse. German anatomists' familiarity with the use of bodies of the executed before 1933 may have contributed to the ease with which they accepted the "opportunities" presented to them by the nearly unlimited access to these bodies provided by the NS regime. The few anatomists who ventured into work with the "future dead," thus introducing a new and vicious paradigm in anatomy, never questioned the legality or ethics of their practice. German postwar anatomy returned to the original paradigm, but was built in part on the bodies of NS victims. All anatomists who had worked in German anatomy during the war and represented the discipline thereafter had been complicit in this appalling work, independent of their own political convictions. This personal involvement explains in part German anatomy's long resistance against a public evaluation of its history during National Socialism and to the general silence on the topic.[7]

Anatomists assisted the NS regime's goal to eradicate all those perceived to be harming the German people. Not only were their lives to be taken, but their bodies were to be utterly destroyed. This complete annihilation included the delivery of victims' bodies to the anatomical institutes and many other places of medical activity, and ended with the murder and industrial use of the bodies of millions of Holocaust victims in the concentration camps.[8] And whereas the anatomists' complicity remained mostly unmentioned in the postwar world, the sensational press reports of the despicable use of prisoners' skin as accessories for SS officers became one of the most haunting symbols of the evils perpetrated by National Socialists on millions of victims, especially the Jews.[9] Their lives, their bodies, and their names were to be entirely expunged and forgotten.

That was the plan. Instead, German National Socialism was defeated in 1945, and many forces since then have actively retrieved the memory of all the victims, from Yad Vashem in Israel to the ever-increasing number of historic memorial sites in Germany and other European countries, and to Holocaust Memorial Museums in the rest of the world. Anatomy has only recently started to recognize its duty to aid in restoring the memory and biographies of its victims. The most appropriate way to honor them lies in accurately retelling their history, and to continue work on ethical guidelines and their practice in modern anatomy and medicine in general. The callous disrespect shown to the last wishes of NS victims facing death should become a reminder to every physician and health worker to listen better to each patient.

A Professional Ethics of Care

While National Socialism led to specific forms of transgressions within anatomy, the underlying forces behind these transgressions are not specific to the Third Reich, but are still active in modern anatomy and other areas of medicine. As the scientific method remains central to new medical insights, medicine's focus, and that includes anatomy, has to reintegrate scientific inquiry with the ethical and political dimensions of human nature and care for the individual. Physicians and those whose research includes living and dead human beings have to reconsider their personal responsibility for the well-being of those who are not only their "research subjects" but are also "in their care," that is, "to be cared for." Marcum recognized this need for medicine to change and identified "two humanistic values that inform the ethical and moral stance or attitude of physicians—emotionally detached concern and empathetic care. [...] For the humanistic or humane practitioner, however, scientific medicine is embedded with empathetic care that includes the patient's and the physician's emotional state."[10]

Anatomy is uniquely suited to become a model for medical education, to foster the socialization of physicians in the professional role of the humane practice of medicine. Just as Marcum identifies gross anatomy as "one of the first steps towards detachment,"[11] anatomy can also become the first step toward a humane approach in medical education. The discipline has at its heart the balance of clinical detachment and empathy and takes into account the care for the person whose body is being investigated as well as the emotions of the dissector. Future physicians can learn to acknowledge the development of clinical detachment and to balance it with empathy, to become caring practitioners of medicine. Christine Montross formulated this potential learning outcome:

> The emotional challenge of anatomy class might be partially intended to prepare classmates and me for the rigors of tending patients whose bodies are sick and maimed. Yet I also believe that the lesson of anatomy is that we do not need to overcome all of our emotion or conquer all difficulty in order to be good clinicians. In fact, in light of the important balance that clinical detachment requires, I should perhaps feel encouraged by my inability to always emotionally disengage.[12]

A new professional ethics of care in anatomy addresses both, the dissector and the dissected, because they are connected.[13] The bodies of the dead matter to the living, even those who died many years ago. And if Antigone was willing to give her life for the proper burial of her brother, then contemporary anatomy must at least make an effort to commemorate the dissected victims of National Socialism in a dignified manner. Further work is necessary to identify these persons—their stories must be told. If, according to Timothy Snyder, the National Socialists turned people into numbers, then German anatomists of the time turned them into tissues and cells: "It is for us scholars, to seek these numbers [tissues and cells] and put them into perspective. It is for us humanists to turn the numbers [tissues and cells] back into people."[14]

The history of anatomy in National Socialism also holds another lesson: it illustrates the need to question societal expectations, as well as institutional and governmental authorities, at any given time. Practices in medicine do not become ethical simply by virtue of being legal or accepted by current society. Bodies of the executed and those that go unclaimed are still legally used for anatomical purposes in many countries around the world, and physicians are officially involved in executions and torture throughout various political systems.[15] The benefit for the individual must remain at the center of medical ethics, not the potential benefit for the society as a whole. In that respect the medical practitioner will always have to take a political stance. The "apolitical" attitude used so often as a defense by physicians who remained in Germany during National Socialism does not exist.[16] When societal taboos are challenged by medicine, scientists need to discourse with the public. Such an open dialogue is essential for the prevention of similar iniquities.[17]

Apart from the convicted murderers Kremer and Hirt, the story of anatomy in the Third Reich has many shades of gray. Anatomists became guilty of ethical transgressions to varying degrees. And even those who opposed the NS regime with all their hearts, if not actions, became collaborators of the authorities in using the bodies of the victims for anatomical work. Therein lies another lesson from this history: "It is the fateful character of our human existence, that even the subjectively honest human being, who acts with the best intention, can become objectively guilty."[18] This is no excuse, but a reminder for the need to stay vigilant.

"The Deeds of Your Parents Cannot be Forgotten"

On returning to Bergen-Belsen fifty years after his liberation from the concentration camp as a child, Rabbi Joseph Polak told his German hosts,

"The deeds of your parents cannot be forgotten [...] you are doomed to be their representatives, and your hands will be stained with blood that you yourselves may not have spilled."[19] Against this background, the words "remembrance" and "commemoration" take on a double meaning. While the victims and survivors of this history have all too often been forgotten by the rest of the world, they themselves cannot forget what happened to them. Even if they desperately want to do so, they are forced to remember, along with their children and grandchildren.[20] Both the "forgetting" and the "not being able to forget" are decisive reasons for an effective and lasting commemoration of the victims of National Socialism, including those whose bodies were used for anatomical purposes. Commemoration is a duty first and foremost for the descendants of the perpetrators—in other words, the current German medical professionals—but ultimately also a privilege and obligation for the whole world community of medicine.

Notes

1. Lippman 2009, 157.
2. Sophocles 442 BCE.
3. Jones 1998, 101–2.
4. Lippman 2009.
5. Proctor 2000; Caplan 1992.
6. Fuchik 1949.
7. Seidelman 2012.
8. E.g., Kühl 2009.
9. Jacobson 2010; Neander 2009.
10. Marcum 2008, 259.
11. Ibid., 262.
12. Montross 2007, 287.
13. On a philosophy of connectedness see Loewy 1997.
14. Snyder 2010, 408.
15. E.g., Annas 2005; Grodin et al. 2013; Gutmann 2014, 287–308.
16. Schagen 2010, .
17. Prüll 2000.
18. Graus 1969.
19. Polak 1995, 26.
20. Kellermann 2009.

Bibliography

Annas, George J. D. 2005. "Unspeakably Cruel: Torture, Medical Ethics, and the Law." *New England Journal of Medicine* 352(20): 2127–32.
Caplan, Arthur L, ed. 1992. *When Medicine Went Mad: Bioethics and the Holocaust.* Totowa, NJ: Humana Press.
Fuchik, Julius. 1949. *Notes from the Gallows.* Salt Lake City: Peregrine Smith Books.

Graus, František. 1969. "Geschichtsschreibung und Nationalsozialismus." *Vierteljahreshefte für Zeitgeschichte* 17(1): 87–95.

Grodin, Michael A., Daniel Tarantola, George J. Annas, and Sofia Gruskin. 2013. *Health and Human Rights in a Changing World*. New York: Routledge.

Gutmann, Ethan. 2014. *The Slaughter: Mass Killings, Organ Harvesting, and China's Secret Solution to Its Dissident Problem*. Amherst, MA, and New York: Prometheus Book.

Jacobson, Mark. 2010. *The Lampshade: A Holocaust Detective Story from Buchenwald to New Orleans*. New York: Simon and Schuster.

Jones, David G. 1998. "Anatomy and Ethics: An Exploration of Some Ethical Dimensions of Contemporary Anatomy." *Clinical Anatomy* 11: 100–105.

Kellermann, Natan P. F. 2009. *Holocaust Trauma: Psychological Effects and Treatment*. New York and Bloomington, IN: iUniverse.

Kühl, Richard. 2009. "Haare, Zähne, Lampenschirme: Die Ausbeutung und 'Verwertung' von Häftlingsleichen in den nationalsozialistischen Konzentrations- und Vernichtungslagern." In: *Die dienstbare Leiche. Der tote Körper als medizinische, soziokulturelle und ökonomische Ressource*, ed. Domink Gross, 85–92. Kassel: Kassel University Press.

Lippman, Laura. 2009. *Life Sentences*. New York: HarperCollins Publishers.

Loewy, Erich H. 1997. *Moral Strangers, Moral Acquaintance, and Moral Friends*. Albany: State University of New York Press.

Marcum, James A. 2008. *Humanizing Modern Medicine: An Introductory Philosophy of Medicine*. Dordrecht: Springer Science + Business Media B.V.

Montross, Christine. 2007. *Body of Work: Meditations on Mortality from the Human Anatomy Lab*. New York: Penguin Press.

Neander, Joachim. 2009. "A Strange Witness to Dachau Human Skin Atrocities: Anton Pacholegg a.k.a. Anton Baron von Guttenberg a.k.a. Antoine Charles de Guttenberg." *theologie.geschichte.de Band 4*. Accessed 29 May 2014. http://universaar.uni-saarland.de/journals/index.php/tg/article/viewArticle/472/511#83.

Polak, Joseph A. 1995. "The Lost Transport." *Commentary* 100(3): 24–27.

Proctor, Robert N. 2000. "Nazi Science and Nazi Medical Ethics: Some Myths and Misconceptions." *Perspectives in Biology and Medicine* 43(3): 335–46.

Prüll, Cay-Rüdiger. 2000. "Der Umgang mit der menschlichen Leiche in der Medizin: die historische Perspektive." In: *Zum Umgang mit der menschlichen Leiche in der Medizin*, ed. Hans-Konrat Wellmer and Gisela Bockenheimer-Lucius, 59–69. Lübeck: Schmidt-Römhild.

Schagen, Udo. 2010. "Walter Stoeckel (1871–1961) als (un)politischer Lehrer: Kaiser der deutschen Gynäkologen." In: *Geschichte der Berliner Universitäts-Frauenklinik: Strukturen, Personen und Ereignisse in und ausserhalb der Charité*, ed. Matthias David and Andreas Ebert, 200–218. Berlin: de Gruyter.

Seidelman, William E. 2012. "Dissecting the History of Anatomy in the Third Reich—1989–2010: A Personal Account." *Annals of Anatomy* 194: 228–36.

Snyder, Timothy. 2010. *Bloodlands: Europe between Hitler and Stalin*. New York: Basic Books.

Sophocles. 442 BCE. "Antigone." Accessed 18 March 2015, http://classics.mit.edu/Sophocles/antigone.html.

Appendix

Table 1. Political Affiliations of German Anatomists, 1933–45

Name	Position	University	NSDAP	SA	SS	Other
Allmer, Konrad[1]	assistant	Wien	Yes, 1932			
Altschul, Rudolf[2]	assistant	Prague	D			
Alverdes, Kurt[3]	professor	Greifswald, Halle, Jena	Yes, 1937			Yes
Auge, Helmut[4]	assistant	Frankfurt	?			
Aunap, E[5]	professor	Dorpat/Tartu	?			
Bachmann, Rudolf[6]	assistant	Leipzig	Yes, 1937	Yes		
Bargmann, Wolfgang[7]	assistant	Frankfurt, Leipzig, Königsberg	Yes, 1933			Yes
Barth, Karl-Heinz[8]	assistant	Leipzig	?			
Bauer, Franz[9]	assistant	Vienna	?			
Bauer, Karl Friedrich[10]	assistant	Berlin, Munich	Yes?	Yes?		
Bautzmann, Hermann[11]	assistant	Hamburg, Kiel	?			
Becher, Hellmut[12]	chairman	Giessen, Marburg, Münster	Yes, 1937	Yes	Yes	Yes
Benninghoff, Alfred[13]	chairman	Kiel, Marburg	Yes, 1941			
Berek, Kurt[14]	assistant	Prague	?			
Berg, Walter[15]	chairman	Königsberg	D			
Bergmann, Louis[16]	assistant	Vienna	D			
Biermann, Olga[17]	assistant	Greifswald	Yes, 1937			

Name	Position	University	NSDAP	SA	SS	Other
Blechschmidt, Erich[18]	chairman	Freiburg, Giessen, Göttingen	Yes, 1937	Yes		
Blotevogel, Wilhelm[19]	chairman	Breslau	Yes, 1933	Yes		Yes
Blume, Werner[20]	assistant	Göttingen	Yes, 1923		Yes	Yes
Bluntschli, Hans[21]	chairman	Frankfurt	D			
Boenig, Horst[22]	assistant	Berlin	Yes			Yes
Boerner-Patzelt, Dora[23]	professor	Graz	Yes, 1939			
Böker, Hans[24]	chairman	Jena, Cologne	Yes, 1934	Yes		Yes
Bracco, Ernst[25]	assistant	Prague	?			
Brandt, Walter[26]	assistant	Cologne	D			
Brockmann, August W.[27]	assistant	Freiburg	Yes, 1930	·	Yes	
Brodersen, Johannes[28]	professor	Hamburg	D			
Bulowa, Johanna[29]	assistant	Vienna	D?			
Burian, Ernst[30]	assistant	Vienna	D?			
Clara, Max[31]	chairman	Leipzig, Munich	Yes, 1935			Yes
Dabelow, Adolf[32]	chairman	Leipzig, Marburg, Munich	Yes, 1937	Yes		
Donalieb, Gerhard[33]	assistant	Berlin	?			
Dragendorff, Otto[34]	professor	Greifswald	No			Yes
Drechsel, Josef[35]	assistant	Vienna	Yes, 1932			
Ebert, Hermann[36]	assistant	Greifswald	No	Yes		
Ehmann, Friedrich[37]	Kustos Museum	Vienna	Yes, 1940			Yes
Eggeling, Heinrich von[38]	chairman	Breslau	No			
Elze, Curt[39]	chairman	Giessen, Rostock, Würzburg	Yes, 1940			Yes
Fahrenholz, Curt Otto[40]	professor	Berlin, Leipzig	?			Yes
Feneis, Heinz[41]	assistant	Marburg, Munich, Tübingen	Yes			
Ferber, Adele[42]	assistant	Graz	Yes, 1939			

Name	Position	University	NSDAP	SA	SS	Other
Ferner, Helmut[43]	assistant	Leipzig, Munich, Prague	Yes			
Feustel, Robert[44]	assistant	Munich	?			
Fick, Rudolf[45]	chairman	Berlin	?			
Fischel, Alfred[46]	Chairman histology	Vienna	D			
Fleischer, Horst[47]	assistant	Greifswald	No			
Freerksen, Enno[48]	chairman	Giessen, Kiel, Marburg, Rostock	Yes, 1932	Yes	Yes	Yes
Friedel, Arthur[49]	assistant	Berlin	?			
Froboese, Hans[50]	assistant	Halle	Yes, 1933	Yes		Yes
Fuchs, Hugo[51]	chairman	Göttingen	Yes, 1933			
Gajsek, Alfred[52]	assistant	Graz	Yes, 1933	Yes		
Gartler, Erich H.[53]	assistant	Graz	Yes, 1938	Yes		
Gehlen, Hans von[54]	assistant	Heidelberg, Kiel	?			
Gieseler, Wilhelm[55]	assistant	Tübingen	Yes, 1933	Yes	Yes	
Gieschen, Karl-Ludwig[56]	assistant	Freiburg	D			
Gisel, Alfred[57]	assistant	Vienna	No			
Glees, Paul[58]	assistant	Bonn	D			
Gluecksmann, Alfred[59]	assistant	Heidelberg	D			
Goldhamer, Karl[60]	assistant	Vienna	D			
Göppert, Ernst[61]	chairman	Marburg	No			
Goerttler, Kurt[62]	chairman	Hamburg, Heidelberg	Yes			
Graeper, Ludwig[63]	chairman	Jena	?	Yes		
Grosser, Otto[64]	chairman	Prague	Yes, 1939			
Groth, Werner[65]	assistant	Greifswald	Yes			
Grueneberg, Hans[66]	assistant	Freiburg	D			
Gruenwald, Peter[67]	assistant	Vienna	D			
Hafferl, Anton[68]	chairman	Graz	Yes, 1938			Yes

Name	Position	University	NSDAP	SA	SS	Other
Hagen, Emmi AK[69]	assistant	Bonn	?			Yes
Haller von Hallerstein, Viktor[70]	chairman	Berlin, Halle	Yes, 1932			
Hasselwander, Albert[71]	chairman	Erlangen	?			
Harting, Kurt EJ[72]	assistant	Bonn	Yes, 1933	Yes		Yes
Hartmann, Adele[73]	assistant	Munich	No			
Hayek, Heinrich von[74]	assistant	Rostock, Shanghai, Würzburg	Yes, 1938	Yes		Yes
Hausknecht, Werner[75]	assistant	Berlin	?			
Heidenhain, Martin[76]	chairman	Tübingen	D?			
Heiderich, Friedrich[77]	chairman	Münster	Yes			
Heidsieck, Erich[78]	assistant	Breslau	?			
Heiss, Robert[79]	chairman	Königsberg	Yes			
Held, Hans[80]	chairman	Leipzig	?			
Henneberg, Bruno[81]	chairman	Giessen	?			
Herrath, Ernst von[82]	assistant	Berlin, Bonn, Giessen	Yes			
Herrlinger, Robert[83]	assistant	Posen	Yes, 1937	Yes		Yes
Hertwig, Günther[84]	assistant	Berlin, Halle, Rostock	D			
Hett, Johannes[85]	professor	Erlangen, Halle	Yes, 1933			Yes
Hirt, August[86]	chairman	Frankfurt, Greifswald, Strasbourg	Yes, 1937		Yes	Yes
Hoepke, Hermann[87]	professor	Heidelberg	D			
Hoffmann, Auguste[88]	assistant	Berlin	?			Yes
Horstmann, Ernst[89]	assistant	Heidelberg	?	Yes		
Jacobj, Walter[90]	professor	Tübingen	No			

Name	Position	University	NSDAP	SA	SS	Other
Jacobson, Werner[91]	assistant	Bonn	D			
Jakobshagen, Eduard[92]	professor	Marburg	Yes			Yes
Jeschek, Wolf[93]	assistant	Graz	Yes, 1934			
Kadanoff, Dimitri[94]	assistant	Würzburg	D			
Kallius, Erich[95]	chairman	Heidelberg	?			
Keller, Ludwig[96]	assistant	Freiburg	Yes, 1937	Yes		
Kemme, Alfred[97]	assistant	Berlin	?			
Kempermann, C. Th.[98]	assistant	Cologne	D			
Kiesselbach, Anton[99]	assistant	Frankfurt, Greifswald, Strasbourg	Yes	Yes		
Kirchner, Egon[100]	assistant	Graz	Yes, 1938			Yes
Kleinschmidt, Adolf[101]	assistant	Berlin	?			
Körner, Fritz[102]	assistant	Jena	Yes			Yes
Kohn, Alfred[103]	chairman histology	Prague	D			
Kopsch, Friedrich[104]	professor	Berlin	?			
Kossmann, Hans Leo[105]	assistant	Berlin	?			
Krantz, Hilde[106]	assistant	Marburg	?			
Krause, Rudolf[107]	chairman	Berlin	?			
Krause, Walter[108]	assistant	Vienna	D			
Kremer, Johann Paul[109]	professor	Münster	Yes, 1932		Yes	
Kügelgen, Alkmar von[110]	assistant	Heidelberg, Kiel	?			Yes
Kugler, Hans[111]	assistant	Vienna	?			
Kühtz, Helmuth[112]	assistant	Frankfurt, Greifswald	No			
Kuhlenbeck, Hartwig[113]	assistant	Breslau	D			
Kull, Harry[114]	professor	Dorpat	?			

Name	Position	University	NSDAP	SA	SS	Other
Kurz, Eugen[115]	professor	Münster	Yes, 1933			
Lange, Bernhard[116]	assistant	Breslau	?			
Langegger, Paul[117]	assistant	Vienna	?			
Lanz, Titus von[118]	professor	Munich	D			
Lembach, ?[119]	assistant	Cologne	?			
Loewy, Adolf[120]	professor emeritus	Berlin	D			
Loidl, Ladislaus[121]	assistant	Graz	Yes, 1936	Yes		
Lohmann, Heinz[122]	assistant	Breslau, Hamburg	Yes?			Yes
Lubosch, Wilhelm[123]	professor	Würzburg	D			
Maerk, Walter[124]	assistant	Innsbruck	Yes, 1932	Yes		
Marcus, Gaston[125]	assistant	Vienna	D			
Marcus, Harry[127]	professor	Munich	D			
Mair, Rudolf[126]	assistant	Halle, Innsbruck	Yes, 1937	Yes		Yes
Mathis, Juerg[128]	chairman	Innsbruck histology	Yes, 1933	Yes		
Mattuschka, Josef[129]	assistant	Vienna	Yes	Yes		Yes
Möllendorff, Wilhelm von[130]	chairman	Freiburg, Zurich	D			
Mollier, Georg[131]	assistant	Heidelberg	Yes?	Yes		
Mollier, Siegfried[132]	chairman	Munich	?			
Moser, Alois[133]	assistant	Graz	Yes, 1938			Yes
Münter, Heinrich[134]	assistant	Heidelberg	D			
Nagel, Arno[135]	chairman	Halle	Yes, 1933			Yes
Nauck, Ernst Th.[136]	chairman	Freiburg, Marburg	Yes, 1933		Yes	Yes
Neubert, Kurt KF[137]	chairman	Rostock, Würzburg	Yes, 1933	Yes		Yes
Niessing, Klaus[138]	assistant	Kiel, Leipzig	Yes, 1937	Yes		Yes

Name	Position	University	NSDAP	SA	SS	Other
Novotny, Otto[139]	assistant	Vienna	?			
Oertel, Otto[140]	chairman	Tübingen	?			
Ortmann, Rolf[141]	assistant	Cologne	?			
Palugyay, Josef[142]	assistant	Vienna	D			
Passarge, Edgar[143]		Rostock	?			
Patzelt, Viktor[144]	chairman	Vienna (Hist)	Yes, 1944			
Pernkopf, Eduard[145]	chairman	Vienna	Yes, 1933	Yes		
Peter, Karl[146]	chairman	Greifswald	No			
Petersen, Hans[147]	chairman	Würzburg	Yes, 1938			
Petry, Gerhard[148]	assistant	Halle	Yes, 1937	Yes		
Pfuhl, Wilhelm[149]	chairman	Frankfurt, Greifswald	Yes, 1933	Yes		
Pichler, Alexander[150]	professor	Vienna	Yes		Yes	Yes
Pick, Joseph[151]	assistant	Vienna	D			
Pischinger, Alfred[152]	chairman	Graz histology	Yes, 1933	Yes		
Plenk, Hanns[153]	assistant	Vienna	?			
Politzer, Georg[154]	assistant	Vienna	D			
Poll, Heinrich[155]	chairman	Hamburg	D			
Pommer, Charlotte[156]	assistant	Berlin	D			
Ponhold, Johann A.[157]	assistant	Graz	Yes, 1938			
Popp, Otto[158]	assistant	Berlin, Greifswald	Yes, 1939			
Pratje, Andreas[159]	assistant	Erlangen	Yes, 1935	Yes		Yes
Quast, Paul[160]	assistant	Bonn	D			
Rabl, Hans[161]	chairman	Graz histology	?			
Raffler, Karl[162]	assistant	Graz	Yes			
Reibmayr, Ilse[163]	assistant	Graz	Yes, 1936			
Reiter, Alfred[164]	assistant	Prague	?		Yes	Yes
Rhein, Georg[165]	assistant	Greifswald	No			

Name	Position	University	NSDAP	SA	SS	Other
Rintelen, Paul W.[166]	assistant	Berlin	?			
Rintelen, Walter[167]	assistant	Berlin	?			
Roethig, Paul[168]	assistant	Berlin	D			
Rolshoven, Ernst[169]	assistant	Greifswald, Marburg, Muenster	Yes, 1933	Yes		Yes
Romeis, Benno[170]	assistant	Munich	No			
Saller, Karl[171]	professor	Göttingen	D			
Sauser, Gustav[172]	chairman	Innsbruck, Vienna histology	D			
Schäuble, Johann[173]	professor	Freiburg	Yes, 1937	Yes		
Scheffler, A.[175]	assistant	Königsberg	?			
Schmeidel, Gustav[176]	assistant	Vienna	Yes			
Schneider, Hans[177]	assistant	Innsbruck	?			
Schreiber, Hans[178]	chairman	Frankfurt	Yes, 1933	Yes		
Schumacher, Sigmund[179]	chairman	Innsbruck histology	?			
Schwarz-Kasten, H. von[174]	assistant	Graz	Yes, 1934			Yes
Seemann, Fritz[180]	assistant	Graz	Yes, 1939	Yes		
Sekol, Egon[181]	assistant	Graz	?	Yes		
Sicher, Harry[182]	assistant	Vienna	D			
Sieglbauer, Felix[183]	chairman	Innsbruck	Yes, 1939			
Simon, Kurt[184]	assistant	Berlin	?			
Simon, St.[185]	assistant	Vienna	?			
Singer, Rudolf[186]	assistant	Vienna	?			
Sobotta, Johannes[187]	chair	Bonn	No			Yes
Sommer, Alfred[188]	professor	Dorpat	?			
Sommer, Hermann[189]	assistant	Greifswald	Yes, 1931		Yes	Yes

Name	Position	University	NSDAP	SA	SS	Other
Spängler, Hans[190]	assistant	Vienna	Yes, 1940			
Spanner, Rudolf M.[191]	chairman	Danzig, Jena, Kiel	Yes, 1936	Yes		Yes
Spitzer, Alexander[192]	assistant	Vienna	D			
Spuler, Arnold[193]	assistant	Erlangen	?			
Stadtmüller, Franz[194]	chairman	Göttingen, Cologne	Yes, 1933			Yes
Stieve, Hermann[195]	chairman	Berlin, Halle	No			Yes
Starck, Dietrich[196]	assistant	Cologne	Yes, 1939			
Stöhr Jr., Philipp[197]	chairman	Bonn	No			Yes
Strauss, Fritz[198]	assistant	Freiburg	D			
Strecker, Friedrich[199]	assistant	Breslau	D			
Tonutti, Emil[200]	assistant	Breslau	Yes, 1937			
Veit, Otto[201]	chairman	Cologne	D			
Voit, Max[202]	professor	Göttingen	No			Yes
Vogt, Walther[203]	chairman	Munich	No			
Volkmann, Rüdiger von[204]	chairman	Jena, Würzburg	Yes			
Voss, Hermann[205]	chairman	Leipzig, Posen	Yes, 1937			
Wagenseil, Ferdinand[206]	chairman	Bonn, Giessen	No			Yes
Waldeyer, Anton[207]	professor	Berlin	Yes, 1934			
Walla, Maximilian[208]	assistant	Greifswald	No	Yes		
Wallraff, Joseph[209]	assistant	Breslau	?			
Wassermann, Fried.[210]	professor	Munich	D			
Watzka, Maximilian[211]	chairman	Prague (Hist)	Yes, 1938		Yes	
Wegner, Richard[212]	assistant	Frankfurt	D?			
Weichherz, Edmund[213]	assistant	Prague	D			

Name	Position	University	NSDAP	SA	SS	Other
Weidenreich, Franz[214]	chair	Heidelberg, Frankfurt	D			
Weinberg, Ernst[215]	professor	Dorpat	?			
Weiss, Harry[216]	assistant	Vienna	D?			
Weissberg, Harry[217]	assistant	Cologne	D			
Weissenberg, Richard[218]	professor	Berlin	D			
Wetzel, Georg[219]	professor	Greifswald	Yes, 1938			Yes
Wetzel, Robert[220]	chairman	Giessen, Würzburg, Tübingen	Yes	Yes		
Weuste, ?[221]	assistant	Cologne	?			
Wiesner, Elisabeth[222]	assistant	Vienna	?			
Wimmer, Karl[223]	assistant	Frankfurt, Strasbourg, Greifswald	Yes	Yes		
Winterstein, Joseph[224]	assistant	Halle	?			
Wirtinger, Wilhelm[225]	assistant	Vienna	Yes, 1938			
Wodtke, Jesko[226]	assistant	Berlin	?			
Wolf-Heidegger, G.[227]	assistant	Bonn	D			
Wolf, Horst[228]	assistant	Greifswald	Yes		Yes	
Wollmann-Kozlik, Fritz[229]	assistant	Danzig	Yes, 1938		Yes	Yes
Zawisch-Ossenitz, Carla[230]	assistant	Vienna	D			
Zeiger, Karl[231]	chairman	Frankfurt, Hamburg, Königsberg	Yes, 1933			Yes
Ziesche, K Th[232]	assistant	Königsberg	?			
Zitzlsperger, Sigfrid[233]	assistant	Kiel, Berlin	?			

Legend

Yes?: membership has to be assumed given the information in source
histology: institute for histology

Archival Sources

Universitätsarchiv Freiburg:
- UAF B24/397
- UAF B24/3200
- UAF B24/1647
Universitätsarchiv Marburg:
- Estate (*Nachlass*) Benninghoff

Bibliography

Alvermann, Dirk. 2015. "'Praktisch begraben'—NS-Opfer in der Greifswalder Anatomie 1935–1947." In: "*... die letzten Schranken fallen lassen": Studien zur Universität Greifswald im Nationalsozialismus*, ed. Dirk Alvermann, 311–51. Köln Weimar Wien: Böhlau Verlag.

Alvermann, Dirk, and Jan Mittenzwei. In press. "The Anatomical Institute at the University of Greifswald during National Socialism: The Procurement of Bodies and Their Use for Anatomical Purposes." *Annals of Anatomy*.

Aly, Götz. 1987. "Das Posener Tagebuch des Anatomen Hermann Voss." In: *Biedermann und Schreibtischtäter: Materialien zur deutschen Täter-Biographie*, ed. Götz Aly, Peter Chroust, and Christian Pross, 15–66. Berlin: Rotbuch Verlag.

Anatomischer Anzeiger. 1933. "Persönliches." *Anatomischer Anzeiger* 76(12–13): 240.

Angetter, Daniela. 1997a. "Krause Interview Version 10/29/1997." Transcript, document held at the Institut für Zeitgeschichte Wien.

———. 1997b. "Gisel Interview First Version, 7/22/1997." Transcript, document held at the Institut für Zeitgeschichte Wien.

———. 1997c. "Interview with Elisabeth Wiesner 11/28/1997." Transcript, document held at the Institut für Zeitgeschichte Wien.

Arias, Ingrid. 2004. "Entnazifizierung an der Wiener Medizinischen Fakultät: Bruch oder Kontinuität? Das Beispiel des Anatomischen Institutes." *Zeitgeschichte* 6(31): 339–69.

Aumüller, Gerhard, Kornelia Grundmann, Esther Krähwinkel, Hans H. Lauer, and Helmuth Remschmidt. 2001. *Die Marburger Medizinische Fakultät im "Dritten Reich."* München: K. G. Saur.

Bargmann, Wolfgang, and A. Scheffler. 1943. "Zur Frage der parthogenetischen Furchung menschlicher Ovarialeizellen." *Anatomischer Anzeiger* 94: 97–100.

Becker, Cornelia, Christine Feja, Wolfgang Schmidt, and Katharina Spanel-Borowski. 2005. *Das Institut für Anatomie in Leipzig. Eine Geschichte in Bildern.* Leipzig: Sax-Verlag.

Beushausen, Ulrich, Hans-Joachim Dahms, Thomas Koch, Almut Massing, and Konrad Obermann. 1998. "Die medizinische Fakultät im Dritten Reich." In: *Die Universität Göttingen unter dem Nationalsozialismus*, ed. Heinrich Becker, Hand Joachim Dahms, and Cornelia Wegeler, 183–286. München: K. G. Saur.

Brehm, Thomas, Horst-Werner Korf, Udo Benzenhöfer, Christof Schomerus, and Helmut Wicht. 2015. "Notes on the History of the Dr. Senckenbergische Anatomie in Frankfurt/Main. Part II. The Dr. Senckenbergische Anatomie during the Third Reich and its Body Supply." *Annals of Anatomy* 201:111-119.

Buddrus, Michael, and Sigrid Fritzlar. 2007. *Die Professoren der Universität Rostock im Dritten Reich. Ein biographisches Lexikon.* München: K. G. Saur.

Catalogus professorum halensis. 2015. "Hans Froboese." Accessed 18 March 2015. http://www.catalogus-professorum-halensis.de/froboesehans.html.

———. 2015. "Haller von Hallerstein." Accessed 18 March 2015. http://www.catalogus-professorum-halensis.de/indexb1933.html.

———. 2015. "Johannes Hett." Accessed 18 March 2015. http://www.catalogus-professorum-halensis.de/indexb1933.html.

————. 2015. "Rudolf Mair." Accessed 18 March 2015. http://www.catalogus-professo rum-halensis.de/indexb1933.html.

————. 2015. "Arno Nagel." Accessed 19 March 2015. http://www.catalogus-professo rum-halensis.de/indexb1933.html.

Drabek, Alexander. 1988. *Die Dr. Senckenbergische Anatomie von 1914 bis 1945*. Hildesheim: Georg Olms AG.

Drews, Ulrich. 1992. "Die Zeit des Nationalsozialismus am Anatomischen Institut in Tübingen: Unbeantwortete ethische Fragen damals und heute." In: *Menschenverachtung und Opportunismus, Tübingen: Zur Medizin im Dritten Reich*, ed. Jürgen Peiffer, 93–107. Tübingen: Attempto.

Eberle, Henrik. 2002. *Die Martin-Luther-Universität in der Zeit des Nationalsozialismus 1933–1945*. Halle (Saale): Mdv Mitteldeutscher Verlag.

Ebert, Christoph. 1971. *Die Personalbiographien der Ordinarien und Extraordinarien der Anatomie mit Histologie und Embryologie, der Physiologie und der Physiologischen Chemie an der Medizinischen Fakultät der Julius-Maximilians-Universität Würzburg*. Medizinische Dissertation, Erlangen-Nürnberg.

Eckart, Wolfgang Uwe. 2012. *Medizin in der NS-Diktatur. Ideologie, Praxis, Folgen*. Wien: Böhlau Verlag.

Eckart, Wolfgang Uwe, Volker Sellin, Eike Wolgast. 2006. *Die Universität Heidelberg im Nationalsozialismus*. Heidelberg: Springer.

Feja, Christine, and Ortrun Riha. 2014. *"Hier hilft der Tod dem Leben": Das Leipziger Institut für Anatomie und das Leichenwesen 1933–1989*. Aachen: Shaker Verlag.

Forsbach, Ralf. 2006. *Die Medizinische Fakultät der Universität Bonn im "Dritten Reich."* München: R. Oldenbourg Verlag.

Freie Universität Berlin. 2013. "Dokumentation: Ärztinnen im Kaiserreich. Hoffmann, Auguste." Accessed 18 March 2015. http://geschichte.charite.de/aerztinnen/HTML/rec00517c3.html.

Gisel, Alfred. 1988. "Julius Tandler." In: *Vertriebene Vernunft II: Emigration und Exil österreichischer Wissenschaft*, ed. Friedrich Stadler, 815–18. Wien: Verlag Jugend und Volk.

Glettler Monika, and Alena Mišková. 2001. *Prager Professoren 1938–1945. Zwischen Wissenschaft und Politik*. Essen: Klartext Verlag.

Grüttner, Michael. 2004. *Biographisches Lexikon zur nationalsozialistischen Wissenschaftspolitik*. Heidelberg: Synchron Wissenschaftsverlag der Autoren.

Heiber, Helmut. 1991. *Universität unterm Hakenkreuz. Teil I: Der Professor im Dritten Reich. Bilder aus der akademischen Provinz*. München: K. G. Saur.

Heindel, Wolfgang. 1971. "Personalbiographien von Professoren und Dozenten des Histologisch-Embryologischen Institutes der Universität Wien im ungefähren Zeitraum von 1848–1968." Medizinische Dissertation Erlangen-Nürnberg.

Hildebrandt, Sabine. 2012. "Anatomy in the Third Reich: Careers Disrupted by National Socialist Policies." *Annals of Anatomy* 184: 251–56.

————. 2013a. "Wolfgang Bargmann (1906–1978) and Heinrich von Hayek (1900–1969): Careers in Anatomy Continuing through German National Socialism to Postwar Leadership." *Annals of Anatomy* 195: 283–95.

————. 2013b. "Anatomische Gesellschaft from 1933 to 1950: A Professional Society under Political Strain—The Benninghoff Papers." *Annals of Anatomy* 195: 381–92.

————. 2013c. "The Case of Robert Herrlinger: A Unique Postwar Controversy on the Ethics of the Anatomical Use of Bodies of the Executed during National Socialism." *Annals of Anatomy* 195: 11–24.

Hlaváčková, Ludmila, and Petr Svobodny. 1998. *Biographisches Lexikon der Deutschen Medizinischen Fakultät in Prag 1883–1945*. Praha: Karolinum.

Humboldt Universität Berlin. 2015. "Auguste Hoffmann." Accessed 18 March 2015. http://www2.gender.hu-berlin.de/ausstellung/Infocomputer/Biographien/N_Hoffmann.html.

Kaiser, Stephanie. 2013. "Tradition or Change? Sources of Body Procurement for the Anatomical Institute of Cologne in the Third Reich." *Journal of Anatomy* 223: 410–18.
Klinische Wochenschrift. 1933. "Hochschulnachrichten." *Klinische Wochenschrift* 43: 1712.
Kreft, Gerald. 2008. "'…nunmehr judenfrei…'. Das Neurologische Institute 1933 bis 1945." In: *Frankfurter Wissenschaftler zwischen 1933 und 1945*, ed. Jörn Kobes and Jan-Otmar Hesse, 125–56. Göttingen: Wallstein Verlag.
Landgericht Münster. 1960. "Das Urteil gegen Dr. Johann Paul Kremer." In: *Justiz und NS-Verbrechen*. Band XVII. Accessed 16 December 2013. http://web.archive.org/web/20081207153240/http://www1.jur.uva.nl/junsv/Excerpts/Kremer.htm.
Lohff, Brigitte. 2005. "'… die Grundgedanken des Nationalsozialismus aufsaugen und verarbeiten'. Die politisch-ideologische Funktion der Medizinischen Fakultät der Christian-Albrechts-Universität zu Kiel 1933–1945." *Jahrbuch für Universitätsgeschichte* 08/05: 211–34
Míšková, Alena. 2007. Die Deutsche (Karls-) Universität vom Münchener Abkommen bis zum Ende des Zweiten Weltkrieges. Prag: Verlag Karolinum.
Mörike, Klaus D. 1988. *Geschichte der Tübinger Anatomie*. Tübingen: JCB Mohr (Paul Siebeck).
München Vorlesungs- und Personalverzeichnisse WS 1932/33–WS 1944/45.
Mussgnug, Dorothee. 1988. *Die vertriebenen Heidelberger Dozenten: Zur Geschichte der Ruprecht-Karls-Universität nach 1933*. Heidelberg: Carl Winter.
Noack, Thorsten. 2008. "Begehrte Leichen. Der Berliner Anatom Hermann Stieve (1886–1952) und die medizinische Verwertung Hingerichteter im Natinoalsozialismus." *Medizin, Gesellschaft und Geschichte. Jahrbuch des Instituts für Geschichte der Medizin der Robert Bosch Stiftung* 26: 9–35.
Oberkofler, Gerhard, and Peter Goller, eds. 1999. *Die medizinische Fakultät Innsbruck. Faschistische Realität (1938) und Kontinuität unter postfaschistischen Bedingungen (1945). Eine Dokumentation*. Innsbruck: Universitätsarchiv.
Oehler-Klein, Sigrid. 2007. *Die Medizinische Fakultät der Universität Giessen im Nationalsozialismus und in der Nachkriegszeit: Personen und Institutionen, Umbrüche und Kontinuitäten*. Stuttgart: Franz Steiner Verlag.
Ortmann. Rolf. 1986. *Die jüngere Geschichte des anatomischen Instituts der Universität Köln 1919–1984*. Köln: Böhlau Verlag.
Pátek Philipp. 2010. "Die Entwicklung der Anatomie in Düsseldorf: Von der Akademie für praktische Medizin zur Universität Düsseldorf." Dissertation, Medical Faculty University of Düsseldorf.
Peiffer, Jürgen. 1998. "Die Vertreibung deutscher Neuropathologen." *Nervenarzt* 69: 99–109.
Personal- und Vorlesungsverzeichnisse der Friedrich-Wilhelms Universität Berlin 1932/33–1944/45. Berlin: University Press.
Potthast, Thomas, and Uwe Hossfeld. 2010. "Vererbungs- und Entwicklungslehre in Zoologie, Botanik und Rassenkunde/Rassenbiologie: Zentrale Forschungsfelder der Biologie an der Universität Tübingen im Nationalsozialismus." In: *Die Universität Tübingen im Nationalsozialismus*, ed. Urban Wiesing, Klaus-Rainer Brintzinger, Bernd Grün, Horst Junginger, and Susanne Michl, 435–83. Stuttgart: Franz Steiner Verlag.
Professorenkatalog der Universität Leipzig. 2010. "Jakobshagen." Accessed 14 October 2010. http://www.uni-leipzig.de/unigeschichte/professorenkatalog/leipzig/jacobshagen_433.
———. 2013. "Fahrenholz." Accessed 29 May 2013. http://www.uni-leipzig.de/unigeschichte/professorenkatalog/leipzig/Fahrenholz_404.
———. 2015. "Adolf Dabelow." Accessed 18 March 2015. http://www.uni-leipzig.de/unigeschichte/professorenkatalog/leipzig/Dabelow_597/.
Scharer, Philipp. 2010. "Robert F. Wetzel (1898–1962)- Anatom, Urgeschichtsforscher, Nationalsozialist. Eine biographische Skizze." In: *Die Universität Tübingen im National-*

sozialismus, ed. Urban Wiesing, Karl-Rainer Brintzinger, Bernd Grün, Horst Junginger, and Susanne Michl, 809–31. Stuttgart: Franz Steiner Verlag.

Scheiblechner, Petra. 2002. *"…Politisch ist er einwandfrei…" Kurzbiographien der an der Medizinischen Fakultät der Universität Graz in der Zeit von 1938 bis 1945 tätigen Wissenschaftlerinnen.* Graz: Akademische Druck-u.Verlagsanstalt.

Schneider, Hans. 1944. "Zur Anatomie des Bewegungsapparates." *Zeitschrift für Anatomie und Entwicklungsgeschichte* 113(1): 187–203.

Schuhmacher, Gert-Horst, and Heinzgünther Wischhusen. 1970. *Anatomia Rostochiensis: Die Geschichte der Anatomie an der 550 Jahre alten Universität Rostock.* Berlin: Akademieverlag.

Schütz, Mathias, Jens Waschke, Georg Marckmann, and Florian Steger. 2013. "The Munich Anatomical Institute under National Socialism: First Results and Prospective Tasks of an Ongoing Research Project." *Annals of Anatomy* 195: 296–302.

Seidler, Eduard, and Karl-Heinz Leven. 2007. *Die Medizinische Fakultät der Alberts-Ludwigs-Universität Freiburg im Breisgau. Grundlagen und Entwicklungen.* Freiburg: Verlag Karl Alber.

Szabó, Aniko. 2000. *Vertreibung, Rückkehr, Wiedergutmachung: Göttinger Hochschullehrer im Schatten des Nationalsozialismus.* Göttingen: Wallstein Verlag.

Tartu Ülikooli Ajaloo Museum. 2013. "Anatoomikumide Narrative." Accessed 6 June 2013. http://www.ajaloomuuseum.ut.ee/752240.

Tomkiewicz, Monika, and Piotr Semków. 2013. *Soap from Human Fat: The Case of Professor Spanner.* Gdynia, Poland: Wydawnictwo Róża Wiatrów.

Ude-Koeller, Susanne, Wilfried Knauer, and Christoph Viebahn. 2012. "Anatomical Practice at Göttingen University since the Age of Enlightenment and the Fate of Victims from Wolfenbüttel Prison under Nazi Rule." *Annals of Anatomy* 194: 304–13.

Volbehr, Friedrich L. C., and Richard Weyl. 1956. *Professoren und Dozenten der Christian-Albrechts-Universität zu Kiel 1665–1954.* Kiel: Ferdinand Hirt.

Vorlesungs- und Personalverzeichnis Breslau 1933–1939. Breslau: University Press.

Voswinckel, Peter, ed. 2002. *Fischer, Isidor: Biographisches Lexikon der hervorragenden Ärzte der letzten fünfzig Jahre. Band III.* Hildesheim: Georg Olms Verlag.

Waibel, Harry. 2011. *Diener vieler Herren: Ehemalige NS-Funktionäre in der SBZ/DDR.* Frankfurt am Main: Peter-Lang.

Wechsler, Patrick. 2005. "La Faculté des Medicine de la 'Reichsuniversität Strassburg' (1941–1945) a l'heure nationale-socialiste." Dissertation. Faculté de Medicine de Strasbourg.

Wendehorst, Alfred. 1993. *Geschichte der Friedrich-Alexander-Universität Erlangen-Nürnberg 1743–1993.* München: Verlag C. H. Beck.

Wiederanders, Bernd, and Susanne Zimmermann. 2004. *Buch der Docenten der Medicinischen Fakultät zu Jena.* Golmsdorf b. Jena: Jenzig-Verlag Gabriele Köhler.

Wien: Öffentliche Vorlesungen an der Universität zu Wien. Vorlesungs- und Personalverzeichnisse 1932–1937/38.

Winkelmann, Andreas. 2015 "The Anatomische Gesellschaft and National Socialism: An Analysis Based on Newly Available Archival Material." *Annals of Anatomy* 201: 17–30.

Winkelmann, Andreas, and Thorsten Noack. 2010. "The Clara Cell: A 'Third Reich Eponym'?" *European Respiratory Journal* 36: 722–27.

Wittern, Renate. 1993. "Aus der Geschichte der medizinischen Fakultät." In: *250 Jahre Friedrich-Alexander-Universität Erlangen-Nürnberg*, ed. Henning Kössler, 315–420. Festschrift. Erlangen: Verlagsdruckerei Schmidt.

Ziesche, Karl T. 1943. "Zur Histologie des Tuber cinereum des Menschen." *Zeitschrift für Zellforschung* 33(1–2): 143–50.

Zimmermann, Susanne. 2000. *Die medizinische Fakultät der Universität Jena während der Zeit des Nationalsozialismus.* Berlin: Verlag für Wissenschaft und Bildung.

Zwilling, Thomas. 2004. "Leben und Werk des Anatomen Georg Wetzel 1871–1951." Medizinische Dissertation Greifswald.

Notes

1. Arias 2004.
2. Peiffer 1998.
3. Waibel, 2011.
4. Drabek 1988.
5. Klinische Wochenschrift 1933.
6. Feja and Riha 2014.
7. Hildebrandt 2013a.
8. Becker et al. 2013.
9. Angetter 1997a.
10. Schütz et al. 2013.
11. Estate Benninghoff.
12. Aumüller et al. 2001.
13. Ibid.
14. Hlaváčková and Svobodny 1998.
15. Voswinckel 2002.
16. Hildebrandt 2012.
17. Alvermann 2015.
18. Ude-Koeller et al. 2012.
19. Grüttner 2004.
20. Ibid.
21. Hildebrandt 2012.
22. Noack 2008.
23. Scheiblechner 2002.
24. Kaiser 2013.
25. Hlaváčková and Svobodny 1998.
26. Hildebrandt 2012.
27. UAF B24/397.
28. Hildebrandt 2012.
29. Angetter 1997b.
30. Gisel 1988.
31. Winkelmann and Noack 2010.
32. Professorenkatalog der Universität Leipzig 2015.
33. Personal- und Vorlesungsverzeichnisse der Friedrich-Wilhelms Universität Berlin 1932/33– 1944/45.
34. Personal communication, electronic mail from Dr. Dirk Alvermann, 2 April 2014.
35. Arias 2004.
36. Alvermann 2015.
37. Arias 2004.
38. Winkelmann 2015.
39. Buddrus and Fritzlar 2007.
40. Professorenkatalog der Universität Leipzig 2013.
41. Drews 1992.
42. Scheiblechner 2002.
43. Personal communication, Mag. Mathias Schütz.
44. München Vorlesungs- und Personalverzeichnisse WS 1932/33–WS 1944/45.
45. Personal- und Vorlesungsverzeichnisse der Friedrich-Wilhelms Universität Berlin 1932/33– 1944/45.
46. Hildebrandt 2012
47. Alvermann and Mittenzwei in print.
48. Aumüller et al. 2001.

49. Personal- und Vorlesungsverzeichnisse der Friedrich-Wilhelms Universität Berlin 1932/33– 1944/45.

50. Catalogus professorum halensis 2015.

51. Beushausen et al. 1998

52. Scheiblechner 2002.

53. Ibid.

54. Eckart et al. 2006.

55. Potthast and Hossfeld 2010.

56. Hildebrandt 2012.

57. Arias 2004.

58. Hildebrandt 2012.

59. Ibid.

60. Ibid.

61. Aumüller et al. 2001.

62. Eckart et al. 2006.

63. Wiederanders 2004.

64. Míšková, 2007.

65. Alvermann and Mittenzwei in print.

66. Hildebrandt 2012.

67. Ibid.

68. Scheiblechner 2002.

69. Forsbach 2006.

70. Catalogus professorum halensis 2015.

71. Wendehorst 1993.

72. Forsbach 2006.

73. Personal communication, Mag. Mathias Schütz.

74. Hildebrandt 2013a.

75. Personal- und Vorlesungsverzeichnisse der Friedrich-Wilhelms Universität Berlin 1932/33–1944/45.

76. Hildebrandt 2013b.

77. Pátek 2010.

78. Vorlesungs- und Personalverzeichnis Breslau 1933–1939.

79. Ebert 2009.

80. Becker et al. 2005.

81. Oehler-Klein 2007.

82. Ibid.

83. Hildebrandt 2013c.

84. Hildebrandt 2012.

85. Catalogus professorum halensis 2015.

86. Wechsler 2005.

87. Mussgnug 1988.

88. Freie Universität Berlin 2013.

89. Eckart et al. 2006.

90. Drews 1992.

91. Hildebrandt 2012.

92. Aumüller et al. 2001.

93. Scheiblechner 2002.

94. Hildebrandt 2012.

95. Eckart 2012.

96. UAF B24/1647

97. Personal- und Vorlesungsverzeichnisse der Friedrich-Wilhelms Universität Berlin 1932/33–1944/45.

98. Hildebrandt 2012.
99. Alvermann 2015.
100. Scheiblechner 2002.
101. Personal- und Vorlesungsverzeichnisse der Friedrich-Wilhelms Universität Berlin 1932/33–1944/45.
102. Zimmermann 2000.
103. Hlaváckovná and Svobodny 1998.
104. Personal- und Vorlesungsverzeichnisse der Friedrich-Wilhelms Universität Berlin 1932/33–1944/45.
105. Ibid.
106. Aumüller et al. 2001.
107. Personal- und Vorlesungsverzeichnisse der Friedrich-Wilhelms Universität Berlin 1932/33–1944/45.
108. Arias 2004.
109. Landgericht Münster 1960.
110. Lohff 2005.
111. Wien: Öffentliche Vorlesungen an der Universität zu Wien.
112. Alvermann 2015.
113. Hildebrandt 2012.
114. Tartu Ülikooli Ajaloo Museum 2013.
115. Pátek 2010.
116. Vorlesungs- und Personalverzeichnis Breslau 1933–1939.
117. Wien: Öffentliche Vorlesungen an der Universität zu Wien.
118. Hildebrandt 2012.
119. Ortmann 1986.
120. Hildebrandt 2012.
121. Scheiblechner 2002.
122. Heiber 1991.
123. Hildebrandt 2012.
124. Oberkofler and Goller 1999.
125. Hildebrandt 2012.
126. Catalogus professorum halensis 2015.
127. Hildebrandt 2012.
128. Oberkofler and Goller 1992.
129. Arias 2004.
130. Hildebrandt 2012.
131. Eckart et al. 2006.
132. Schütz et al. 2013.
133. Scheiblechner 2002.
134. Hildebrandt 2012.
135. Catalogus professorum halensis 2015.
136. Aumüller et al. 2001.
137. Buddrus and Fritzlar 2007.
138. Aumüller et al. 2001.
139. Wien: Öffentliche Vorlesungen an der Universität zu Wien.
140. Mörike 1988.
141. Ortmann 1986.
142. Hildebrandt 2012.
143. Schuhmacher 1970.
144. Arias 2004.
145. Ibid.
146. Alvermann and Mittenzwei in print.

147. Personal communication, Mag. Mathias Schütz.
148. Eberle 2002.
149. Alvermann and Mittenzwei in print.
150. Arias 2004.
151. Hildebrandt 2012.
152. Scheiblechner 2002.
153. Wien: Öffentliche Vorlesungen an der Universität zu Wien.
154. Hildebrandt 2012.
155. Ibid.
156. Ibid.
157. Scheiblechner 2002.
158. Alvermann and Mittenzwei in print.
159. Wittern 1993.
160. Hildebrandt 2012.
161. Heindel 1971.
162. Scheiblechner 2002.
163. Ibid.
164. Míšková 2007.
165. Alvermann and Mittenzwei in print.
166. Personal- und Vorlesungsverzeichnisse der Friedrich-Wilhelms Universität Berlin 1932/33–1944/45.
167. Ibid.
168. Hildebrandt 2012.
169. Aumüller et al. 2001.
170. Schütz et al. 2013.
171. Hildebrandt 2012.
172. Arias 2004.
173. CV and Fragebogen Schaeuble 1937, UAF B24/3200.
174. Scheiblechner 2002.
175. Bargmann and Scheffler 1943.
176. Angetter 1997b.
177. Schneider 1944.
178. Brehm et al. 2015.
179. Oberkofler and Goller 1999.
180. Scheiblechner 2002.
181. Ibid.
182. Hildebrandt 2012.
183. Oberkofler and Goller 1999.
184. Personal- und Vorlesungsverzeichnisse der Friedrich-Wilhelms Universität Berlin 1932/33–1944/45.
185. Wien: Öffentliche Vorlesungen an der Universität zu Wien.
186. Ibid.
187. Forsbach 2006.
188. Anatomischer Anzeiger 1933.
189. Alvermann 2015.
190. Arias 2004.
191. Tomkiewicz and Semków 2013.
192. Hildebrandt 2012.
193. Wendehorst 1993.
194. Kaiser 2013.
195. Noack 2008, and personal communication.
196. Kreft 2008.

197. Forsbach 2006.
198. Hildebrandt 2012.
199. Ibid.
200. Estate Benninghoff.
201. Hildebrandt 2012.
202. Szabó 2000.
203. Schütz et al. 2013.
204. Zimmermann 2000.
205. Aly 1987.
206. Aumüller et al. 2001.
207. Waibel 2011.
208. Alvermann 2015.
209. Vorlesungs- und Personalverzeichnis Breslau 1933–1939.
210. Hildebrandt 2012.
211. Míšková 2007.
212. Hildebrandt 2012.
213. Ibid.
214. Eckart et al. 2006.
215. Klinische Wochenschrift 1933.
216. Gisel 1988.
217. Hildebrandt 2012.
218. Ibid.
219. Zwilling 2004.
220. Scharer 2010.
221. Ortmann 1986.
222. Angetter Daniela 1997c.
223. Alvermann 2015.
224. Hildebrandt 2013b.
225. Arias 2004.
226. Personal- und Vorlesungsverzeichnisse der Friedrich-Wilhelms Universität Berlin 1932/33–1944/45.
227. Hildebrandt 2012.
228. Alvermann 2015.
229. Tomkiewicz and Semków 2013.
230. Hildebrandt 2012.
231. Hildebrandt 2013a.
232. Ziesche 1943.
233. Volbehr 1956.

Table 2. Scholars of Anatomy Whose Careers Were Disrupted by National Socialist Policies

Name/dates	Place/position before	Year	Reason	Place/position after
Altschul, Rudolf (1901–63)	Prague, assistant histology	1939	Jewish desc	University of Saskatchewan, Canada, professor of anatomy
Benesi, Oscar (1878–1956)	Vienna, professor of ENT	1941	Jewish desc	Detroit, research in anatomy; New York, 1944 hospital aide
Berg, Walther (1878–1945?)	Königsberg, emeritus professor anatomy	1934/35?	Jewish desc	no documentation
Bergmann, Louis (1907–92)	Vienna, assistant anatomy	1938	Jewish desc	New York University, professor of anatomy
Bielschowsky, Max (1869–1940)	Berlin, director KWI neurobiology	1933	Jewish desc	1933–36 research Holland; 1936–39 Berlin; 1939 London
Bluntschli, Hans (1877–1962)	Frankfurt, chairman anatomy	1933	political	chairman of anatomy, University of Bern, Switzerland
Brandt, Walter (1889–1971)	Cologne, professor of anatomy	1936	political and wife Jewish desc	lecturer of anatomy, Birmingham, UK
Brodersen, Johannes (1878–1970)	Hamburg, professor of anatomy	1941	political	1951 professor emeritus, Hamburg, Germany
Cohn, Ludwig (1873–1935)	Bremen, zoologist, Municipal Museum	1935	Jewish desc	1935 death of natural causes
DeBruyn, Peter (1910–90s?)	Leyden, Holland, assistant	1941	?	professor of anatomy and histology, Chicago
Driesch, Hans (1868–1941)	Leipzig, chairman philosophy	1933	political	1941 death of natural cause
Elias, Hans (1907–85)	Giessen, student zoology	1934	Jewish desc	professor of anatomy, Chicago

Name/dates	Place/position before	Year	Reason	Place/position after
Erdmann, Rhoda (1870–1935)	Berlin, chair, experimental cytology	1933	political	reinstated in new institute 1934
Fischel, Alfred (1868–1938)	Vienna, emeritus chairman of histology	1939*	Jewish desc	1938 death of natural causes
Flesch, Max (1852–1943)	Frankfurt, gynecologist, professor emeritus anatomy Bern	1939*	Jewish desc	1943 death in Terezin concentration camp
Florian, Jan (1897–1942)	Brno, chairman embryology	1939	political	tried to work privately; 1941 prison; 1942 executed in Mauthausen
Frankenberger, Zdenek (1892–1966)	Prague, director histology/ embryology	1939	political	Senior assistant Vinohrady hospital; postwar reinstated
Gieschen, Karl–Ludwig (1905–1945)	Freiburg, assistant anatomy	1933	Jewish descent	Died in December 1945
Glees, Paul (1909–99)	Bonn, assistant anatomy	1936	political, "voluntary"	1936–39 anatomist Holland; 1939–61 anatomist England; 1961–78 professor of anatomy Göttingen
Glücksmann, Alfred (1904–1985)	Heidelberg, assistant anatomy	1933	Jewish desc	Strangeway Research Lab, Cambridge, UK
Goldhamer, Karl / Golthamer, Charles R. (1896–1963)	Vienna, assistant anatomy, director anatomical radiology	1938	Jewish desc	6 weeks KZ Dachau; radiologist California
Grueneberg, Hans (1907–82)	Freiburg, assistant anatomy	1933	Jewish desc	professor of genetics, Uni College London
Gruenwald, Peter (?)	Vienna, assistant histology	?	Jewish desc?	after 1942 professor of histology and pathology, United States

Name/dates	Place/position before	Year	Reason	Place/position after
Harms, J.-Wilhelm (1884–1956)	Tübingen, Jena, chairman zoology	1938	political, "voluntary"	reinstated as professor in Jena shortly after
Hatschek, Berthold (1854–1941)	Vienna, chairman emeritus zoology	1938	Jewish desc	1941, death of natural causes
Heringa, Gerard Carel (1891–1972)	Amsterdam, chairman histology	1941	political	removed by NS authorities; KZ with wife, who died there; reinstated after the war
Herrmann, Heinz (1911–2009)	Vienna, student	1936?	Jewish desc?	research Copenhagen; professor of cell biology University of Connecticut
Hertwig, Günther (1888–1970)	Rostock, professor of anatomy	1937	political, "voluntary"	1937–45 research Berlin; 1945 chairman anatomy Rostock
Hoepke, Hermann (1889–1983)	Heidelberg, professor of anatomy	1937	Jewish desc of wife	1940–45 private practice, then reinstated
Hoyer Jr., Henryk (1864–1947)	Krakow, chair comparative anatomy, emeritus	1939	political	imprisoned "Sonderaktion Krakau," survived KZ; reinstated postwar
Jacobsohn-Lask, Louis (1863–1940)	Berlin, neurologist and neuroanatomist	1936	Jewish desc, "voluntary"	research Sewastopol, USSR
Jacobson/ Jacobsen, Werner (1906–?)	Bonn, assistant anatomy	1933	Jewish desc	Strangeway Research Lab, Cambridge, UK
Joseph, Heinrich (1975–1941)	Vienna, zoology	1938	Jewish desc	1941 committed suicide with wife
Kadanoff, Dimitri (1900–82)	Würzburg, assistant anatomy	1933	Political, "voluntary"?	1933–39 private practice, Bulgaria; then chairman anatomy, University of Sofia/Bulgaria

Name/dates	Place/position before	Year	Reason	Place/position after
Kempermann, Carl Theodor (?)	Cologne, assistant anatomy	1935	Jewish desc	?
Kohn, Alfred (1867–1959)	Prague, chairman emeritus histology	1938*	Jewish desc	survived KZ Terezin
Kostanecki, Kazimierz von (1863–1940)	Krakow, chairman anatomy	1939	political	prison, perished in KZ Sachsenhausen, "Sonderaktion Krakau"
Krause, Walter (1910–2007)	Vienna, assistant anatomy	1938	political	prison, 1940–41; soldier; postwar professor anatomy Vienna University
Kuhlenbeck, Hartwig (1897– 1984)	Breslau, assistant anatomy	1933	political, "voluntary"	professor of neuroanatomy, Philadelphia, United States
Lachman, Ernest (1901–79)	Berlin, Virchow Hospital radiology	1933	?	professor of anatomy/ radiology, Oklahoma City, United States, until 1979
Lanz, Titus von (1897–1967)	Munich, professor anatomy	1938	wife Jewish desc	stipend from NDW; postwar chairman Munich University
Levi, Guiseppe (1872–1965)	Turin, chairman anatomy	1938	Jewish desc	research Belgium, Italy; postwar reinstatement Turin
Loewenthal, Karl (1892–1944)	Berlin, pathologist	1933	Jewish desc	1933–38 professor of embryology/ histology University of Istanbul; 1939 Rhode Island, United States
Loewy, Adolf (1862–1937)	Berlin, emeritus professor of anatomy and physiology	1933	Jewish desc	since 1922 director Swiss Institute of Physiology
Lubosch, Wilhelm (1875–1938)	Würzburg, professor anatomy	1935	Jewish desc	Died 1938

Name/dates	Place/position before	Year	Reason	Place/position after
Marcus, Gaston (1908–1967)	Vienna, assistant anatomy, under Tandler, then surgeon	1938	Jewish desc?	Medical career in military
Marcus, Harry (1880–1976)	Munich, professor anatomy	1934/38	Jewish desc	1938 via Italy to Bolivia; 1939–54 professor anatomy Bolivia; 1954 returned to Munich
Martin-Oppenheim, Stephanie (?)	Munich, anthropologist	?	Jewish desc	Prison Memmingen Holland; KZ Terezin; postwar in Germany
Meyer, Robert (1864–1947)	Berlin, pathology, gynecology, embryology	1935	political and Jewish desc	until 1939 work in Berlin; 1939 immigration to US, research and consulting
Moellendorff, Wilhelm von (1887–1944)	Freiburg, chairman anatomy	1935	political, "voluntary"	chairman anatomy Zurich
Muenter, Heinrich (1883–1957)	Heidelberg, professor of anatomy and anthropology	1934	political	position in London
Münzer, Franz Theodor (1895–1944)	Prague, professor neurology	1939	Jewish desc?	Private practice 1939/40; 20 Nov 1942 Terezin; 23 Oct 1944 Auschwitz, where he perished
Palugyay, Josef (1890–1953)	Vienna, anatomical radiology	?	wife of Jewish desc	Continued teaching, Vienna radiology, not anatomy
Peterfi, Tibor (1883–1953)	Berlin, department head KWI biology	1934	political and Jewish desc	1932–39 research Cambridge; Kopenhagen, 1939–45; professor embryology/histology Istanbul; 46–48 chair anatomy Budapest
Pick, Joseph (1908–68)	Vienna, assistant anatomy	1937/38?	Jewish desc, "voluntary"	professor anatomy, New York University Medical School

Name/dates	Place/position before	Year	Reason	Place/position after
Pinkus, Felix (1868–1947)	Berlin, professor dermatology	1933	Jewish desc	1939 emigration, private practice Michigan
Politzer, Georg (1898–1956)	Vienna, assistant anatomy	1937	Jewish desc, "voluntary"	radiologist in India; 51 reinstated in Vienna
Poll, Heinrich (1877–1939)	Hamburg, chairman anatomy	1933	Jewish desc	private research Hamburg; immigration to Sweden, 1939
Pommer, Charlotte (1914–2004)	Berlin, assistant anatomy	1943	political, "voluntary"	war, hospital work; postwar, private practice
Quast, Paul (1894–?)	Bonn, assistant anatomy	?	?	unplaced 1936
Roethig, Paul (1874–1940)	Berlin, neuroanatomist	1933	Jewish desc	nervous breakdown, institutionalized
Saller, Karl (1902–69)	Göttingen, professor anatomy/ anthropology	1935	political	private practice; postwar chairman anthropology, Munich
Sauser, Gustav (1899–1968)	Vienna, chairman histology	1938	political	private practice; postwar chairman histology, Innsbruck
Scharrer, Berta (1906–95)	Frankfurt, neuroanatomist, Edinger Institute	1937	political, "voluntary"	career as neuroanatomist, chairman, Einstein University New York
Scharrer, Ernst Albert (1905–65)	Frankfurt, neuroanatomist, Edinger Institute	1937	political, "voluntary"	career as neuroanatomist, chairman, Einstein University New York
Sicher, Harry (1889–1974)	Vienna, assistant anatomy	1938	Jewish desc	professor dental anatomy, Loyola University Chicago
Slonimski sen., Piotr W. (1893–1944)	Warsaw, histology	1939	political	taught in underground university; killed in Warsaw Uprising
Spitzer, Alexander (1868–1943)	Vienna, emeritus chairman neuroanatomy	1938	Jewish desc	1942 transport to Terezin; died there in 1943

Name/dates	Place/position before	Year	Reason	Place/position after
Stahr, Hermann (1868–1947)	Danzig, director pathology	1933	Jewish desc	private research
Strauss, Fritz (1907–1994)	Freiburg, assistant anatomy	1933	Jewish desc	career in anatomy, Switzerland
Strecker, Friedrich (1879–1959)	Breslau, assistant anatomy; private practice	1933	political	imprisoned 1933; private practice until 1945; 1946 chairman anatomy Rostock
Studnicka, Frantisek K. (1870–1955)	Brno, chairman zoology/ histology	1939	political	private research
Terni, Tullio (1888–1946)	Padua, professor of anatomy	1938	Jewish desc	reinstated 1945; removed again 1946; suicide
Veit, Otto (1884–1972)	Cologne, chairman anatomy	1937	Jewish desc	postwar reinstated
Vogt, Cécile (1875–1962)	Berlin, KWI Brain research	1937	political	new institute in Black Forest
Vogt, Oskar (1870–1959)	Berlin, KWI Brain research	1937	political	new institute in Black Forest
Wallenberg, Adolf (1862–1949)	Danzig, emeritus director internal/ psychiatric medicine, municipal hospital, neuroanatomist	1938	Jewish desc	research in Oxford, England; 1943 immigration to Chicago, lecturer
Wassermann, Friedrich (1884–1969)	Munich, chairman histology	1936	Jewish desc	anatomical career in United States
Wegner, Richard (1884–1967)	Frankfurt, professor anatomy	1935	political?	private practice; 1948 chairman anatomy Greifswald
Weichherz, Edmund (1901/3–36?)	Berlin, assistant anatomy	1933	Jewish desc?	possibly Prague
Weidenreich, Franz (1873–1948)	Frankfurt, chairman anthropology	1934	Jewish desc	anthropological research; 1938 United States/China

Name/dates	Place/position before	Year	Reason	Place/position after
Weissberg, Harry (?)	Cologne, assistant anatomy	1938	Jewish desc	work in Cologne hospital
Weissenberg, Richard (1882–1974)	Berlin, professor anatomy	1933	Jewish desc	research in England, Italy, United States
Wermuth, Erich G. (1914–86)	Vienna, research student anatomy?	1938	political, "voluntary"	England, Canada, private practice
Woerdeman, M. W. (1892–1990)	Amsterdam, chairman anatomy	1941	political	prison for one month; reinstated thereafter
Wolf–Heidegger, Gerhard (1910–86)	Bonn, assistant anatomy	1935	Jewish desc	anatomical career Basel
Zawisch-Ossenitz, Carla (1888–1961)	Vienna, assistant histology 1938 political	1938	political	imprisoned, immigration to US, 1947 chairman histology Graz

Revised from Hildebrandt, Sabine. 2012a. "Anatomy in the Third Reich: Careers Disrupted by National Socialist Policies." *Annals of Anatomy* 184: 251–56.

Legend

Name/dates: Name of scholar with years of birth and death
Place/position before: place and position of employment before dismissal and/or emigration
Year: year of dismissal and/or emigration
*: membership Leopoldina struck
Reason: Reason for dismissal or "voluntary" leaving of position and/or emigration:
Jewish desc: Jewish descent—dismissal due to anti-Semitic legislation
Political: political dissent with regime (including religious dissent)
Also: actions of occupying German forces against universities and individuals
Place/position after: place and position after dismissal and/or emigration
KWI: *Kaiser-Wilhelm-Institut*
NDW: *Notgemeinschaft Deutscher Wissenschaftler* (emergency committee of German scientists; aid organization)
KZ: concentration camp
ENT: ear, nose, and throat specialist

Table 3. Body Supply of Anatomical Departments of German Universities, 1933–45

University	Body-register	U	Ho	Su	Do	Ps	Pr	CC	FL	PO	GS	CO	Total bodies	#exec	ID exec	ID others
Berlin	No	X					X						?	>182	Some	No
Bonn	Yes	X	X			X	X		X	X	X	X	1025	191	Yes, all	Some
Breslau	?	X					X						?	?	No	No
Cologne	<1942	X	X				X	X	X	X		X	>774	85	Some	?
Danzig	No	X					X	X	?		X		>147	?	No	No
Dorpat	?												?	?		
Erlangen	?	X					X		X				?	?	No	No
Frankfurt	?	X					X						?	?	Some	No
Freiburg	No	X					X		X	X			?	?	No	Some
Giessen	Yes	X					X	X	X	X	X		405	53	Some	No
Göttingen	No	X	X		X	X	X	X	X		X	X	517	233	?	No
Graz	Yes	X	X			X	X					X	?	>95	Some	Some
Greifswald	Lost 1947	X	X			X	X		X	X	X		>432	>33	Some	Some
Halle	<1937	X	X			X	X		X	X	X	X	?	~108	No	No
Hamburg	Yes	X	X			X	X	X	X	X	X	X	?	>315	Some	No
Heidelberg	Yes	X	X			X	X		X	X	X	X	970	280	Some	Many
Innsbruck	Yes	X		X			X		X	X		X	?	>62	Some	Some
Jena	Yes	X	X	X		X	X	X	X		X	X	2224	200	Yes, most	Some
Kiel	No	X	X			X	X		X				?	>73	Some	No
Königsberg	?	X					X						?	?	Some	No

University	Body-register	U	Ho	Su	Do	Ps	Pr	CC	FL	PO	GS	CO	Total bodies	#exec	ID exec	ID others
Leipzig	?	X					X					X	?	>75	57	No
Marburg	Yes	X	X	X		X	X	X	X	X			~800–900	?	Some	No
Munich	Some	X	X				X						?	>34	Some	No
Münster	?	X					X						?	?	No	No
Posen	No	X					X	X			X		?	?	No	No
Prague	?	X					X						?	?	No	No
Rostock	?	X					X				X		?	?	No	No
Strasbourg	No	X						X		X			300–400	>86	86	No
Tübingen	Yes	X	X			X	X	X	X	X	X	X	1077	~285	Some	Many
Vienna	No	X	X		X	X	X				X	X	>5341*	1377	Yes **	Yes
Würzburg	Yes	X	X			X	X	X	X	X	X	X	944	120	Some	No

Legend

University: location of anatomical department

Body register: body register found or not

Sources of bodies: U: unclaimed in general; Ho: hospital; Su: suicide; Do: donation; Pr: prisoners dead of natural causes and executed; CC: concentration camp; FL: forced-labor camp; PO: prisoner of war camp; GS: directly delivered by GeStapo (*Geheime Staatspolizei*; secret police) or police or forensic medicine; Co: communal institutions, e.g., retirement homes

Total bodies: overall number of bodies received during time period studied, generally 1933 to 1945 or time of NS occupation

exec: number of bodies of victims of executions following civilian and military trials

ID exec: number of victims of execution identified

ID others: identification of any other groups of victims

* plus 7,000 children, miscarriages and stillbirths

** names identified, but not published because of data privacy protection

Table 4. Stieve's List

Nr.	Last Name	First Name	Age	Birth Date	Death Date	Nationality
1	Jünemann	Charlotte	24	11/23/1910	8/26/1935	G
2	Schröter	Bruno	53	9/11/1938	7/7/1937	G
3	Wittke	August	21	5/28/1915	7/25/1936	G
4	Georger	Marie	41	5/26/1896	5/25/1937	G
5	Kneup	Katharina	39	7/7/1899	10/4/1938	F
6	Kuhlmann	Helmuth	23	10/26/1914	7/19/1938	G
7	Seyfarth	Anna	36	11/24/1902	11/12/1938	G
8	Schwitzer	Georg	34	8/1/1903	6/15/1938	G
9	Schwitzer	Anna	41	1/8/1897	6/15/1938	G
10	Gässner	Arno	40	11/16/1899	12/7/1939	G
11	Gose	Arthur	19	2/13/1920	3/2/1939	G
12	Diecker	Marie	33	7/29/1906	7/29/1940	G
13	Mertyn	Janine	20	5/15/1921	8/21/1941	P
14	Tyschenski	Paula	41	4/13/1899	3/21/1941	G
15	Schubert	Ruth	20	11/22/1920	2/1/1941	G
16	Augustyniak	Veronika	42	11/30/1899	8/15/1942	P
17	Ball	Julie	61	4/24/1880	3/19/1942	G
18	Baum	Marianne	30	2/9/1912	8/18/1942	G
19	Budach	Else	30	7/3/1911	5/17/1942	G
20	Czubakowska	Bronisława	29	7/9/1916	8/15/1942	P
21	Fiedermann	Anna	23	9/3/1918	6/20/1942	G
22	Götze	Ursula	26	3/29/1916	8/5/1943	G
23	Golembiewski	Sophie	35	2/8/1907	8/7/1942	P
24	Grossvogel-Pesant	Jeanne	42	9/16/1901	7/6/1944	B
25	Hanusch	Erika	22	4/25/1920	7/17/1942	G
26	Jadamowitz	Hildegard	26	2/12/1916	8/15/1942	G
27	Kächele	Juliette	21	11/20/1920	10/2/1942	F
28	Kochmann	Sala	30	6/7/1912	8/18/1942	G
29	Korsing	Frieda	53	4/17/1889	6/5/1942	G
30	Lambert	Elisabeth	45	10/15/1897	12/14/1942	G
31	Laetsch	Ulla	31	8/22/1910	7/8/1942	G
32	Redepenning	Elfriede	31	10/25/1910	7/30/1942	?
33	Reichmann	Luisa	38	4/25/1904	12/18/1942	C
34	Saarow	Lieselotte	19	5/16/1923	6/5/1942	G
35	Sadowska	Josefa	27	4/14/1915	5/4/1943	P
36	Schejner	Marianne	64	3/20/1878	8/15/1942	P

Nr.	Last Name	First Name	Age	Birth Date	Death Date	Nationality
37	Schulze-Boysen	Libertas	29	11/20/1913	12/22/1942	G
38	Schumacher	Elisabeth	38	4/28/1904	12/22/1942	G
39	Tucholla	Käthe	32	1/10/1910	9/28/1942	G
40	Walther	Irene	23	1/23/1919	8/18/1942	G
41	Zbierska	Leokadia	25	3/22/1917	8/22/1942	P
42	Bach	Margarete	59	8/1/1883	5/4/1943	G
43	Baumgartner	Gertrud	22	8/26/1920	4/15/1943	G
44	Beek	Cato Bontjes van	22	11/14/1920	8/5/1943	G
45	Behnke	Marta	68	12/22/1874	7/29/1943	G
46	Böller	Frieda	32	11/21/1911	12/29/1943	G
47	Berkowitz	Liane	20	8/7/1923	8/5/1943	G
48	Biesenack	Emilie	43	10/13/1899	5/17/1943	G
49	Briese	Hildegard	20	1/9/1923	5/17/1943	G
50	Buch	Eva	22	1/31/1921	8/5/1943	G
51	Castek	Jaroslava	34	6/5/1908	5/25/1943	C
52	Comelli	Jeanne	19	1/17/1924	9/24/1943	F
53	Coppi	Hilde	34	5/13/1909	8/5/1943	G
54	Delacher	Helene	39	8/25/1904	11/12/1943	A
55	Dietrich	Gisela	23	6/10/1920	12/21/1943	G
56	Dvorak	Emilie	23	9/27/1919	3/29/1943	C
57	Dymski	Monika	25	4/28/1918	6/25/1943	P
58	Engler	Stephanie	32	11/18/1910	6/25/1943	A
59	Erlik	Georgette	29	10/26/1913	8/20/1943	F
60	Froese	Toska	40	12/9/1903	11/12/1943	G
61	Gasczak	Marianne	28	12/28/1914	9/28/1943	P
62	Hampel	Elise	39	10/27/1903	4/8/1943	G
63	Hanke	Helene	26	3/1/1917	3/16/1943	G
64	Henkel	Elfriede	42	8/19/1901	6/25/1943	G
65	Hirsch	Hella	22	3/6/1921	3/4/1943	G
66	Hitzuber	Frieda	43	2/20/1900	7/23/1943	G
67	Hofmann	Rosa	23	5/27/1919	3/9/1943	A
68	Horstbrink	Frieda	47	9/4/1896	11/2/1943	G
69	Joachim	Marianne	21	11/5/1921	3/4/1943	G
70	Joch	Charlotte	37	2/22/1906	6/25/1943	G
71	Jezierska	Wieslawa	30	10/13/1912	3/9/1943	P
72	Kocowa	Mirosława	35	1/31/1908	2/16/1943	P
73	Koniczna	Halina	21	8/25/1921	2/16/1943	P
74	Kosin	Martha	39	5/9/1904	9/30/1943	G

Nr.	Last Name	First Name	Age	Birth Date	Death Date	Nationality
75	Kummerow	Ingeborg	31	8/23/1912	8/5/1943	G
76	Krsnak	Maria	43	1/22/1900	4/12/1943	C
77	Lederer	Vera	31	3/25/1912	7/1/1943	C
78	Lefevre	Madeleine	36	1/2/1907	3/25/1943	?
79	Lindner	Herta	22	11/3/1920	3/29/1943	G
80	Loewy	Hildegard	20	8/4/1922	3/4/1943	G
81	Maiereder	Karoline	36	4/28/2007	4/5/1943	G
82	Makowiak	Helene	41	?	11/2/1943	?
83	Marivet	Marguerite	34	8/13/1909	8/20/1943	F?
84	Harnack	Mildred von	40	9/16/1902	2/16/1943	U
85	Modrzewski	Hedwig von	21	2/20/1922	10/13/1943	G
86	Naumann	Else	33	5/15/1914	7/29/1943	G
87	Oesterreich	Ruth	49	6/6/1894	6/25/1943	G
88	Pheter	Simone	26	3/8/1917	8/20/1943	B?
89	Piotrzkowski	Elli	33	9/18/1909	6/1/1943	G
90	Prokop	Olga	20	7/22/1922	3/9/1943	P
91	Rachel	Marta	39	5/16/1904	11/2/1943	G
92	Rohde	Anna	38	10/13/1905	12/9/1943	G
93	Rink	Maria	41	3/22/1901	2/16/1943	P
94	Rognant	Lucienne	22	10/7/1920	9/24/1943	F
95	Rubal	Wilhelmine	42	2/6/1901	3/29/1943	C
96	Rychly	Franziska	35	11/30/1907	3/4/1943	C
97	Slach	Eleonore	23	10/14/1919	4/12/1943	C
98	Springer-Velaerts	Flora	34	1/25/1909	8/20/1943	F?
99	Szaidel	Emma	26	2/18/1917	11/12/1943	?
100	Scheffczyk	Pelagia	28	3/8/1915	10/5/1943	P
101	Scheitza	Elisabeth	40	11/30/1902	7/1/1943	G
102	Schlösinger	Rose	36	10/5/1907	8/5/1943	G
103	Brockdorff	Erika von	32	4/29/1911	5/13/1943	G
104	Schottmüller	Oda	38	2/9/1905	8/5/1943	G
105	Scholz	Elfriede	40	3/25/1903	12/16/1943	G
106	Tassin	Lucienne	20	6/21/1923	10/13/1943	F
107	Tanton	Renee	21	11/1/1921	9/24/1943	F
108	Terwiel	Marie	33	6/7/1910	8/5/1943	G
109	Veith	Henryka	28	12/19/1914	6/25/1943	P
110	Völkner	Käthe	37	4/12/1906	7/28/1943	G
111	Węgierska	Wanda	24	1/31/1919	6/25/1943	P
112	Altermann	Maria Anna	23	10/10/1921	10/20/1944	G

Nr.	Last Name	First Name	Age	Birth Date	Death Date	Nationality
113	Arnould	Rita	30	4/11/1914	8/20/1943	B
114	Bejd	Antonie	33	9/30/1911	9/8/1944	C
115	Bernasova	Libuse	22	3/9/1922	11/30/1944	C
116	Bethge	Emma	52	7/13/1891	3/9/1944	G
117	Beutler	Liesbeth	43	2/3/1901	8/11/1944	G
118	Brinkmeyer	Therese	41	7/2/1903	7/27/1944	G
119	Ceroinka	Vera	39	8/27/1905	11/24/1944	G
120	Danhofer	Agnes	55	12/18/1888	12/6/1944	G
121	Dörffel	Gertrud	54	8/15/1889	7/27/1944	G
122	Dubsky	Marie	28	5/27/1916	7/20/1944	C
123	Dziallas	Elfriede	32	4/30/1912	12/29/1944	G
124	Francke	Erna	46	11/23/1897	4/25/1944	G
125	Frank-Schultz	Ehrengard	59	3/23/1885	12/8/1944	G
126	Gast	Gertrud	36	11/16/1907	8/25/1944	G
127	Göttmann	Elsa	41	3/23/1903	11/10/1944	G
128	Günther	Wilhelmine	27	7/18/1917	6/9/1944	G
129	Hallmann	Erna	27	7/16/1916	5/26/1944	G
130	Henin	Marie-Luise	45	12/9/1898	6/9/1944	B
131	Hilbz	Margarete	36	12/4/1907	6/29/1944	G
132	Hoffmann	Lina	47	3/18/1897	9/29/1944	G
133	Hölzlsauer	Anna Maria	41	6/30/1902	4/19/1944	A
134	Jahn	Hedwig	53	12/24/1890	3/2/1944	G
135	Jaster	Martha	45	3/19/1899	6/21/1944	G
136	Kallenbach	Wanda	42	6/13/1902	8/18/1944	G
137	Kazlauskaite	Filumena	29	1/101915	5/19/1944	S
138	Klute	Marie	39	5/3/1905	12/15/1944	G
139	Korth	Helene	33	4/17/1911	10/6/1944	G
140	Kosch	Marta	32	7/27/1911	1/27/1944	G
141	Kreulich	Marie	54	10/5/1889	3/19[17]/1944	G
142	Lombaerts	Christine	40	1/24/1904	1/27/1944	B
143	Latoschinski	Marianne	41	9/12/1903	9/29/1944	G
144	Liersch	Maria	38	11/24/1906	10/27/1944	G
145	Linke	Ursula	21	7/15/1923	12/8/1944	G
146	Liptow	Marie	43	2/16/1901	9/29/1944	G
147	Litzenberg	Charlotte	20	10/8/1923	5/26/1944	G
148	Lucas	Andree	20	8/1/1924	11/24/1944	F?
149	Melzen	Frieda	61	11/29/1882	10/6/1944	G
150	Mlynarz	Genovewa	41	12/29/1902	7/7/1944	P

Nr.	Last Name	First Name	Age	Birth Date	Death Date	Nationality
151	Möller	Helene	37	4/30/1907	5/9/1944	G
152	Le Muzie	Annemarie	24	7/21/1920	9/15/1944	F
153	Obolensky	Vera	33	6/24/1911	8/4/1944	G?
154	Olejak	Hedwig	37	9/25/1906	2/18/1944	G
155	Parmentier	Madeleine	20	9/3/1923	3/2/1944	F
156	Petit	Maurice	21	1/14/1923	11/24/1944	F
157	Pilan	Amanda	62	7/27/1881	5/9/1944	G
158	Prehn	Ilse	24	7/17/1920	9/15/1944	G
159	Rainaud	Therese	21	8/25/1923	9/29/1944	F
160	Lavíčková	Krista	27	12/15/1917	8/11/1944	C
161	Stentzel	Frieda	39	11/3/1905	10/20/1944	G
162	Styma	Anna	18	3/2/1926	9/1/1944	G
163	Svatos	Levcena	44	7/12/1889	4/25/1944	C
164	Swierzek	Gertrud	22	12/22/1921	9/8/1944	P
165	Sztwiertnia	Elisabeth	39	3/26/1905	8/11/1944	C?
166	Szyriajew	Maria	20	9/26/1924	10/20/1944	?
167	Schmenkel	Hulda	49	9/23/1894	5/9/1944	G
168	De Smedt	Josef	23	6/16/1921	11/10/1944	B?
169	Schneider	Anna	43	12/6/1900	6/9/1944	A
170	Thillmann	Herta	27	6/30/1917	1/21/1944	G
171	Toews	Hedwig	23	5/22/1920	4/6/1944	G
172	Tygör	Elfriede	40	10/10/1903	8/25/1944	G
173	Uriga	Elisabeth	41	10/4/1913	9/15/1944	G
174	Warret-Carion	Berthe	40	6/6/1904	9/1/1944	F
175	Westeroth	Ida	60	2/10/1884	8/11/1944	G
176	Wieczorek	Maria	58	8/4/1886	9/22/1944	G
177	Wegener	Margarete	30	5/15/1913	1/7/1944	G
178	Wittke	Frieda	49	5/16/1895	5/26/1944	G
179	Wosikowski	Irene	34	2/9/1910	10/27/1944	G
180	Zagora	Anna	29	9/16/1915	9/22/1944	P
181	Zehden	Emmi	44	3/28/1900	6/9/1944	G
182	Ziegeler	Hildegard	25	10/22/1918	9/15/1944	G

Revised from Hildebrandt, Sabine. 2013. "The Women on Stieve's List: Victims of National Socialism Whose Bodies Were Used for Anatomical Research." *Clinical Anatomy* 26: 3–21.

Legend

Dates are given as Month/Day/Year
Nationality: G = German; F = French; P = Polish; B = Belgium; C = Czech; A = Austrian, S = Soviet; U = American (US)

Table 5. Identification of NS Victims Used by Clara and His Group

Abbreviated name, age, sex	Name	Nationality	Date of Execution (Month/Day/Year)	Co
Fa 47y m	Fast, Gottfried		5/14/1935	
Ri 42y m	Richter, Willi		5/25/1940	
	Riedel, Arthur	German	1/26/1937	X
Pe 30y f	Pechatz, Emilie	Czech	8/12/1938	X
Pa 52y f	Parakenings, Maria	German	10/10/1937	X
Ri 18y m	Richter, Paul	German	4/17/1940	
Op 20y m	Opitz, Erich			
Hou 23y m	Housa, Rudolf	Czech	4/23/1940	
Kuz 25y m	Kuzdas, Karl	Czech	2/26/1941	
Ha 27y m	Hanzl, Jaroslaus	Czech	8/6/1940	X
Bau 27y m	Bauer, Anton	German	8/17/1940	X
Kar 28y m	Karbaum, Werner	German	5/11/1940	X
Chlu 29y m	Chlupaty, Friedrich	Czech	2/28/1941	
Schmi 29.5y m	Schmidt, Erwin	German	5/27/1937	
Odv 31y m	Odvarka, Josef	Czech	10/3/1941	
Schi 31y m	Schiessl, Max	German	1/22/1940	
Gue 32y m	Guenther, Max	German	4/26/1940	
Stez 33 y m	Steinmetz, Herbert	German	2/26/1941	
Svo 33y m	Svoboda, Johann	Czech	10/3/1941	
Cyr 33y m	Cyranek, Ludwig	German	7/4/1941	
Bej 34y m	Bejr, August	Czech	1/18/1941	
Jar 34y m	Jarolim, Albert	Czech	8/1/1941	
Schey 35y m	Schejbal, Eduard	Czech	8/22/1941	
Hol 36y m	Holub, Josef	Czech	8/22/1941	
Jez 38y m	Jezek, Josef	Czech	8/6/1940	X
Glei 39 y m	Gleisner, Otto	German	9/14/1938	X
Nie 39y m	Nietschmann, Richard	German	10/22/1935	
Hai 41y m	Haiboeck, Heinrich	Austrian	9/22/1941	
Bart 42y m	Bartos, Josef	Czech	7/12/1941	
Eng 43y f	Engler, Marie	German	8/1/1941	X
Muel 43y m	Mueller, Erich	German	4/26/2015	X
Duer 44y m	Duerschner, Thomas	German	8/10/1940	
Kra 45y m	Kraus, Wenzel	Czech	3/21/1941	
Ce 46y m	Cech, Josef	Czech	5/7/1940	
Drn 47y m	Drnek, Cenek	Czech	6/28/1941	
Vet 51y m	Vetter, Max Oskar	German	8/17/1940	X

Abbreviated name, age, sex	Name	Nationality	Date of Execution (Month/Day/Year)	Co
Ung 73y f	Unger, Anna	German	1/13/1942	
Ko 18y m	Koschnik, Willi	German	1/13/1942	
Ste 19y m	Stecyk, Bronislav	Polish	12/22/1941	
Cze 20y m	Czernek, Tadeusz	Polish	12/22/1941	
Ba 20.5y m	Baron, Stefan	Polish	12/22/1941	
Mo 22y m	Moravec, Alois	Czech	1/13/1942	
Deuz 23 y m	Deuzenberg, Alexander	German	7/10/1940	X
Wos 25 y m	Woszynski, Andrzey	Polish	11/21/ 1941	
Kur 25.5.y m	Kurc, Johann	German	12/6/1941	
Mach 28y m	Machac, Alois	Czech	1/13/1942	
Mat 31y m	Matousek, Karl	Czech	1/13/1942	
Fi 31.5y m	Fiedler, Rudolf	German	10/21/1942	
Vse 31.5y m	Vsejansky, Johann	Czech	11/28/1941	
Sem 33y m	Semerak, Ulrich	Czech	11/28/1941	
Gloe 37y m	Gloeckle, Wilhelm	Czech	11/28/1941	
Cho 37y m	Choronzak, Bronislav	Polish	1/16/1942	
Sla 37.5y m	Sladek, Metodej	Czech	1/13/1942	
Klo 39.5ym	Kloss, Fritz	German	10/8/1941	
Pro 41y m	Prokes, Fritz	Czech	1/13/1942	
Mate 42y m	Matejicek, Bohumil	Czech	12/19/1941	
Al 44y m	Albig, Ernst	German	11/28/1941	
Kli 42y m	Klimes, Karl	Czech	11/28/1941	

Revised from Hildebrandt, Sabine. 2013. "Current Status of Identification of Victims of the National Socialist Regime Whose Bodies Were Used for Anatomical Purposes." *Clinical Anatomy* 27: 514–36.

Legend

y = years of age

m = male

f = female

Co: X = Anatomy transport confirmed by research doctor Birgit Sack, memorial site Dresden

Table 6. List of Forced Laborers Whose Bodies Were Transported from Camp Breitenau to the Department of Anatomy at the University of Marburg

Name	Nationality	Date of Birth	Date of Death	Cause of Death
Kaczurek, Henryk	Polish	12/12/ 1922	8/20/1941	accident
Bielozobodow, Basil	Lithuanian	5/6/1906	4/21/1943	seizures
Wesolewski, Tadeusz	Polish	5/21/1921	9/20/1941	suicide/ hanging
Maciol, Johann	*			
Gorzinski, Theodor	*			
Kudelko, Clemens	*			
Szperna, Heinrich	Polish	7/30/1918	6/17/1941	execution, (sudden cardiac death)
Wisniewski, Stanislaw	Polish	10/19/1918	6/17/1941	execution, (sudden cardiac death)
Jurkiewicz, Josef	Polish	3/14/1909	1/26/1942	execution
Polednik, Albert	Polish	1898	3/19/1942	execution
? Knapik, Josef	Polish	1922	2/22/1942	execution
Nowak, Johann	Polish	2/6/1912	5/6/1942	execution
Witecki, Ignaz	Polish	1910	6/16/1942	execution
Wypych, Maryjan	Polish	12/13/1921	7/9/1942	execution
? Luba, Stefan	Polish	4/14/1916	7/17/1942	execution
? Pecka, Bronislaw	Polish	1916	10/26/1942	execution
? Bafja, Antoni	Polish	1918	11/21/1942	execution
? Dytrich, Jan	Polish	1924	11/21/1942	execution
? Cieply, Anton	Polish	8/31/1911	12/19/1942	execution
? Janicki, Anton	Polish	1924	12/19/1942	execution
? Kolczynski, Mieczyslaw	Polish	1918	12/19/1942	execution
? Orlowski, Marian	Polish	1913	12/19/1942	execution
? Wojcik, Jan	Polish	1921	12/19/1942	execution
? Stephan, Kasimir	Polish	1924	12/19/1942	execution

Based on information from Richter, Gunnar. 2009. *Das Arbeitserziehungslager Breitenau (1940–1945). Ein Beitrag zum nationalsozialistischen Lagersystem.* Kassel: Verlag Winfried Junior.

Legend

? = evidence of transport to the anatomy department at Marburg without documentation in body register

(sudden cardiac death) = cause of death given on death certificate

* = documentation for transport to anatomy department at Marburg, no further information

INDEX

www.ingramcontent.com/pod-product-compliance
Lightning Source LLC
Chambersburg PA
CBHW070901030426
42336CB00014BA/2275